Surface Acoustic Wave Filters

T0348665

Surface Acoustic Wave Filters
With Applications to Electronic Communications and Signal Processing

David Morgan

Impulse Consulting, Northampton, UK

AMSTERDAM • BOSTON • HEIDELBERG • LONDON • NEW YORK • OXFORD
PARIS • SAN DIEGO • SAN FRANCISCO • SINGAPORE • SYDNEY • TOKYO
Academic Press is an imprint of Elsevier

Academic Press is an imprint of Elsevier
Linacre House, Jordan Hill, Oxford OX2 8DP, UK
84 Theobald's Road, London WC1X 8RR, UK
30 Corporate Drive, Suite 400, Burlington, MA 01803, USA

First edition 1985. Second edition 2007

Copyright © 2007 Elsevier Ltd. All rights reserved

Published by Elsevier Ltd. All rights reserved
The right of *David Morgan* to be identified as the author of this work has been
asserted in accordance with the Copyright, Designs and Patents Act 1988

British Library Cataloguing in Publication Data
A catalogue record for this book is available from the British Library

Library of Congress Cataloging-in-Publication Data
A catalog record for this book is available from the Library of Congress

ISBN: 978-0-1237-2537-0

For information on all Academic Press publications visit
our web site at http://books.elsevier.com

Typeset by CharonTec Ltd (A Macmillan Company), Chennai, India,
www.charontec.com.

Working together to grow
libraries in developing countries

www.elsevier.com | www.bookaid.org | www.sabre.org

ELSEVIER   BOOK AID International   Sabre Foundation

Transferred to Digital Printing 2010

CONTENTS

Chapter 6   Bandpass filtering using non-reflective transducers   **157**

Chapter 7   Correlators for pulse compression radar and communications   **183**

Chapter 8   Reflective gratings and transducers   **225**

PREFACE

'The history of science teems with examples of discoveries which attracted little notice at the time, but afterwards have taken root downwards and borne much fruit upwards'

Lord Rayleigh
(Presidential Address to the British Association, Montreal, 1884)

In the context of surface acoustic waves, this statement could hardly be more prophetic. In 1885 Rayleigh described an acoustic wave motion which plays an important part in seismology, and recently this has taken on a quite different significance as the basis for a huge range of electronic devices. The potential for electronics applications was first suggested in the 1960s, noting that the waves could provide substantial signal delays which would be inconvenient to obtain by conventional methods. They also give substantial versatility because transducers for generation and reception can be located anywhere in the propagation path, and lithographic fabrication techniques can provide almost arbitrary geometries with high precision. Consequently, a wide variety of devices have been developed, and they have found their way into many branches of electronics, including signal processing for radar and communications systems. The most notable application area is in bandpass filtering for communications, including the ubiquitous mobile telephone systems.

A previous book on this subject, *Surface-Wave Devices for Signal Processing*, was published by Elsevier in 1985, coincidently the centenary of Rayleigh's 1885 paper. At that time, in 1985, many devices had been established in practical electronic systems, including signal processing devices and bandpass filters for communications and television systems. These devices remain in widespread service now. However, the 1980s were something of a watershed for surface waves. The rise of communications systems such as mobile telephony demanded new capabilities, particularly for low-loss bandpass filters satisfying exacting specifications. In response, a considerable variety of novel devices emerged. A common factor here was the use of reflecting structures, either in the form of reflective transducers or as components in resonators. Reflectivity is employed in various types of single-phase unidirectional transducers for bandpass filtering at intermediate frequencies. Resonators are found in various low-loss bandpass filters for radio frequencies, as well as in surface-wave oscillators. In parallel, many new surface-wave materials were established, some using new types of surface wave. The theory of surface-wave generation and propagation, particularly in reflective structures, has developed substantially and become much more complex. The continual vigor of the subject is well illustrated by the constant stream of publications, as shown by the many recent references quoted here.

This new book expands the coverage of the earlier one to include the recent developments. The earlier material is described in Chapters 2 to 7. This includes various acoustic waves (Chapter 2), surface excitation (Chapter 3), propagation effects and materials (Chapter 4), quasi-static transducer theory (Chapter 5) and non-reflective bandpass filters (Chapter 6). Devices for correlation, used in pulse-compression radar and spread-spectrum communications, are in Chapter 7. These accounts are similar to those of the earlier book, but to allow space for new material, there is some compression and some topics have been omitted. Some results for transducer analysis, derived from the earlier sections, are summarized in Section 5.3 of Chapter 5.

New areas include filters using unidirectional transducers (Chapter 9), waveguiding and transversely-coupled resonators (Chapter 10), and resonator filters (Chapter 11). Preceding these, Chapter 8 describes the theory of reflective transducers and gratings, including analysis using the Reflective Array Model (RAM) and Coupling-of-Modes (COM) theory. At the beginning, Chapter 1 gives a survey of the whole subject. This is intended to be readable independently of the rest of the book.

The book is written at a post-graduate level, assuming some familiarity with topics such as matrix algebra and the Y- and S-matrix descriptions for linear devices. However, much of the material should also be comprehensible at an undergraduate level, particularly the survey in Chapter 1. A prior knowledge of acoustic waves is not necessary, since this topic is summarized in Chapter 2. In fact, much of the theory in Chapters 3 and 5 follows simply from the assumption, often valid, that a piezoelectric substrate supports surface-wave propagation with velocity dependent on whether the surface is free or metallized. This approach is adequate for many devices, and it requires little further knowledge of acoustic waves. The theoretical developments make much use of Fourier analysis, and the required relations are summarized in Appendix A. Since the original book was published, it has become common to describe the behavior of unapodized transducers and gratings in terms of a scattering matrix called the P-matrix, so the present book makes use of this form. Appendix D considers the reciprocity and power-conservation constraints on this matrix, and also cascading techniques for analyzing devices with several components such as resonators. This Appendix also considers the all-important topic of multiple-transit signals in surface-wave devices.

An initial reading of the book might start with the survey in Chapter 1 and then skip to Section 5.3 of Chapter 5, which summarises properties of non-reflective transducers as given by the quasi-static theory. This leads on to the use of non-reflective transducers for bandpass filtering (Chapter 6) and matched filtering in radar and communications (Chapter 7). The topic of internal reflections is then introduced in Chapter 8.

The book has benefitted from interactions with many colleagues, particularly in the Nippon Electric Company, the University of Edinburgh and Plessey Research (Caswell). Many of the ideas in the 1985 book arose from work in the Caswell group, consisting of R. Allen, R. Almar, R. Arnold, R.E. Chapman, R.K. Chapman, J. Deacon, R. Gibbs, W. Gibson, J. Heighway, J. Jenkins, P. Jordan, B. Lewis, J. Metcalfe, R. Milsom, J. Purcell, D. Selviah and D. Warne. There was also much interaction with E.G.S. Paige and M.F. Lewis at RSRE Malvern (now part of Qinetiq). The Foreword to the 1985 book was contributed by Dr. J. Bass, Director of Plessey Research (Caswell). In 1991 a soft-cover edition was published with a Foreword by E.G.S. (Ted) Paige, who sadly passed away in 2004. Many of the surface-wave workers in the UK in the 1970s and 1980s, including the Caswell group, owe a debt to Ted for the inspiration he showed as director of the surface-wave group in RSRE Malvern and sponsor of much of the UK research effort in this field.

Other colleagues with whom discussions have been helpful include B. Abbott, I. Avramov, S. Biryukov, C. Campbell, D.-P. Chen, X. Chen, J. Collins, M. DaCunha, C. Hartmann, K.-Y. Hashimoto, J. Heighway, V. Kalinin, J. Koskela, C.S. Lam, M. Lewis, C. Liang, D. Malocha, R. Milsom, R. Peach, V. Plessky, C. Ruppel, M. Salomaa, M. Sharif, K. Shibayama, J.-B. Song, T. Thorvaldsson, M. Weinacht, K. Yamanouchi and S. Zhgoon.

For the present book, I am grateful to colleagues for providing some figures, namely B. Abbott (Fig. 1.4), C. Ruppel (Fig. 1.14) and M. Solal (Fig. 9.6). S. Zhgoon and A. Shvetsov provided data for leaky-wave dispersion (Figs. 11.22 and 11.23) calculated using finite-element analysis. K.-Y. Hashimoto is thanked for making sophisticated surface-wave software available publicly, and for its use here in Figs. 3.2, 11.19, 11.22 and 11.23. Helpful comments on initial drafts of the book have been made by B. Abbott, V. Plessky, D. Malocha , C. Ruppel, M. Solal and S. Zhgoon.

Finally, it is a great pleasure to thank Professor Sir Eric Ash for providing the Foreword to this book. In the early days, Eric was one of the few who recognized the potential future of the subject and initiated its development, and my own work in this field started as one of his students at University College London in the 1960s. I have spent nearly all of my working life in this fascinating subject, and this Foreword now brings it to a full circle. Or perhaps I should say, since the subject continues apace, a further turn of a helix. I hope that readers will find the book both helpful and enlightening.

David Morgan
Northampton, 2007

FOREWORD TO SECOND EDITION

It is a digital world – and has been for a long time. The transformation toward digital electronics which started in the middle of the last century gathered increasing momentum in the early 1970s with the arrival of the microcomputer. Thereafter analog electronics, except where it is intended to interface directly with our human senses, was doomed to an ever diminishing role. Game, set, match for digital? Actually not quite! Just at that moment, there appeared a new and completely unanticipated technology, that of ultrasonic signal processing, and specifically the use of surface acoustic waves – the subject of this book.

Surface acoustic waves were hardly new – they had been discovered by Lord Rayleigh in 1885 theoretically, and their reality confirmed by their appearance in seismic records. The fact that these sub-one Hertz waves, when translated to frequencies six to nine orders of magnitude higher, could perform useful tasks in signal processing came as a great surprise. So what is so good about surface acoustic waves? Firstly for a given frequency their wavelength is very small. As compared, for example, with microwaves one can achieve a great deal in a very small space. The second merit is that surface waves hug the surface; they can be influenced by electrodes placed on the surface. The third merit is the existence of single-crystal piezoelectric materials in which propagation losses are very small, and which provide the opportunity for efficient transformation, in both directions, between electrical and acoustic signals. Surface-acoustic-wave signal processing is therefore a *planar* technology, based on the use of photolithography to define the structures. It shares this feature with semiconductor microelectronics, but differs in that it normally requires just a single photolithographic mask. The common use of masks has enabled surface-wave devices to benefit from the huge advances in mask technology made by the semiconductor industry.

The first and still dominant use of surface acoustic waves is for the realization of bandpass filters, followed secondly by resonators. David Morgan has been a very major contributor to the whole field since its inception. The book provides a brilliant exposition of the subject – in this following the much admired clarity of his earlier book of 20 years ago. Much has happened in the interim. Morgan's book illuminates the whole field, including the rapid developments in the theory of the basic wave types involved and the conceptual advances in wave manipulation. These are advances which have led to performance achievements which could not be approached two decades ago.

Surface-acoustic-wave devices play a very large role in modern electronics. Oddly enough this is known only to those very directly involved – it is almost a secret society! Few people appreciate that a mobile phone will contain at least two and sometimes as many as six distinct surface-acoustic-wave filters. These filters have specifications on passband shape, stop-band rejection, temperature performance which are as rigorous as have *ever* been attempted in telecommunication systems. Most television receivers have two such filters. In addition surface-acoustic-wave devices are used extensively in advanced radar systems for pulse expansion and compression. The total annual production of surface-acoustic-wave filters is at least four billion and possibly nearer seven billion per annum.

Surface-acoustic-wave devices are sensitive animals. They are affected by stress, pressure, proximity of chemical substances, temperature, all of which must be carefully considered in the design of filters and resonators to ensure that when exposed to the real world, they remain within specification. But the other side of that coin is that these elements can act as sensors, with applications for a wide range of sensing needs. They can be used for stress and torque measurements. One notable developing interest is the possibility of their use as selective bio-sensors. There are intriguing possibilities of incorporating them in car tyres to provide a permanent record of pressure – a potentially very large market.

For all those engaged in advancing the science, the art and the applications of surface acoustic waves, David Morgan's book will provide an invaluable and coherent guide to what is known, and to how current knowledge can be applied toward the further advances in this fascinating area of applied science.

Eric Ash
London, 2007

FOREWORD TO PREVIOUS EDITION (1991)

It is a source of amazement that, given a mask with which to perform the photo-lithography, all you need is a single crystal with a thin metal film deposited on its surface in order to make a surface acoustic wave device with outstanding performance. It can function as a band-pass filter, passing most of the signal in the pass band but providing 60 dB or more rejection out-of-band. It can be made into a matched filter capable of extracting a signal for noise when the strength of the signal is four orders of magnitude below the prevailing noise. How is it that such a seemingly simple device can achieve so much? An important part of the answer lies in the fact that it is dependent for its performance on *intrinsic* properties of the single crystal-surface acoustic wave velocity and piezo-electric constant. But a vital ingredient is the design of the mask for here is where the subtlety and sophistication enters. It is the design of the mask which makes the difference between a SAW device having either excellent or mediocre performance; the design of the lithography mask can be equated to the design of the device.

A major strength of Dr. Morgan's book is the logical and coherent development of those topics which underpin or are relevant to the design of SAW devices. This book provides a clear and careful treatment of topics ranging from the basic theory of bulk and surface acoustic waves, the electrical excitation and detection of surface waves, material properties, through to the design, performance and application of a range of key SAW devices. Though originally planned as a reference book for the SAW device design engineer, the development of the material is well suited to undergraduate and MSc courses. Students will appreciate not only the clear exposition of the central subject matter but also the linkage with such topics as the application of Fourier Transform techniques and signal processing as applied to radar, and to video and audio telecommunication systems.

Dr. Morgan's book, making its first appearance exactly one century after Rayleigh's classic paper recording his discovery of elastic surface waves, is now five years old. It says much for the original choice and treatment of the subject that virtually all is as relevant today as the day it was written.

E.G.S. PAIGE
November 1990

1

BASIC SURVEY

Acoustic devices have been used in electronics for almost a century, as described in the historical introduction of Royer and Dieulesaint's book [1]. In 1915 the first transducers were developed for sonar in submarines, making use of the piezoelectric effect in quartz to generate acoustic waves in the sea. A piezoelectric plate can serve as an electrically coupled acoustic resonator, and the use of this for stable oscillators began development around 1920. Today, the quartz resonator is of familiar and widespread use for regulating oscillator frequencies. Among the attractions of acoustic waves are the low velocities (giving a compact device for a given frequency) and, in suitable materials, low losses leading to good resonator Q-values.

In the 1960s it was first suggested that *surface* acoustic waves (SAWs) might also be useful. This type of wave motion, guided along the surface of a solid material, introduces the possibility of accessing the wave within its propagation path, and it enormously increases the potential versatility of the devices. A key requirement is some means for generating and detecting the waves, and many methods were known at the time though they were not suitable for electronics applications because of clumsiness or inefficiency. The key starting point was the advent of the interdigital transducer (IDT) in 1965. With this component, the surface-wave device becomes a suitably shaped metallic thin film deposited on the surface of a piezoelectric crystal such as quartz or lithium niobate. This development immediately changed the practicality of the subject, because such devices could be made easily and cheaply by lithographic techniques borrowed from semiconductor manufacture. In addition to the IDT, a range of other components were developed, particularly reflecting gratings which are widely used in surface-wave resonators.

In the subsequent years, a very wide variety of devices have been developed, including delay lines, bandpass filters, resonators, oscillators and matched filters, all having a variety of forms. The devices are found in many practical

1

systems, including professional radar and communications equipment and consumer applications such as domestic television and the high-volume area of mobile telephones. Worldwide device production is in the region of 3 billion (3×10^9) devices annually.

This chapter gives an overview of the subject, without going into too much depth technically. Most of the topics are described in more detail in later chapters. Note that we are concerned with devices for electronics, excluding other areas such as seismology and non-destructive testing in which surface waves are involved. Some earlier books on the subject are listed as references [1–11]. The history of the subject is described elsewhere [12].

1.1 ACOUSTIC WAVES IN SOLIDS

In a solid material, an acoustic (or 'elastic') wave involves changes to the relative positions of the atoms, described quantitatively in terms of *strains*. The position changes are also specified in terms of displacement from the equilibrium state, and the displacement is generally a function of position. In the presence of strains, the material generates internal forces which tend to return the material to its equilibrium, unstrained, state. The forces are expressed in terms of the *stresses*. Acoustic waves are propagating phenomena involving stresses and strains. For example, a localized displacement, such as that due to a hammer blow, will cause strains which in turn generate stresses. The stresses generate further strains at more distant points. In this way, the disturbance propagates away from the point of the blow in the form of acoustic waves, which can travel over the whole extent of the solid. Waves that propagate freely inside a solid material, with a character unaffected by boundaries, are called *bulk* waves.

We initially consider isotropic materials, so that the relevant material properties are independent of the material orientation. Anisotropic materials will be considered later. In an isotropic solid the simplest forms of acoustic wave are plane waves, that is, waves in which all the variables are constant over a plane called the wavefront. The propagation direction is normal to the wavefront. In contrast to free-space electromagnetic waves, acoustic waves are of two basic types. Firstly, longitudinal waves are such that the displacement is parallel to the propagation direction. Secondly, transverse (or 'shear') waves have displacements in any direction parallel to the wavefront, normal to the propagation direction. These two types of motion are illustrated in Fig. 1.1. They both can propagate independently of boundaries and are non-dispersive, so that the velocity is independent of frequency. Typical velocities are 6000 m/s for longitudinal waves and 3000 m/s for transverse waves.

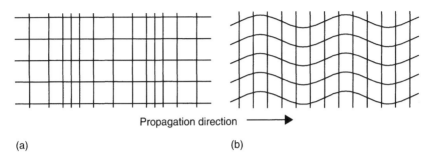

Propagation direction ⟶

(a)   (b)

Figure 1.1 Displacement of a rectangular grid during propagation of bulk acoustic waves. (a) Longitudinal wave and (b) transverse (shear) wave. In both cases, the wave propagates to the right.

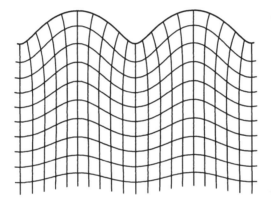

Figure 1.2 Displacement of a rectangular grid during propagation of a Rayleigh wave. The propagation direction is to the right.

The existence of a surface acoustic wave (SAW) was shown in 1885 by Lord Rayleigh [13]. The wave, often called a Rayleigh wave, propagates along the plane surface of an isotropic solid half-space, with amplitude decaying exponentially away from the surface. The solution needs to satisfy the condition that there is no traction on the surface, and to do this it has two components corresponding to bulk shear and longitudinal waves. The wave is non-dispersive, with a velocity a little less than that of the bulk shear wave. The motion of the material is in the *sagittal plane*, that is, the plane containing the surface normal and the propagation direction. Figure 1.2 shows the instantaneous distortion of the material, with the displacements much exaggerated. The motion of individual atoms is elliptical.

Rayleigh pointed out that the wave explained some seismological signals not previously understood. Other scientists developed this topic, particularly Love who found a purely shear wave, with displacements normal to the sagittal plane, existing in a half-space covered with a layer of slower material. Sezawa found that Rayleigh-type waves, with displacements in the sagittal plane, could also exist in a layered system and, like Love waves, they were dispersive and gave a series of modes with different velocities. Rayleigh, Love and Sezawa waves all occur in surface-wave devices, though Rayleigh-type waves are dominant. All of these waves are considered to be included in the topic of surface-wave devices.

Properties of acoustic waves are discussed in Chapter 2, which covers a variety of relevant waves in both isotropic and anisotropic materials. The latter case is much more complex but it is important because piezoelectricity, a crucial factor for surface-wave devices, can occur only in anisotropic materials.

Rayleigh waves have been used for non-destructive testing, since they are reflected by defects such as cracks in the material. Consequently, there are many methods for generating the waves [14]. Commonly, a plate transducer of a piezoelectric material such as PZT (lead zirconium titanate) is used to generate a bulk wave, with an arrangement to convert this to a surface wave.

The interest in *electronics* applications was initially spurred by requirements for pulse compression radar [15]. The radar transmitter is envisaged to generate a 'chirp' pulse, that is, a pulse whose frequency varies with time. For a point target, an attenuated and delayed copy of this pulse is received. In the radar receiver, a 'pulse compression' filter has a corresponding delay-frequency dependence, but reversed such that all the frequency components of the signal arrive at the output at the same time. This results in substantial shortening of the pulse and the signal-to-noise ratio is much increased. The effect of this is to increase the range capability of the radar.

Various technologies were considered for the pulse compression filter [16], particularly acoustic devices because they give velocities much less than those of electromagnetic waves (typically 3000 m/s instead of 3×10^8 m/s). The delays needed, typically 10 μs or more, would thus be obtained very compactly. The devices included 'wedge' delay lines using bulk acoustic waves. To obtain the required dispersion, arrays of sources were arranged with varied spacing, such that the waves took paths of different lengths for different frequencies. From this, it was a small step to envisage using surface waves for pulse compression. This principle was proposed independently in 1963 by Mortley and by Rowen [17].

In 1965, White and Voltmer [18] put forward the interdigital transducer (IDT) as a source or receiver of surface waves. The transducer consists of a sequence of metal electrodes, usually of aluminum, alternately connected to two bus bars. Figure 1.3 shows a basic surface-wave device using two transducers, one for

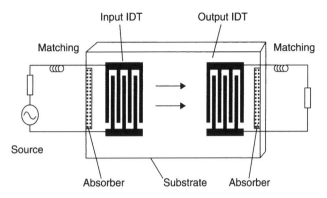

Figure 1.3 Basic surface-wave device.

generation and the other for reception of the waves. In addition to the novel transducer structure, this device is also novel in that it uses a substrate material that is *piezoelectric*, meaning that electric and mechanical fields are coupled within the material. Piezoelectricity occurs in many crystalline materials naturally, without any special intervention. Typical materials used in this context are quartz and lithium niobate. When a voltage is applied, the input transducer on the left in Fig. 1.3 generates a periodic electric field, and a corresponding elastic stress is set up because of the piezoelectric effect. Surface waves are generated strongly for frequencies such that the wavelength is similar to the transducer pitch. At the output, the receiving transducer acts in a reciprocal manner, converting incident waves to an output voltage. Thus, the device converts an electrical input signal into an electrical output signal. Typically, a transducer might have 20–1000 electrodes. The electrode width is a quarter of the center-frequency wavelength, and the minimum possible width, determined by the fabrication technology, limits the frequency obtainable. In commercial production, the minimum width is about 0.5 μm. For a typical velocity of 3000 m/s this gives a maximum operating frequency of 1.5 GHz, though special techniques can extend this to 5 GHz. The lowest frequencies used are in the region of 20 MHz, simply because at low frequencies the device size becomes inconvenient.

The device of Fig. 1.3 can be regarded as a delay line, giving a delay corresponding to the distance between transducer centers, typically about 3 μs for each cm of path length. Alternatively, we note that the transducers are effective mainly at a frequency where the transducer pitch corresponds to the surface-wave wavelength. Hence the response is strongest at this frequency and weaker at other frequencies, so the device can also be regarded as a bandpass filter. This is a prominent application for surface waves, though practical bandpass filters are usually much more sophisticated than this. Note that absorbing material

Figure 1.4 Surface-wave bandpass filter for a mobile phone basestation (courtesy of B. Abbott, Triquint).

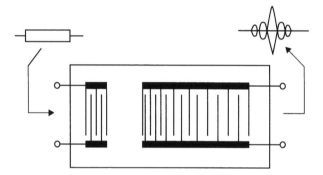

Figure 1.5 Interdigital pulse compression filter.

is applied at each end to eliminate unwanted reflections from the ends of the substrate. These absorbers are present in many surface-wave devices, but for simplicity they are not usually shown on the diagrams.

Figure 1.4 shows a bandpass filter for a mobile phone basestation, with a center frequency of 100 MHz. The electrodes are not visible individually but they can be seen as the darker central area. The device uses withdrawal weighting, explained later. This is an unusually large device, with length 76 mm.

The transducer is a very convenient surface-wave component. It exploits piezo-electricity, so that much of the work is done by the crystal substrate. It is also simple to fabricate using lithographic techniques, similar to those used for integrated circuits but simpler because only one layer is involved. A crucial advantage is that the transducer geometry can be varied in an almost arbitrary manner, giving enormous versatility. This is well illustrated by the huge range of surface-wave devices that have been developed subsequently.

An early surface-wave device was the pulse compression filter described by Tancrell *et al.* [19] in 1969, using the principle shown in Fig. 1.5. Here the right transducer has graded periodicity, and for a given frequency it couples to the waves mainly where the pitch corresponds to the surface-wave wavelength. Hence the delay of the output signal varies with frequency, as required for pulse compression radar. In the radar system the transmitted signal is swept in frequency and the filter compresses this to a short pulse, as indicated in the figure. Tancrell's device used a lithium niobate substrate and had a center frequency of 60 MHz, bandwidth of 20 MHz and a time dispersion of 1 μs. This type of pulse compression filter soon became entrenched in radar systems, making pulse compression a practical reality.

1.2 PROPAGATION EFFECTS AND MATERIALS

Following the demonstration of basic devices, a prime requirement was to search for suitable materials to be used as substrates. This will be considered in Chapter 4. A key requirement is piezoelectricity. This property can only be present if the material is *anisotropic*, that is, its properties vary with direction in relation to the internal structure. Nearly always, a single-crystal material is chosen because such materials can also give low losses. The requirements for surface-wave devices have led to a very substantial search for suitable materials, a complex topic because for each potential material it is necessary to assess the relevant properties for all crystal orientations. Important properties are the wave velocity, piezoelectric coupling, temperature effects, diffraction, attenuation and the level of unwanted bulk-wave generation. It was found that on some crystals the surface-wave propagation was almost ideal, even for frequencies up to 1 GHz. However, even small imperfections can be relevant when exacting requirements are to be met, and temperature effects nearly always need to be considered.

Since the material properties vary with direction, it is essential to quote the orientation when specifying a material. For example, a common material is $Y–Z$ lithium niobate (LiNbO$_3$), where the notation indicates that the surface is normal to the crystal Y-axis and the wave propagates in the crystal Z-direction. Another case is $128°Y–X$ lithium niobate. This is a rotated Y-cut. The surface normal makes an angle $128°$ with the crystal Y-axis, and the wave propagates in the crystal X-direction. Both of these cases have strong piezoelectric coupling, but the temperature stability is poor. The reverse is the case for $ST–X$ quartz (SiO$_2$). This is a $42.7°$ rotated Y-cut, with wave propagation along X. It has good temperature stability, as quantified by the temperature coefficient of delay (TCD). The delay is maximized at $21°$C, where the TCD is zero, and for practical purposes the delay is a quadratic function of temperature. The center frequency of a filter has the same temperature coefficient, but with a change of sign.

Many of the surface-wave properties of a material can be deduced by calculating the wave velocity, a complex calculation because it involves anisotropy and piezoelectricity. The wave velocities for a free surface and a metallized surface are denoted by v_f and v_m, respectively. In the latter case, the surface has an idealized metal coating which shorts out the parallel component of electric field at the surface, but is too thin to have any mechanical effect. The fractional difference between these velocities, $\Delta v/v$, characterizes the piezoelectric coupling to the wave. It is also common to define a coupling constant K^2 as twice this value, so we have

$$\Delta v/v \equiv (v_f - v_m)/v_f \equiv K^2/2. \tag{1.1}$$

The above lithium niobate orientations have strong $\Delta v/v$ values, and this is suited to wide-band filters. For ST–X quartz, $\Delta v/v$ is weak and the material is generally limited to narrow-band devices. On the other hand, such devices often need good temperature stability, which quartz provides.

Diffraction of surface waves is important when assessing a material. In some respects, this is similar to the familiar diffraction of light in air. Close to the source, there is a near-field region where the diffraction has relatively little effect. Farther away there is a far-field region, where the waves spread out radially. In most devices the receiving transducer is within the near field of the launching transducer, so the diffraction does not have much effect. However, diffraction often needs to be considered if a demanding requirement is to be met. Anisotropy introduces several novelties. In general beam steering can occur, that is, the energy flow direction is not perpendicular to the wavefronts. The wavefronts are parallel to the electrodes of the launching transducer. Beam steering is usually avoided by the choice of the material orientation. In some cases, the anisotropy is such that diffraction spreading is substantially reduced.

A low-diffraction case is Y–Z lithium niobate, already mentioned. This is excellent for high-performance filters, but it has an important limitation – when using a transducer designed to generate surface waves, there is also unwanted coupling to bulk waves. In a bandpass filter, this gives unwanted signals in the upper stop band. A common cure is to use a set of parallel metal strips, known as a *multistrip coupler*, to couple two parallel tracks. This will be seen later in Fig. 1.14. Surface waves are coupled preferentially, thus reducing the bulk-wave component in comparison. Alternatively, the $128°Y$–X orientation of lithium niobate gives lower bulk-wave generation but it is not a minimal-diffraction case. This avoids the need for a multistrip coupler and hence reduces the device size, though diffraction limits the performance somewhat.

Table 1.1 summarizes the properties of selected materials, and others are included in Chapter 4. The parameter ε_∞ is defined such that the capacitance

Table 1.1 Common surface-wave materials.

	Y–Z lithium niobate (LiNbO$_3$)	128° Y–X lithium niobate (LiNbO$_3$)	*ST*–*X* quartz (SiO$_2$)	36° Y–X lithium tantalate* (LiTaO$_3$)
v_f (m/s)	3488	3979	3159	4212
$\Delta v/v$	2.4%	2.7%	0.06%	2.4%
$\varepsilon_\infty/\varepsilon_0$	46	56	5.6	50
TCD (ppm/°C)	94	75	0	32
Advantage	Low diffraction, strong coupling	Low bulk waves, strong coupling	Small TCD	Strong coupling, moderate TCD
Disadvantage	Large TCD, strong bulk waves	Large TCD	Weak coupling	
Suitability	Wide-band filters, RACs, convolvers	Wide-band filters	Narrow-band filters, resonators, pulse compression	Low-loss filters, RF filters

RAC: reflective array compressor.
*Leaky wave, special considerations apply – see text.

of a unit-aperture single-electrode transducer, per period, is simply ε_∞. This applies when the electrode widths equal the interelectrode gap widths. The table includes data for 36° Y–X lithium tantalate. This material gives coupling similar to lithium niobate, but with a better temperature stability. It has been widely used in radio frequency (RF) filters for mobile phones, where narrow skirt width requirements have to be met over a wide temperature range. The wave involved is a 'leaky' wave, rather different from the usual Rayleigh wave, but further explanation will be deferred to later, in Chapter 11.

1.3 BASIC PROPERTIES OF INTERDIGITAL TRANSDUCERS

1.3.1 Transducer reflectivity and the triple-transit signal

The basic two-transducer surface-wave device has an important limitation arising from reflections of the waves by the transducers, as shown in Fig. 1.6. Unwanted reflected waves reach the output after three transits of the device, giving a 'triple-transit' signal. Of course, there are also higher-order reflections, called multiple-transit signals. In the device frequency response, these signals cause ripples in the amplitude and phase. In most devices the ripples must be well controlled, and this implies that the triple-transit signal must be minimized either by accepting a large insertion loss or by using a radically different device structure. The problem arises basically because a transducer has three ports. In a two-port device, such as a conventional filter, reflections are eliminated by matching the ports electrically, and this also minimizes the insertion loss.

Consider first a transducer with no internal reflections, a condition described by the term *non-reflective*. It can be shown (Appendix D) that such transducers are

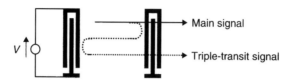

Figure 1.6 Origin of unwanted triple-transit signal.

always *bidirectional*. This means that they generate waves of equal amplitude in the two directions when a voltage is applied. The *efficiency* is maximized when the transducer is electrically matched to a source, and in this case the power loss is ideally zero. When a voltage is applied, half of the available input power emerges as surface waves in each direction, so the *conversion loss* is 3 dB. The same loss factor applies for a transducer receiving surface waves, so a two-transducer device will have an insertion loss of 6 dB. For the same matching, the reflection coefficient for a receiving transducer is −6 dB. This result, shown in Appendix D, follows from reciprocity and power conservation. The unwanted triple-transit signal is then 12 dB below the main signal, much too large for most applications.

The reflection coefficient of a non-reflective transducer is ideally zero if it is shorted, that is, when it is connected to an electrical load with zero impedance. However, reflections appear when the electrical load is finite, as it must be in practice because otherwise there could be no transfer of power to the load. The reflections arise by a simple physical process. A surface wave incident on the transducer causes a voltage to appear across the bus bars. This voltage causes excitation of secondary surface waves (in both directions), just as it would if there were no incident surface wave. One of these waves is of course in the direction opposite to the incident wave, and this is the reflected wave. The same phenomenon applies for both transducers. The triple-transit signal can be controlled by ensuring that the load impedance is small enough, and this implies a relatively large insertion loss, typically 20 dB or more. Although this loss is inconvenient, many surface-wave devices are designed to accept it because non-reflective transducers can give excellent performance.

In *reflective* transducers there is an additional internal reflection mechanism due to the electrodes. Each electrode reflects only weakly, but the reflections become significant if they are added in phase. In a 'single-electrode' transducer the center-frequency wavelength is $\lambda_0 = 2p$. The reflections add in phase at this frequency, as shown in Fig. 1.7a. A common method for eliminating this problem is to use a *double-electrode* transducer shown in Fig. 1.7b. Here the electrodes are connected to the two bus bars in pairs. The electrode pitch p is

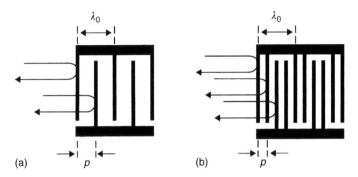

Figure 1.7 Internal reflection of surface waves. (a) Single-electrode transducer (shown with reduced-width electrodes) and (b) double-electrode transducer.

now $\lambda_0/4$. At the center frequency, reflections from adjacent electrodes have phases differing by $180°$, so that they cancel. In contrast, the original transducer of Fig. 1.7a is called a *single-electrode* transducer. Thus, double-electrode transducers are non-reflective, while single-electrode transducers are generally reflective. In a single-electrode transducer the electrode reflectivity not only causes the transducer reflections, but it also causes substantial distortion of the transducer frequency response. The distortion is a complex phenomenon which is inconvenient to deal with, and it limits the performance obtainable. Consequently, double-electrode transducers are often chosen, though the reduction of electrode pitch from $\lambda_0/2$ to $\lambda_0/4$ limits the operating frequency.

When electrically matched, a reflective transducer with *symmetrical* geometry gives a conversion loss of 3 dB and a reflection coefficient of −6 dB, as for a non-reflective transducer, so the triple-transit problem remains (Appendix D).

If a reflective transducer is connected to a source or load with finite impedance, we now have two reflection mechanisms: the load-dependent reflection and the electrode reflectivity. These two mechanisms interact in a complex manner. It is possible to design a non-symmetric transducer such that these mechanisms cancel, so that there is no overall reflection even for a finite electrical load. Such transducers are known as single-phase unidirectional transducers (SPUDTs), and they are considered further in Section 1.7.3.

1.3.2 Non-reflective transducers: delta-function model

A basic theory for non-reflective transducers is the *quasi-static* method given in Chapter 5, developed from some fundamentals in Chapter 3. This is a rather complicated theory applicable to a wide variety of transducers. However, for

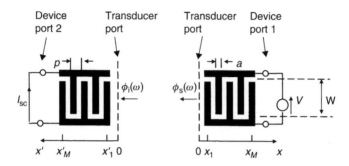

Figure 1.8 Surface-wave device using two uniform transducers.

limited types of transducer there are some simple results, and these are summarized here without giving the theoretical background. Uniform transducers are considered here, and they are assumed to be non-reflective, as in the case of double-electrode transducers. Propagation conditions are assumed to be ideal, ignoring propagation loss and diffraction. The results are closely related to the *delta-function model* of Tancrell and Holland [20], which was one of the first methods for analyzing transducers. Each transducer is assumed to have *regular* electrodes, that is, the electrodes all have the same width a and pitch p. The metallization ratio a/p is taken to be $1/2$, a common value.

Figure 1.8 shows a two-transducer device, with uniform single-electrode transducers. The transducer on the right has an applied voltage V. It has M electrodes, centered at $x = x_1, x_2, \ldots, x_M$. The x values are relative to the transducer port shown as the broken line. This is just outside the transducer, but its precise location is not important. We consider the surface wave generated in the $-x$-direction at frequency ω. Of course, there is also a wave generated to the right, but this is not considered here.

A simple approach is just to regard each electrode as a source of surface waves which travel through the transducer with phase velocity v and wavenumber $k = \omega/v$. The waves are assumed to be unaffected by any electrodes that they pass under. To a good approximation they are non-dispersive, so that v is independent of frequency. The amplitude of the wave due to electrode m has the form $\exp[jk(x - x_m)]$, where $j \equiv \sqrt{-1}$. For the right transducer in Fig. 1.8 we take the lower bus bar to be grounded and apply this formula to each live electrode, adding the waves generated. Live electrodes are identified by defining a polarity $\hat{P}_m$ for electrode m, such that $\hat{P}_m = 1$ for a live electrode (connected to the upper bus in the figure) and $\hat{P}_m = 0$ for a grounded electrode. Thus, the wave generated by electrode m becomes $\phi_{sm}(x, \omega) = VE(\omega)\hat{P}_m \exp[jk(x - x_m)]$, where the factor $E(\omega)$ allows for physical processes. The system is known to be linear, so the total wave amplitude is obtained by summing these contributions. We define $\phi_s(\omega)$ as the amplitude at $x = 0$, which is regarded as the transducer port. Thus,

$$\phi_s(\omega) = \sum_{m=1}^{M} \phi_{sm}(0, \omega) = VE(\omega) \sum_{m=1}^{M} \hat{P}_m \exp(-jkx_m). \tag{1.2}$$

It is assumed here that all variables are proportional to $\exp(j\omega t)$, with this factor omitted. As is common practice, the actual amplitude is to be obtained by multiplying eq. (1.2) by $\exp(j\omega t)$ and then taking the real part. The amplitude $\phi_s(\omega)$ is interpreted as the potential accompanying the wave, at the surface of the substrate. $E(\omega)$ is an element factor, representing the response of each individual electrode. This factor varies slowly with ω, and often it may be taken to be approximately constant. If $E(\omega)$ is taken as a constant, eq. (1.2) shows that each electrode acts as a localized source of waves situated at the electrode center. Mathematically, this is a delta-function source in the x-domain, hence the term 'delta-function model'.

Equation (1.2) applies for any set of electrode polarities $\hat{P}_m$. For a single-electrode transducer the polarities have the sequence $\hat{P}_m = 0, 1, 0, 1, \ldots$, while a double-electrode transducer has $\hat{P}_m = 0, 0, 1, 1, 0, 0, 1, 1, \ldots$ Both of these cases are covered by eq. (1.2). For a single-electrode transducer, the center frequency ω_{s0} occurs when the electrode pitch p equals $\lambda/2$, giving $\omega_{s0} = \pi v/p$. At this frequency it is found that $E(\omega_{s0}) = 1.694 \, j\Delta v/v$, and we have $E(\omega) \approx E(\omega_{s0})$ for frequencies near ω_{s0}. For a double-electrode transducer the center frequency ω_{d0} occurs when $p = \lambda/4$, giving $\omega_{d0} = \pi v/(2p)$. Here, we have $E(\omega_{d0}) = 1.247 \, j\Delta v/v$. In Chapter 5 it will be shown that $E(\omega) = j(\Delta v/v)\overline{\rho}_f(k)/\varepsilon_\infty$, with $\overline{\rho}_f(k)$ defined in eq. (5.59).

The transducer response can be written in terms of an array factor $A(\omega)$, which gives the frequency variation excluding the element factor. This is defined as

$$A(\omega) = \sum_{m=1}^{M} \hat{P}_m \exp(-jkx_m). \tag{1.3}$$

Equation (1.2) gives $\phi_s/V = E(\omega)A(\omega)$. The transducer response $H_t(\omega)$ is defined by the expression

$$H_t(\omega) = -jE(\omega)A(\omega)\sqrt{\omega W \varepsilon_\infty/(\Delta v/v)}. \tag{1.4}$$

Using this expression, the potential of the wave generated can be written as

$$\phi_s/V = jH_{tL}(\omega)\sqrt{(\Delta v/v)/(\omega W \varepsilon_\infty)} \tag{1.5}$$

where subscript L has been added to indicate a launching transducer.

A related approach can be used for a receiving transducer, such as that on the left in Fig. 1.8. Mathematically, the process is much the same – an incident wave traveling through the transducer causes electrical excitation at the electrodes one at a time. When a surface wave is incident on port 1 of a shorted transducer, with potential $\phi_i(\omega)$ at this port, the output current I_{sc} can be shown to be given by

$$I_{sc}/\phi_i = -jH_{tR}(\omega)\sqrt{\omega W \varepsilon_\infty/(\Delta v/v)}. \qquad (1.6)$$

Here, the response is again defined as in eq. (1.4), with subscript R added to indicate a receiving transducer.

With these equations, the response of the two-transducer device of Fig. 1.8 is easily found. When a voltage V is applied to the launching transducer, the wave amplitude at the transducer port is given by eq. (1.5). At the receiving transducer the amplitude is the same except for a phase change $-kd$ due to propagation over the distance d between the two-transducer ports. Hence $\phi_i = \phi_s \exp(-jkd)$. The short-circuit current produced by the receiver is given by eq. (1.6), so the ratio of output current to input voltage is

$$I_{sc}/V = H_{tL}(\omega)H_{tR}(\omega)\exp(-jkd). \qquad (1.7)$$

Thus, the device response is simply the product of the transducer responses, with a linear phase change due to the transducer separation. The triple-transit signal is absent from this equation because we have assumed the transducers to be non-reflective. This implies that they do not reflect when shorted or connected to a zero-impedance source.

The function I_{sc}/V is often denoted by Y_{21}, which is one component of the device admittance matrix Y_{ij}. Y_{21} is defined as the short-circuit current at the device port 2 (the left transducer in Fig. 1.8), when unit voltage is applied to port 1 (the right transducer). If we apply the voltage at the left and measure the current at the right, the result gives Y_{12}. From the above approach, it can be seen that $Y_{12} = Y_{21}$. This is an important *reciprocity relation*. It applies for most linear devices, and it applies for all the linear devices described in this book.

Frequency response

To examine the frequency response of a uniform transducer, consider the array factor of eq. (1.3) for a single-electrode transducer, such as those in Figs. 1.7 and 1.8. The electrode centers can be taken to be at $x_m = mp$, and we can take $\hat{P}_m = 0, 1, 0, 1, \ldots$ Equation (1.3) becomes a sum of N_p terms with spacing

$\Delta x = 2p$, where $N_p = M/2$ is the number of periods and M is assumed to be even. This gives

$$A(\omega) = \sum_{n=1}^{N_p} \exp(-2jkp) = \frac{\sin(N_p kp)}{\sin(kp)} \exp[-j(N_p + 1)kp] \qquad (1.8)$$

where the summation has been done by treating the series as a geometric progression. This equation gives a series of peaks at $kp = n\pi$, that is, at $p = n\lambda_0/2$. These correspond to the fundamental response of the transducer (where $n = 1$ and $\lambda_0 = 2p$) and the harmonics ($n > 1$). The array factor has the same magnitude at these peaks, but the transducer response gives different magnitudes because of the influence of the element factor. Electrode reflections have been ignored here. This is sometimes valid when the electrode reflectivity is not too large and the transducer is not too long.

For the fundamental response, kp is close to π. Define $\theta = kp - \pi$, so that $\sin(kp) = -\sin\theta \approx -\theta$. We can express θ as $\theta = \pi(\omega - \omega_0)/\omega_0$, where $\omega_0 = \pi v/p$ is the center frequency of the fundamental response. The magnitude of the array factor is

$$|A(\omega)| \approx N_p \left| \frac{\sin(N_p \theta)}{N_p \theta} \right|. \qquad (1.9)$$

This gives the form of the transducer response approximately, because the element factor $E(\omega)$ is a slowly varying function. The array factor is plotted in Fig. 1.9. The zeros nearest to the center frequency f_0 occur at $f = f_0 \pm f_0/N_p$. The points at $f - f_0 = \pm f_0/2N_p$ are very close to the 4 dB points of the function. If we define $T = N_p/f_0$ as the transducer length in time units, the bandwidth is conveniently written as $\Delta f \approx 1/T$. Uniform double-electrode transducers give a very similar response, with the bandwidth having the same relation to the transducer length.

Transducer admittance

The electrical admittance seen looking into a transducer is an important factor when electrical matching is considered. It is often dominated by a capacitance denoted by C_t. The admittance can be represented by adding further terms in parallel, associated with the acoustic activity. These are a conductance $G_a(\omega)$ and a susceptance $B_a(\omega)$. The total admittance is therefore

$$Y_t(\omega) = G_a(\omega) + jB_a(\omega) + j\omega C_t. \qquad (1.10)$$

The equivalent circuit thus has these three components in parallel, as in Fig. 1.10. For a uniform transducer, the capacitance is approximately proportional to the

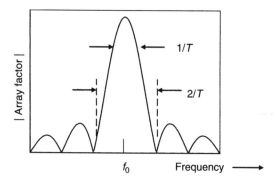

Figure 1.9 Array factor of a uniform transducer in the region of the fundamental response.

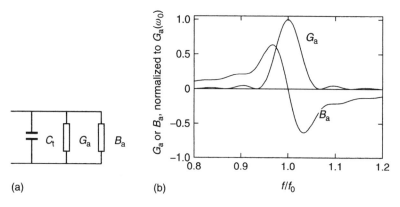

(a) (b)

Figure 1.10 (a) Equivalent circuit of transducer. (b) Acoustic conductance and susceptance for uniform transducer.

number of periods, N_p. Thus, if C_{t1} is the capacitance per period, we have $C_t = N_p C_{t1}$. It will be shown in Chapter 5, Section 5.5 that $C_{t1} = W\varepsilon_\infty$ for single-electrode transducers, and $C_{t1} = W\varepsilon_\infty \sqrt{2}$ for double-electrode transducers.

The conductance $G_a(\omega)$ is the real part of Y_t, and it can be deduced from power conservation. When a voltage V is applied, the power absorbed by the transducer is $P_a = G_a |V|^2/2$. The power of a wave with surface potential ϕ_s can be shown to be $P_s = (1/4)\omega W\varepsilon_\infty |\phi_s|^2/(\Delta v/v)$, and ϕ_s is given by eq. (1.5). Surface waves of equal amplitude are generated in both directions, so the total wave power is $2P_s$. Equating this to P_a, we find the simple relation

$$G_a(\omega) = |H_t(\omega)|^2. \tag{1.11}$$

This equation applies for any unapodized non-reflective transducer, with $H_t(\omega)$ given by eq. (1.4).

If the transducer is also uniform the array factor $A(\omega)$ is given by eq. (1.9), and this gives the form of the response approximately. Hence, for the fundamental response a uniform transducer has

$$G_a(\omega) \approx G_a(\omega_0)[(\sin X)/X]^2 \qquad (1.12)$$

where $X \equiv N_p\theta = N_p(\omega - \omega_0)/\omega_0$. This function is shown in Fig. 1.10b, normalized to the center-frequency value $G_a(\omega_0)$. As for the array factor, the nulls nearest to the center frequency occur at frequency $f_0 \pm f_0/N_p$. In Chapter 5, Section 5.5, it will be shown that $G_a(\omega_0) = \alpha\omega_0\varepsilon_\infty WN_p^2\Delta v/v$, with $\alpha = 2.87$ for single-electrode transducers and $\alpha = 3.11$ for double-electrode transducers. Figure 1.9 also shows the susceptance $B_a(\omega)$ which is related to $G_a(\omega)$ because of causality, as shown in Chapter 5, Section 5.2.2.

A two-transducer device is described electrically by its admittance matrix Y_{ij}. Here Y_{11} is defined as the admittance seen looking into port 1 of the device (here taken as the launching transducer) when port 2 is shorted. The receiving transducer does not reflect in this situation, because we have assumed the transducers to be non-reflective. It follows that Y_{11} is simply the admittance Y_t of the launching transducer, eq. (1.10). Similarly, Y_{22} is the admittance of the receiving transducer, and $Y_{12} = Y_{21}$ are equal to I_{sc}/V, given by eq. (1.7). We now have all four components of the admittance matrix. This is an important part of the analysis because it enables the device response to be calculated allowing for the electrical source and load impedances, as described in Appendix D. In particular, this calculation includes the triple-transit signal, which does not need to be calculated explicitly. For non-reflective transducers the triple-transit signal does not occur in the Y-matrix, but it arises in the device response when the loads are finite. The calculation also gives the insertion loss. The results allow for the circuit effect, which is a small distortion arising from the frequency dependence of the transducer admittances and the finite source and load impedances. The response is often expressed in terms of the scattering matrix S_{ij} ($i, j = 1, 2$), in which $S_{12} = S_{21}$ represents transmission from source to load, and vice versa, and S_{11} and S_{22} are electrical reflection coefficients. It is quite common to include inductors, as in Fig. 1.3, to tune out the transducer capacitances. This can improve the electrical matching to the transducers, thus reducing the insertion loss. However, care is needed when using inductors because this can lead to unacceptable triple-transit signals.

The above analysis refers to non-reflective transducers, such as the double-electrode type. Single-electrode transducers often have significant internal reflections, in which case a different approach is needed. Chapter 8 describes two methods: the reflective array method (RAM) and the coupling-of-modes (COM) method. The COM has long been applied to periodic systems occurring in microwave and optical devices, but for surface waves there is the particular complication of distributed transduction. A notable result is that the reflections distort the conductance G_a of a transducer such that it becomes asymmetric with its peak at a reduced frequency, instead of the symmetric curve of Fig. 1.10.

1.4 APODIZATION AND TRANSVERSAL FILTERING

As seen in Section 1.1 the basic interdigital transducer can be modified by grading the electrode periodicity. Another common transducer variant is *apodization*. This means that the active overlaps of the electrodes are varied along the length of the transducer, as illustrated in Fig. 1.11. By choosing the overlaps suitably the device can be designed to give a specified frequency response, as considered in Chapter 6. This was first demonstrated by Hartemann and Dieulesaint in 1969 [21], and they suggested its use as an intermediate-frequency (IF) filter for domestic television (TV) receivers. This developed into a great success, and all TV receivers now use a surface-wave IF filter. As explained above, the transducers usually have double electrodes in order to minimize electrode reflections. However, diagrams such as Fig. 1.11 are often drawn with single electrodes for clarity.

The principle of this device can be appreciated from its impulse response. This is defined as the output of the device when a short impulse is applied to the input. For present purposes, assume that the left transducer in Fig. 1.11 is relatively short. If a short pulse is applied to this transducer, a corresponding surface-wave pulse is generated, propagating along the surface. As this pulse passes through the right transducer it generates an electrical signal which appears at the bus bars. At any time, the strength of this signal is proportional to the overlap of the transducer electrodes at the location of the pulse. Hence the impulse

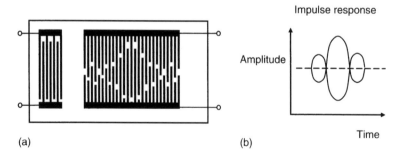

(a)   (b)

Figure 1.11 (a) Basic bandpass filter using one apodized transducer and one unapodized transducer. (b) Corresponding impulse response.

response, a function of time, corresponds to the overlap function of the right transducer, a function of position. In principle, *any* impulse response function can be obtained, simply by using the corresponding apodization. Moreover, the required impulse response is easily calculated from the required frequency response, because these two responses are related by the Fourier transform. This is a very powerful concept which illustrates the great versatility of surface-wave devices.

As an example, a common requirement is for a filter that passes signals over a frequency band of width B, say, and rejects signals outside this band. The frequency response has a rectangular shape. The corresponding impulse response is a modulated carrier, with envelope proportional to $[\sin(\pi Bt)]/(\pi Bt)$. Hence, an apodization of this form gives a rectangular frequency response. Another example is a uniform transducer, such as one of the transducers in Fig. 1.7 or 1.8. This has a flat impulse response, a rectangle of length T, say. Its frequency response has the form $[\sin(\pi \Delta f T)]/(\pi \Delta f T)$, where $\Delta f = f - f_0$ is the frequency deviation from the center frequency f_0. These examples are illustrated in Fig. 1.12. They are strongly related because the Fourier transformation from the time to frequency domains is almost the same as from frequency to time domains. In practice, the filter response will be affected by the length of the short transducer, which was ignored above. Also a $(\sin x)/x$ form for the impulse response cannot be produced in practice because it has infinite length, so it has to be approximated by a finite-length function.

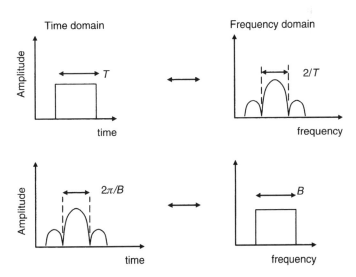

Figure 1.12 Relation between time-domain and frequency-domain functions.

The response of an apodized transducer is obtained by a simple modification of the earlier theory. In the array factor of eq. (1.3), the electrode polarity $\hat{P}_m$ is replaced by u_m/W, where u_m is the distance that electrode m extends from the upper bus bar. The array factor becomes

$$A(\omega) = \sum_{m=1}^{M} (u_m/W) \exp(-jkx_m). \qquad (1.13)$$

To justify this, note that when a voltage is applied electrode m generates a surface-wave beam of width u_m. The short-circuit current generated by the uniform transducer is proportional to this beam width. Alternatively, we can regard the electrode overlaps as the surface-wave sources; it turns out that this makes no difference to the results. When $u_m = 0$ or W we have $u_m/W = 0$ or 1, so eq. (1.13) agrees with eq. (1.3) if the transducer is unapodized. Equation (1.13) is very significant because it corresponds to the response of a *transversal filter*. This is a conceptual device in which a signal is delayed by a series of times $t_m = x_m/v$, and each delayed version is multiplied by a constant coefficient u_m/W before being added to give the total response. The theory of transversal filters shows that, with some provisos, an *arbitrary* frequency response can be synthesized by a suitable choice of the coefficients u_m.

Design details are described in Chapter 6. An important limitation is that the impulse response needs to have finite length, and this leads to a need for optimized design methods. Fortunately, there is a very flexible and optimal method called the Remez algorithm, originally developed for design of digital finite-impulse-response (FIR) filters.

The frequency response of a practical filter for domestic television receivers is shown in Fig. 1.13. The shaded lines indicate the required response, with tolerances, showing that a very complex amplitude variation is needed. The lower part of the figure is the group delay. This needs to vary with frequency, that is, the filter is dispersive. This is needed because the longitudinally coupled (L-C) filter, which the surface-wave filter replaces, is dispersive. The television transmitter introduces distortion to compensate for this, so it needs to be present in the TV receiver. The response shown only just meets the specification, but this should not be regarded as a near failure – the device was designed to minimize area, and hence cost, while still meeting the requirements.

A photograph of an operating TV filter is shown in Fig. 1.14. Here the device is imaged by an electron microscope, which senses the electric fields of the surface wave directly, including those associated with the surface waves.

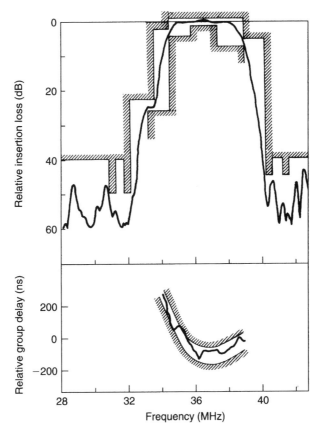

Figure 1.13 Frequency response of a surface-wave bandpass filter for domestic TV receivers (courtesy of Plessey Research).

The photograph has been strobed in phase with the wave so as to display the instantaneous wave amplitude. In this device, the wave is launched by an apodized transducer at the left and then passed to a multistrip coupler, which was mentioned in Section 1.2. The coupler displaces much of the energy sideways so that it impinges on a uniform transducer in a parallel track, discriminating against signals due to bulk waves. However, many filters use a substrate such as $128°\,Y$–X lithium niobate, on which a multistrip coupler is not necessary.

Figure 1.14 is an example of surface-wave *probing*, that is, direct observation of the surface-wave distribution in an operating device. More conveniently, there are several optical techniques for probing and measuring the surface waves, as described in Chapter 6 of the earlier book [4] and in more recent literature [22].

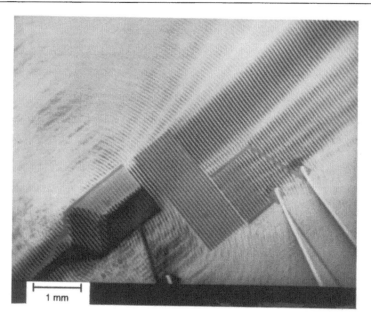

Figure 1.14 Electron micrograph of an operating TV filter (courtesy of C. Ruppel, EPCOS AG).

As an alternative to apodization, a transducer can be modified by withdrawal weighting (Chapter 6, Section 6.3.3). In this technique the transducer is unapodized but weighting is obtained by removing sources selectively, such that the density of the remaining sources mimics the amplitude distribution required. For narrow-band filters, this can give a frequency response similar to that of an apodized transducer, with the advantage of being much less subject to diffraction errors.

1.5   CORRELATION AND SIGNAL PROCESSING

We have already introduced pulse compression, for radar, in Section 1.1. The interdigital pulse compression filter of Fig. 1.5 is a common way of realizing this function. This is an example of a *matched filter*, or *correlator*, used to process a signal accompanied by noise in such a way as to maximize the signal-to-noise ratio. Chapter 7 describes this application, and also the application of matched filtering in spread-spectrum communication systems.

For correlation, the impulse response of the filter needs to be the time-reverse of the ideal input signal. For radar, both of these are normally chirp waveforms. The filter increases the signal-to-noise power ratio (SNR) by approximately the time-bandwidth product, TB, which is typically 50–500. The increase of

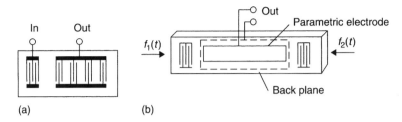

Figure 1.15 Correlation devices. (a) Tapped delay and (b) non-linear convolver.

range capability is equivalent to increasing the peak power of a conventional radar by a factor TB, so the benefit is substantial. An alternative device called a reflective array compressor (RAC) uses reflections from groove arrays. This enables larger TB products to be obtained, up to 16 000 experimentally. Metal strips can also be used as reflectors, in place of grooves, and they are more practical owing to ease of fabrication.

Correlation is also widely used in 'spread-spectrum' communication systems, which use a bandwidth much larger than that of the data. Typically, each data bit becomes a phase shift keyed (PSK) waveform, consisting of a carrier modulated with phase changes of 0° or 180° according to a binary code. The matched filter for this waveform can take the form of a surface-wave tapped delay line, shown in Fig. 1.15a. Each tap is a short transducer, and the taps are connected to the output with a polarity sequence corresponding to the code. When a short pulse is applied to the input transducer at the left, a packet of surface waves travels along the device, exciting the taps in sequence and producing a PSK waveform. It is feasible to connect the taps via electronic switches, so that the coding can be changed electronically. Spread-spectrum systems employ a separate code for each user, with advantages of security and discrimination against unwanted signals due to reflection from buildings, for example.

Another correlation device is the non-linear *convolver* sketched in Fig. 1.15b. In this device, input signals are applied to transducers at each end of a $Y-Z$ lithium niobate crystal, producing surface waves traveling toward each other. A non-linearity in the material produces a field proportional to the product of the two signals, and this is sensed by a metal 'parametric electrode' between the transducers. Mathematically, the output is the convolution of the two input signals. This implies that the device acts like a *linear* filter such as a bandpass filter or chirp filter. However, the 'impulse response' is effectively one of the input waveforms, and this is almost arbitrary. Thus, the convolver is extremely flexible. The basic convolver has very weak non-linearity, but greater efficiency can be obtained by confining the waves using waveguides (Chapter 7).

Semiconductor devices

The search for better efficiency has also prompted investigation of semiconductor interactions. A variety of experimental semiconductor devices are described in Kino's substantial review [23]. A silicon sheet can be held close to the lithium niobate surface, so that non-linearity occurs in the silicon. Alternatively, a zinc oxide film can be deposited on a (non-piezoelectric) silicon substrate. This enables interdigital transducers to be used, and the silicon gives much stronger non-linear effects than lithium niobate. In addition, it is possible to store signals in diodes fabricated on the silicon surface. Thus, a coded waveform can be stored and then read out later, or used as an 'impulse response' for later correlation of a corresponding signal waveform.

Zinc oxide films on silicon can also be used for correlation of PSK waveforms, using the tapped delay line of Fig. 1.15a. This brings in the possibility of constructing circuitry in the silicon, leading to an integrated device combining the surface-wave filter with the circuitry. In particular, voltage-controlled switches have been incorporated, giving independent control of the phases of the contributions from individual taps. In this way, the PSK filter becomes programmable.

Piezoelectric films are also of interest for generating surface waves on other non-piezoelectric substrates. In particular, high-velocity substrates enable high frequencies to be used because an increase of velocity leads to a higher frequency for a given electrode width. For example, piezoelectric zinc oxide or aluminum nitride films can be fabricated on non-piezoelectric diamond or sapphire substrates (Chapter 4, Section 4.4.3).

1.6 WIRELESS INTERROGATION: SENSORS AND TAGS

Surface-wave devices are sensitive to temperature changes, a factor that usually needs to be considered carefully in order to ensure sufficient temperature stability. Alternatively, it can be exploited in order to measure temperature. Many other quantities can also be measured, and related devices can be used as identification tags. Many of these devices are of the basic one-port form shown in Fig. 1.16 [24]. An interdigital transducer is connected directly to an antenna, generating surface waves when an radio frequency (RF) pulse signal is received. A series of reflectors, typically short gratings or transducers, is provided. Reflected waves are received by the transducer, so that the antenna radiates a return signal with characteristics determined by the reflectors. Typically, the 2.4 GHz ISM band is used.

The return signal generated by this device is dependent on environmental factors such as temperature or stress, so these can be measured. It is also characteristic

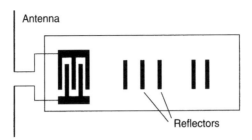

Figure 1.16 Reflective delay line for wireless tag or sensor.

of the device design, and can therefore be used for device identification. Surface-wave devices have the advantage of passive operation, so that a power supply is unnecessary. Also, wireless operation is possible because the device can give relatively long delays, so that the return signal is not swamped by the interrogation signal.

Identity tags can be made by using a sequence of reflectors coded in some way [25]. For example, they might be in a regular sequence except that selected reflectors are omitted. Such devices have potentially wide-ranging applications, including for example identification of cars and trains [24].

Temperature can be measured by comparing the phases of waves from individual reflectors, and if several reflectors are used the range can be increased without ambiguity. A resolution of 0.1°C is feasible. Strain can also be measured [26], and with a device mounted on a sealed cavity this can also indicate pressure. Other devices measure traces of chemicals in air or liquids. This is done by coating the device with a film which absorbs the target chemical, causing a change of density and hence surface-wave velocity [27]. As an alternative to the delay line, it is possible to use one-port resonators such as those described in the Section 1.7 [26].

1.7 RESONATORS AND LOW-LOSS FILTERS

Reflecting gratings can be used as efficient reflectors of surface waves, and two gratings can form a cavity resonator with application to high-stability oscillators. Resonators are also used in several types of low-loss bandpass filter, as described in Chapters 10 and 11. In this chapter, gratings and resonators are covered in Section 1.7.1, followed by RF resonator filters in Section 1.7.2. Section 1.7.3 describes devices used for low-loss IF applications.

1.7.1 Gratings and resonators

Surface waves can be reflected strongly by a reflecting *grating*, consisting of regular array of strips. The strips are often shorted metal electrodes, but they could alternatively be grooves, as in the RAC mentioned earlier. At the Bragg frequency, such that the periodicity equals half the wavelength, reflections from individual strips have the same phase so they add coherently, as in a single-electrode transducer. Strong reflections are obtained when $N|r_s| > 1$, where N is the number of strips and r_s is the reflection coefficient of one strip. Typically, $|r_s|$ might be 2% and N might be 200 or more. The strong reflection coefficient is obtained at a cost of restricted bandwidth – the reflection is strong only over a fractional bandwidth $\Delta f/f_0 \approx 2|r_s|/\pi$. Starting in the 1970s, gratings have received much attention because of their application to resonators, oscillators and filters. They are discussed in Chapters 8 and 11. Chapter 8 gives the analysis using the RAM and COM methods, which can also be applied to reflective arrays used in optical or microwave devices.

Figure 1.17a shows a resonator structure, having a transducer in the cavity formed by two reflecting gratings. This behaves approximately as the device of Fig. 1.17b, where a transducer is located within a Fabry–Perot resonator formed by two idealized mirrors representing the gratings. The mirror separation L is larger than the grating separation, for reasons discussed in Chapter 11. A Fabry–Perot resonator gives resonances when $2L$ is a multiple of the wavelength. In a surface-wave device the length L may be many wavelengths, so one might expect to see many resonances. However, resonances are only produced in the band where the gratings reflect strongly, and the device can be designed so that only one resonance is produced. The substrate is nearly always ST–X quartz, because high-Q devices usually require good temperature stability. For the gratings, shorted metal electrodes can be used, but grooved gratings were

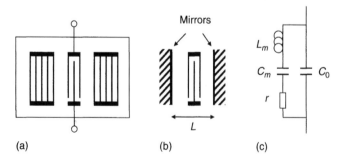

Figure 1.17 (a) One-port resonator. (b) Approximate Fabry–Perot equivalent. (c) Butterworth–van Dyke equivalent circuit.

found to give better Q-values. Experimentally, Q-values can be 10^5 at 100 MHz, or 10^4 at 1000 MHz.

Figure 1.17c shows the well-known Butterworth–van Dyke equivalent circuit. This consists of a series resonant circuit in parallel with a capacitance C_0, which is approximately the transducer capacitance C_t. The inductance L_m and capacitance C_m are 'motional' components associated with the surface-wave resonance. The circuit is approximately valid for many piezoelectric resonators, provided they are designed to have one prominent resonance free from unwanted spurious signals. The admittance is large at the resonant frequency $\omega_r = 1/\sqrt{(L_m C_m)}$. Owing to the static capacitance C_0, there is also an antiresonant frequency ω_a where the admittance is small (zero when $r \to 0$). The Q of the resonator is $1/(\omega_r C_m r)$, where the resistance r accounts for losses.

The resonator can be used as a controlling element for a high-stability oscillator. For this purpose, a 'two-port' resonator can be used, in which there are two transducers between the gratings instead of one. The oscillator is made by connecting one transducer to the other via an amplifying feedback circuit, with the amplifier gain exceeding the loss of the resonator (Chapter 11, Section 11.2). In comparison with oscillators using bulk-wave resonators, the surface-wave device can operated at higher frequencies, up to 500 MHz, but its stability is not so high.

1.7.2 Low-loss filters for RF

As described in Section 1.3.1 above, the basic surface-wave devices are limited by the unwanted triple-transit signal, and consequently the insertion loss is generally quite high. In the 1980s a substantial demand began for filters with low losses for applications in wireless communications, particularly for mobile telephones. Responding to this, a variety of novel surface-wave devices were developed. Several devices use resonator structures, described in Chapter 11.

Bandpass filters based on various resonator technologies have been established for many decades. Conventional L-C bandpass filters are basically combinations of L-C resonators, and the design methods can be adapted to other technologies such as resonant waveguides or dielectric resonators [28]. For piezoelectric resonators the same methods can be applied, but the bandwidth is limited by the static capacitance C_0 (Fig. 1.17c). Because of this, the design principles used in surface-wave filters are rather different from those of other technologies.

The *impedance element filter* (IEF) is a type of ladder filter illustrated in Fig. 1.18a. It has a sequence of resonators connected purely electrically, with

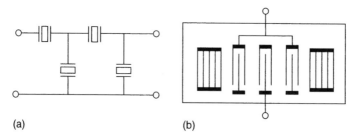

Figure 1.18 Low-loss bandpass filters using resonators. (a) Impedance element filter (IEF). Each resonator is a two-terminal device such as that in Fig. 1.17a. (b) Longitudinally-coupled resonator (LCR) filter.

no acoustic coupling, though they are fabricated on the same substrate. Each resonator is a two-terminal device consisting of a single-electrode transducer with a grating on each side, much like Fig. 1.17a. With a suitable design this can produce a clean acoustic resonance, even though the single-electrode transducer reflects the waves strongly. The series resonators are similar to each other, as are the parallel resonators. The peak admittance (resonance) of the series resonators is arranged to be at the same frequency as the peak impedance (antiresonance) of the parallel resonators. At this frequency, a signal is passed along the filter with little attenuation. At frequencies where there is little acoustic activity the device behaves as a network of capacitors, giving some attenuation, so that there is a stop band. The attenuation here depends on the ratio of capacitances, and it may be increased by increasing the number of resonators.

Such filters are widely used in mobile phone applications, where there is need for strong piezoelectric coupling and moderate temperature stability. To meet these needs, the filters normally use a *leaky wave*. The substrate is a rotated Y-cut of lithium tantalate, with $36° Y–X$ or $42° Y–X$ orientation. Details are given in Chapter 11, Section 11.4. The leaky wave is basically a shear-horizontal (SH) wave, with acoustic displacements normal to the sagittal plane. This choice gives strong piezoelectric coupling with moderate temperature stability. The leaky wave needs the presence of a continuous set of electrodes (gratings or transducers) because these trap the wave at the surface. These devices have given excellent performance, and they are manufactured in enormous quantities. Filters at 950 MHz give insertion losses of 2 dB or less and they handle input powers up to about 2 W, both features important for the phone applications. Filters centered around 2 GHz are also common.

Another resonator device is the *longitudinally-coupled resonator* (LCR) filter, shown in Fig. 1.18b. This technology, also known as the DMS (double-mode SAW) filter, is an alternative method for realizing low-loss filters at RF frequencies. It uses resonances in a different way. The device consists of three transducers in the space between two reflecting gratings. The gratings form a cavity giving a sequence of resonant modes, and the transducers are arranged symmetrically so that they can only respond to symmetric modes. The device is designed such that only two symmetric modes are within the reflection band of the gratings, so that it behaves as a two-pole filter. Some complications arise because single-electrode transducers are used, giving internal reflections, but these do not affect the basic principles. As for the IEF, a leaky wave in lithium tantalate is normally used. The LCR filter tends to give a rather poor roll-off in the high-frequency skirt. In comparison with the IEF described above, the LCR filter can give better stop band rejection though the power handling capability is less.

Another technology for RF filtering is known as the film bulk acoustic wave resonator (FBAR) [29]. Conceptually, this consists of a series of bulk-wave resonators connected in a ladder circuit like that in Fig. 1.18a. The FBAR filter is made compactly by fabricating the resonators on a common silicon substrate. The piezoelectric cavity is usually an aluminum nitride film. FBAR filters at 2 GHz have shown frequency responses quite similar to surface-wave IEFs. They are applicable at higher frequencies than IEFs but, for a given frequency, they tend to be physically larger.

1.7.3 Low-loss filters for IF

In addition to the RF devices described above, there have also been special techniques for low-loss IF filtering. The main examples are the transversely coupled resonator (TCR) filter and a variety of unidirectional transducers known as SPUDTs. These devices normally use quartz substrates, chosen for their temperature stability.

The TCR filter, shown in Fig. 1.19a, is essentially two identical resonators situated close together. Each resonator has a central transducer with a reflecting grating on each side, as in the basic resonator of Fig. 1.17a. For the TCR filter the transducer is a single-electrode type. The key ingredient here is *wave-guiding*, described in Chapter 10. The resonators have narrow apertures, a few wavelengths, and they are located close together, with the result that significant acoustic coupling occurs between them. Through this mechanism the device input is coupled to the output. The overall structure has two waveguide modes, symmetric and antisymmetric, which give the device a two-pole frequency response. This device can give a remarkably small fractional bandwidth, in the region of 0.1%. It is also very small physically, an important factor for mobile phone handsets.

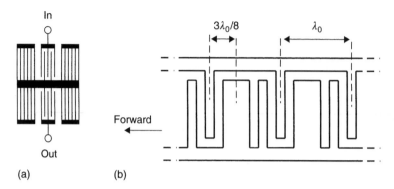

Figure 1.19 (a) Transversely-coupled resonator (TCR) filter. (b) Distributed acoustic reflection transducer (DART).

The TCR filter and the RF resonator filters in Section 1.7.2 above avoid the triple-transit problem by using multiple transits. An alternative approach is to use special transducers in which the reflectivity is canceled. As considered in Section 1.3.1, transducer reflectivity arises from two effects: internal reflections and load-dependent reflectivity. Using special transducer structures, it is possible to arrange for these effects to cancel. It is shown in Chapter 9 that this requires the transducer to have internal reflections and to lack symmetry. The transducers used for this purpose are called *single-phase unidirectional transducers*, or SPUDTs.

A common type of SPUDT is the distributed acoustic reflection transducer, or DART, shown in Fig. 1.19b. Here each period has length equal to λ_0, the center-frequency wavelength, and it contains two narrow electrodes of width $\lambda_0/8$ and one wide electrode of width $3\lambda_0/8$. For a shorted DART, reflections arise mainly from the wide electrodes. When a voltage is applied, the upper narrow electrode acts as a source of waves, at a distance $3\lambda_0/8$ from the center of the wide electrode. This distance is such that the reflector reinforces waves in one direction and reduces waves in the other direction. The effect is reinforced by additional periods of the transducer, so that the directivity increases as the number of periods is increased.

For any directional transducer, it can be shown that the electrical loading can be determined such that there are no acoustic reflections. Thus, the internal reflectivity is canceled by the load-dependent reflectivity. This property can be used to eliminate the unwanted triple-transit signal in a two-transducer device. Note, however, that the signal is canceled only for one specific electrical load, and only at one frequency. Thus, the triple-transit signal is still an important issue.

Bandpass filters consisting of two DARTs have been widely used at IF in mobile phone systems. These devices can employ weighting without apodization. The distribution of transduction can be modified by selectively omitting some of the upper narrow electrodes. For the reflections, a wide reflector electrode can be replaced by two narrow electrodes in order to eliminate its reflectivity. The transducer can thus have cells with differing transduction and reflection parameters. These can be thought of as derived from smooth continuous functions representing the transduction and reflection, with the cell designs produced by a sampling process. With these modifications the DART behavior becomes very complex, but it can be modeled effectively using COM or RAM analysis.

As mentioned earlier (Section 1.4), filters using non-reflective transducers can be designed using established methods developed for digital transversal filtering. However, these methods are not applicable to SPUDT filters, which have a much more complex behavior involving internal reflections. Sophisticated non-linear numerical techniques are used. In a remarkable degree of sophistication, the triple-transit signal can be *included* in the design process. In fact, many filters actually exploit this signal to sharpen the roll-off in the skirts of the response. Typically, these filters have center frequencies of 80–220 MHz and bandwidths around 1.2 MHz. The substrate is usually *ST–X* quartz because the frequency responses have quite narrow skirts, implying that good temperature stability is needed. Apart from the DART, there are many other types of SPUDT, using different electrode structures and different substrate materials.

In addition to DARTs and other types of SPUDT, some other low-loss techniques are described in Chapter 9, Section 9.6.

1.7.4   Performance of bandpass filters

Table 1.2 summarizes the capabilities of the main filter types. The transversal filter, considered in Section 1.4 above, can provide very exacting performance. For example, it can give in-band ripple of less than 1 dB and stop band rejection of 60 dB or more. On the other hand, the need to minimize distortion due to the triple-transit signal leads to relatively high insertion losses. Other devices give lower losses but generally with less stop-band rejection and more in-band amplitude variation. The transversal filter is widely used, particularly for television systems. Other types are widely used for mobile phones – the DART and TCR types are used for IF applications, while the IEF and LCR are used at RF.

From a practical point of view, the starting point for filter development is a device specification. Because of the versatility of surface-wave devices, it is best to

Table 1.2 Typical performance data for surface-wave bandpass filters.

Type	Transversal	DART filter	LCR filter	IEF	TCR filter
Substrate	Various	*ST–X* quartz	42°*Y–X* LiTaO$_3$	42°*Y–X* LiTaO$_3$	*ST–X* quartz
Center frequency (MHz)	50–500	80–500	500–2000	600–3000	50–200
Bandwidth	1–50%	1–3%	1–3%	1–4%	0.05–0.15%
Insertion loss	15–40 dB	6–10 dB	1–3 dB	1–3 dB	5–7 dB
Maximum rejection	60 dB	45 dB	55 dB	45 dB	50 dB
Minimum shape factor*	1.1	1.5	3.0	1.5	2.0

* Shape factor = ratio of 40 and 3 dB bandwidths.

think of this in terms of a template such as that in Fig. 1.13, so that the amplitude and tolerance of the response are specified at all frequencies over some range. The designer needs to tighten this to allow for the shift of the response over the specified temperature range. Bandpass filters are usually supplied in standard metal packages or, particularly at high frequencies, ceramic types. In the 1990s the need to miniaturize the devices for mobile phone handsets has focused attention on the packages [30] as well as the device designs. To reduce package sizes, a 'flip-chip' assembly is sometimes used, in which balls are attached to tracks on the surface-wave device. The latter is inverted and pressed on to conducting tracks on a board in the package, so that bonding wires are not necessary. Even smaller is a 'chip-size SAW package' in which a surface-wave device is made on a substrate in the usual way and the package is then assembled on the same substrate, with a clear area above the active part of the device. This package has a footprint no larger than the surface-wave device itself. As a result, IF filters for the IS-95 CDMA (code division multiple access) system can be only 7 mm long. Sizes of radio frequency filters are even more remarkable, with package devices as small as 2×2 mm being produced [30].

Bandpass filters are sometimes designed assuming that lumped-component matching circuits will be added, compensating for some distortion due to these circuits. At high frequencies a vital factor is the feedthrough, that is, direct electromagnetic coupling from the filter input to output. The device designer takes considerable care to minimize feedthrough within the surface-wave package but there can also be troublesome feedthrough outside the package, and the circuit design (feeds, matching components and so on) needs careful consideration.

Some devices are designed for balanced operation, that is, at each port there are two live terminals designed to receive voltages with equal magnitude but opposite sign. Balanced drives are quite common in communication systems, and many surface-wave devices can have modified designs in order to suit this. In some cases a device may be used in either a balanced or unbalanced mode. It is also possible to use a balanced drive at one port and an unbalanced drive at the other port. In this case, the surface-wave device acts as a balun transformer, converting from balanced to unbalanced drive, in addition to its frequency filtering function.

1.8 SUMMARY OF DEVICES AND APPLICATIONS

Surface-wave technology shows remarkable versatility owing to the ability to fabricate surface structures with almost arbitrary geometries. In addition, analytical techniques are capable of predicting device behavior with high accuracy. In consequence, a wide variety of devices have been developed, with a wide range of applications. Surveys of this topic were given by Williamson [31], Hartmann [32], Lam *et al.* [33] and Campbell [8, 34]. In 1977, Williamson quoted 44 types of surface-wave device, of which 10 were in common practical usage, and 45 government systems were quoted as using these devices. Later, Hartmann [32] quoted similar numbers, with more emphasis on commercial activity. Table 1.3 summarizes the main devices and their main applications.

Table 1.3 Main surface-wave devices and selected applications.

Device	Applications
Delay line	Oscillator, monopulse radar, electronic countermeasures (repeaters), sensors
Tapped delay line (fixed or programmable)	PSK correlation in communications
Tapped delay line, reflective	Sensors (temperature, pressure, chemicals), identity tags
Resonator (one or two port)	Oscillator, sensors
Chirp filter (interdigital or RAC)	Radar pulse compression and expansion, spectral analysis, Fourier analysis, variable delay
Bandpass filter, transversal	Domestic TV receiver, TV equipment, general communications and radar, PSK–MSK conversion, cordless phones, pagers, clock recovery (in fiber-optic repeaters), clutter filters for radar
Bandpass filter bank	Electronic surveillance
Low-loss bandpass filter[a] (IEF or LCR filter)	RF front end filtering[b]
Low-loss bandpass filter[a] (TCR filter)	Narrow-band IF filtering[b]
Low-loss bandpass filter (SPUDT)	IF filtering[b, c]

[a] Resonator filter.
[b] Including mobile phone handset.
[c] Including mobile phone basestation.

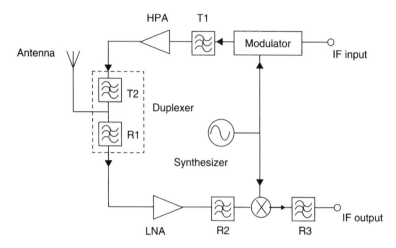

Figure 1.20 Simplified partial circuit of a digital mobile telephone handset for global system for mobiles (GSM). The bandpass filters shown are usually surface-wave devices. LNA: low noise amplifier, and HPA: high-power amplifier. Adapted from Fig. 10.9 of Campbell [8].

As already mentioned, surface-wave filters are widely used in mobile phone systems [8, 33, 35]. Figure 1.20 shows a typical circuit for a mobile phone handset, in which the bandpass filters shown are usually surface-wave devices. The transmitter chain includes two RF filters, T1 and T2. The receiver chain has RF filters R1 and R2, and an IF filter R3. At the antenna, filters T2 and R1 are connected directly, forming a duplexer. The two filters in a duplexer are often contained in one package. The system transmits and receives over separate bands, each with bandwidth typically 25 MHz. Some examples of the bands used are listed in Table 1.4. The transmitter filters T1 and T2 are required to have low loss in the transmitter band and good rejection in the receiver band, and corresponding remarks apply for the receiver filters R1 and R2. A particular requirement is that filter T2 needs to handle quite high power levels, up to 2 W. Filter R1 needs to have very low insertion loss, less than 2 dB, because its loss strongly affects the noise figure of the receiver. In fact, all the filters need to have low loss simply because high losses imply a need for more amplification elsewhere, increasing the battery drain.

The requirements of this and other wireless systems spurred the development of novel low-loss surface-wave filters. In the transmitter, filter T2 is usually the IEF type because this can handle high powers (2 W) with very low insertion losses (2 dB). Filter R1 is also usually an IEF. Filters T1 and R2 are often LCR filters, which can give low loss and good stop-band rejection, though with less power handling capability. Filter R3 is at the IF. For global system for mobiles

Table 1.4 Frequencies for mobile and cordless telephones [8, 33].

System	Transmitter band* (MHz)	Receiver band* (MHz)	RF bandwidth (MHz)	IF bandwidth (kHz)
Analog cellular phone (FDMA)				
AMPS (Americas, Australia)	824–849	869–894	25	30
ETACS (UK)	871–904	916–949	33	25
NMT 900 (Europe)	890–915	935–960	25	12.5
Digital cellular phone (mostly TDMA)				
GSM (Europe)	890–915	935–960	25	200
IS-95 (CDMA)	824–849	869–894	25	1250
PDC (Japan)	940–956	810–826	16	25
PCN/DCS-1800 (Europe)	1710–1785	1805–1880	75	200
Digital cordless phone (TDMA)				
DECT (Europe)	1880–1990	1880–1990	110	1728
PHS (Japan)	1895–1907	1895–1907	12	300

AMPS: advanced mobile phone system;
DECT: digital European cordless telephone;
ETACS: extended total access communication system;
GSM: global system for mobiles;
PCN: personal communication network;
PDC: personal digital cellular and
PHS: personal handyphone system;
*Of handset.

(GSM) the bandwidth needed is very small (Table 1.4), and here the TCR filter has been used. However, other systems such as CDMA have a wider bandwidth, and for this purpose filters using unidirectional transducers such as the DART have been developed.

Figure 1.20 is only one example of a handset circuit. In particular, some handsets are made to handle several different mobile systems, and for this purpose they can require different filters for the different systems. For example, Hikita *et al.* [36] describe an RF handset module for four different systems, containing six RF surface-wave filters.

The success of surface-wave devices can be attributed to a number of factors:

(a) Crystalline materials are available on which the propagation is almost ideal (low loss, diffraction and dispersion), with suitable piezoelectric coupling and with adequate temperature stability.
(b) Long delays, needed for delay lines and chirp filters, are obtained compactly.
(c) Patterns with arbitrary geometry can be made by photolithography, giving very high versatility. Typical substrates are several hundred wavelengths long, giving many degrees of freedom.
(d) Excellent accuracy and repeatability arise from the precision of photomask generation machines and the reproducibility of crystalline substrate materials.

(e) Complex designs, exploiting the versatility, are feasible because of sophisticated numerical analysis and design techniques.
(f) Single-stage photolithography is well suited to mass production of low-cost devices, with many devices per wafer.

REFERENCES

1. D. Royer and E. Dieulesaint. *Elastic Waves in Solids*, Vols. 1 and 2, Springer, 2000.
2. H. Matthews (ed.), *Surface Wave Filters*, Wiley, 1977.
3. A.A. Oliner (ed.), *Acoustic Surface Waves*, Springer, 1978.
4. D.P. Morgan. *Surface-Wave Devices for Signal Processing*, Elsevier, 1985, 1991.
5. S. Datta. *Surface Acoustic Wave Devices*, Prentice-Hall, 1986.
6. M. Feldmann and J. Henaff. *Surface Acoustic Waves for Signal Processing*, Artech House, 1989.
7. S.V. Biryukov, Y.V. Gulyaev, V.V. Krylov and V.P. Plessky. *Surface Acoustic Waves in Inhomogenous Media*, Springer, 1995.
8. C.K. Campbell. *Surface Acoustic Wave Devices for Mobile and Communications Applications*, Academic Press, 1998.
9. K.-Y. Hashimoto. *Surface Acoustic Wave Devices in Telecommunications – Modelling and Simulation*, Springer, 2000.
10. C.C.W. Ruppel and T.A. Fjeldy (eds.), *Advances in Surface Acoustic Wave Technology, Systems and Applications*, World Scientific, 2000 (Vol. 1), 2001 (Vol. 2).
11. G.S. Kino. *Acoustic Waves: Devices, Imaging and Analog Signal Processing*, Prentice-Hall, 1987.
12. D.P. Morgan. 'A history of surface acoustic wave devices', in C.C.W. Ruppel and T.A. Fjeldy (eds.), *Advances in Surface Acoustic Wave Technology, Systems and Applications*, Vol. 1, World Scientific, 2000, pp. 1–50.
13. Lord Rayleigh (J. Strutt). 'On waves propagated along the plane surface of an elastic solid', *Proc. London Math. Soc.*, **17**, 4–11 (1885).
14. R.M. White. 'Surface elastic waves', *Proc. IEEE*, **58**, 1238–1276 (1970).
15. J.R. Klauder, A.C. Price, S. Darlington and W.J. Albersheim. 'The theory and design of chirp radars', *Bell Syst. Tech. J.*, **39**, 745–808 (1960).
16. I.N. Court. 'Microwave acoustic devices for pulse compression filters', *IEEE Trans.*, **MTT-17**, 968–986 (1969).
17. W.S. Mortley. UK Patent 988 102 (1963). Also J.H. Rowen, U.S. Patent 3 289 114 (1963).
18. R.M. White and F.W. Voltmer. 'Direct piezoelectric coupling to surface elastic waves', *Appl. Phys. Lett.*, **7**, 314–316 (1965).
19. R.H. Tancrell, M.B. Schulz, H.H. Barrett, L. Davies and M.G. Holland. 'Dispersive delay lines using ultrasonic surface waves', *Proc. IEEE*, **57**, 1211–1213 (1969).
20. R.H. Tancrell and M.G. Holland. 'Acoustic surface wave filters', *Proc. IEEE*, **59**, 393–409 (1971).
21. P. Hartemann and E. Dieulesaint. 'Acoustic surface wave filters', *Electron. Lett.*, **5**, 657–658 (1969).
22. K. Kokkonen, J.V. Knuuttila, V.P. Plessky and M.M. Salomaa. 'Phase-sensitive absolute-amplitude measurements of surface waves using heterodyne interferometry', *IEEE Ultrason. Symp.*, **2**, 1145–1148 (2003).
23. G.S. Kino. 'Acoustoelectric interactions in acoustic-surface-wave devices', *Proc. IEEE*, **64**, 724–748 (1976).

24. F. Schmidt and G. Scholl. 'Wireless SAW identification and sensor systems', in C. Ruppel and T. Fjeldy (eds.), *Advances in Surface Acoustic Wave Technology, Systems and Applications*, Vol. 2, World Scientific, 2001, pp. 277–325.

25. C.S. Hartmann. 'A global SAW ID tag with large data capacity', *IEEE Ultrason. Symp.*, **1**, 65–69 (2002).

26. J. Beckley, V. Kalinin, M. Lee and K. Voliansky. 'Non-contact torque sensors based on SAW resonators', *IEEE Intl. Freq. Contr. Symp.*, 2002, pp. 202–213.

27. E. Berkenpas, S. Bitla, P. Millard and M. DaCunha. 'Pure shear horizontal SAW biosensor on langasite', *IEEE Trans. Ultrason. Ferroelect. Freq. Contr.*, **51**, 1404–1411 (2004).

28. G. Matthaei, L. Young and E.M.T. Jones. *Microwave Filters, Impedance Matching Networks and Coupling Structures*, Artech House, 1980. Also A.I. Zverev. *Handbook of Filter Synthesis*, Wiley, 1967.

29. M. Ylilammi, J. Ellä, M. Partanen and J. Kaitila. 'Thin film bulk acoustic wave filter', *IEEE Trans. Ultrason. Ferroelect. Freq. Contr.*, **49**, 535–539 (2002).

30. P. Selmeier, R. Grünwald, A. Przadka, H. Krüger, G. Feiertag and C. Ruppel. 'Recent advances in SAW packaging', *IEEE Ultrason. Symp.*, **1**, 283–292 (2001).

31. R.C. Williamson. 'Case studies of successful surface-acoustic-wave devices', *IEEE Ultrason. Symp.*, 1977, pp. 460–468.

32. C.S. Hartmann. 'Systems impact of modern Rayleigh wave technology', in E.A. Ash and E.G.S. Paige (eds.), *Rayleigh-Wave Theory and Application*, Springer, 1985.

33. C.S. Lam, D.S. Stevens and D.J. Lane. 'BAW- and SAW-based frequency control products for modern telecommunication systems and their applications in existing and emerging wireless communications equipment', *Intl. Meeting on Future Trends of Mobile Communication Devices*, Tokyo, January 1996.

34. C.K. Campbell. 'Applications of surface acoustic and shallow bulk acoustic wave devices', *Proc. IEEE*, **77**, 1453–1484 (1989).

35. K. Yamanouchi. 'Mobile communication systems and surface acoustic wave devices', *Electron. Commun. Jpn.*, **76**(3), 43–51 (1993).

36. M. Hikita, N. Shibagaki, K. Yokoyama, S. Matsuda, N. Matsuura and O. Hikino. 'Investigation of merged RX-differential output for multiband SAW front-end module', *IEEE Ultrason. Symp.*, **1**, 393–396 (2003).

2

ACOUSTIC WAVES IN ELASTIC SOLIDS

Many different types of acoustic waves can propagate in solid materials, and here we are particularly concerned with surface waves. A brief account of the analysis and properties of the waves is presented, including waves in piezo-electric materials since these are used in practical devices. More detail is given in other sources [1–4]. The basic nature of the waves is essential to an understanding of the devices. Although the devices make use of piezo-electric materials, many of the waves in such materials have counterparts in isotropic materials, which are non-piezoelectric. Section 2.2 therefore gives some details of the isotropic case, and this helps to illustrate some fundamentals which also relate to the piezoelectric case. The piezoelectric case is described in Section 2.3.

The subject of surface-acoustic-wave (SAW) devices is concerned with all types of surface waves that can propagate on a half-space, making use of piezoelec-tricity. It will be seen that this includes piezoelectric Rayleigh waves, leaky surface waves, surface transverse waves (STWs), Bleustein–Gulyaev waves and, in layered systems, layered Rayleigh waves and Love waves. All of these are included in the subject. The term 'SAW', without further qualification, is often taken to mean a piezoelectric Rayleigh wave. The various properties of the waves need to be considered when selecting materials for practical devices, and this topic is considered in Chapter 4.

2.1 ELASTICITY IN ANISOTROPIC MATERIALS

We first describe the elastic behavior of anisotropic materials, summarizing the development given in more detail elsewhere [3–6]. It is convenient to con-sider the non-piezoelectric case first, and then consider piezoelectric materials later.

2.1.1 Non-piezoelectric materials

Elasticity is concerned with the internal forces within a solid and the related displacement of the solid from its equilibrium, or force-free, configuration. It is assumed here that the solid is homogeneous. The forces will be expressed in terms of the *stress*, T, while the displacements are expressed in terms of the *strain*, S.

We consider first the strain, expressed as the motion of a particle. The term 'particle' is used to refer to an elementary region of the material, much smaller than any characteristic elastic dimension such as wavelength, but much larger than the interatomic distance. Thus, irregularity on the atomic scale is ignored. Suppose that, in the equilibrium state, a particle in the material is located at the point $\mathbf{x} = (x_1, x_2, x_3)$. When the material is not in its equilibrium state, this particle is displaced by an amount $\mathbf{u} = (u_1, u_2, u_3)$, where the components u_1, u_2 and u_3 are in general functions of the coordinates x_1, x_2, x_3. Thus, a particle with equilibrium position $\mathbf{x}$ has been displaced to a new position $\mathbf{x} + \mathbf{u}$. For the present, the displacement $\mathbf{u}$ is taken to be independent of time, t. Now clearly there will be no internal forces if $\mathbf{u}$ is independent of $\mathbf{x}$, since this simply denotes a displacement of the material as a whole. There will also be no forces if the material is rotated. To avoid these cases, the strain at each point is defined by

$$S_{ij}(x_1, x_2, x_3) = \frac{1}{2}\left(\frac{\partial u_i}{\partial x_j} + \frac{\partial u_j}{\partial x_i}\right), \quad i, j = 1, 2, 3. \tag{2.1}$$

With this definition, any displacements or rotations of the material as a whole cause no strain, and the strain is related to the internal forces. The strain is a second-rank tensor and is clearly symmetrical, so that

$$S_{ji} = S_{ij}. \tag{2.2}$$

Thus, only six of the nine components are independent.

The internal forces are described by a stress tensor T_{ij}. To define this, consider the plane $x_1 = x_1'$ within the material, where x_1' is a constant. If the material is strained, the material on one side of the plane exerts a force on the material on the other side. The force may be in any direction, and it may vary with the coordinates x_2, x_3 in the plane. The stress is defined such that the force per unit area has an x_i-component equal to $T_{i1}(x_1', x_2, x_3)$ with $i = 1, 2, 3$. This is the force exerted on the material at $x_1 < x_1'$. The force exerted on the material at $x > x_1'$ is the negative of this. The definition applies for any value of x_1', so we can write the stress as $T_{i1}(x_1, x_2, x_3)$. Similarly, we may consider forces on planes perpendicular to the x_2 and x_3 axes, defining the stresses in the same way, to arrive at the second-rank stress tensor $T_{ij}(x_1, x_2, x_3)$. Although we have only considered planes perpendicular to the coordinate axes, it can be shown that the forces acting on any plane can be deduced from this tensor [3, 4]. The force per unit area on a plane within the material has an x_i component $f_i = \Sigma_j T_{ij} n_j$. Here $\mathbf{n} = (n_1, n_2, n_3)$ is the unit vector normal to the plane and pointing away from the material on which the force acts. The material on the other side of the plane experiences the reversed force, since it has the sign of $\mathbf{n}$ changed. It can also be shown that the stress tensor is symmetric, that is,

$$T_{ji} = T_{ij}. \tag{2.3}$$

For practical purposes, many materials behave in an elastic manner, that is, if the material is deformed by applying a force, the material returns to its original state when the force is released. In addition, the strains are usually proportional to the stresses if the latter are small, so that the relation is linear. This is a generalization of Hooke's law, which states that stress is proportional to strain for the one-dimensional case. Unless stated otherwise, it is assumed throughout this book that the material is elastic and linear, and hence each component of stress is given by a linear combination of the strain components. The coefficients needed are given by the *stiffness tensor* c_{ijkl}, defined such that

$$T_{ij} = \sum_k \sum_l c_{ijkl} S_{kl}, \quad i,j,k,l = 1,2,3. \tag{2.4}$$

In such equations it is common to use Einstein's summation convention, in which summation signs are omitted for integers repeated on the right and absent on the left. Thus, both summation signs can be omitted in eq. (2.4). However, this convention is not used here.

The stiffness is a fourth-rank tensor with 81 elements. However, many of these elements are related. The symmetry of S_{ij} and T_{ij}, eqs (2.2) and (2.3), implies that the stiffness is unaltered if i and j are interchanged, or if k and l are interchanged, so that

$$c_{jikl} = c_{ijkl} \quad \text{and} \quad c_{ijlk} = c_{ijkl}. \tag{2.5}$$

Thus, only 36 of the 81 elements are independent. It can also be shown, from thermodynamic considerations, that the second pair of indices can be interchanged with the first pair, so that

$$c_{klij} = c_{ijkl}. \tag{2.6}$$

This reduces the number of independent elements to 21. These elements are of course physical properties of the material under consideration, so that the number of independent elements may well be reduced further by the symmetry of the material. For example, a crystalline material with cubic symmetry has only three independent elements. It should be noted that the coordinate axes x_1, x_2, x_3 will not in general be parallel to the axes of the crystal lattice.

Equation of motion

If the stress and strain are functions of time as well as position, the motion is subject to Newton's laws in addition to the above equations, and these constraints can be combined in the form of an equation of motion. Consider an elementary cube of material, centered at $\mathbf{x}' = (x_1', x_2', x_3')$. The edges are parallel to the x_1, x_2 and x_3 axes, and each edge is of length δ. The material surrounding the cube exerts forces on all six faces. For the faces at $x_1 = x_1' \pm \delta/2$, the components of force in the x_i direction are $\pm \delta^2 T_{i1}(x_1' \pm \delta/2, x_2', x_3')$. The forces on the faces normal to the x_2 and x_3 axes are obtained in the same way, and we add these to obtain the total force on the cube. Taking δ to be small, the total force has an x_i component

$$\delta^3 \left[\sum_j \frac{\partial T_{ij}}{\partial x_j} \right]_{\mathbf{x}'}.$$

This must be equal to the acceleration $\partial^2 u_i(\mathbf{x}')/\partial t^2$, multiplied by the mass $\rho \delta^3$, where ρ is the density. This is valid for all points $\mathbf{x}'$, and hence

$$\rho \frac{\partial^2 u_i}{\partial t^2} = \sum_j \frac{\partial T_{ij}}{\partial x_j}, \quad i, j = 1, 2, 3, \tag{2.7}$$

which is the equation of motion.

2.1.2 Piezoelectric materials

Piezoelectricity is a phenomenon which, in many materials, couples elastic stresses and strains to electric fields and displacements. It occurs only in anisotropic materials, whose internal structure lacks a center of symmetry. This includes many classes of single crystals, but the effect is often weak, so that it has little influence on the elastic behavior. However, here we are concerned with devices that make crucial use of piezoelectricity, so it is necessary to take account of it in the analysis. Only insulating materials will be considered here.

In a homogeneous piezoelectric insulator, the stress components T_{ij} at each point are dependent on the electric field $\mathbf{E}$ (or, equivalently, the electric displacement $\mathbf{D}$) in addition to the strain components S_{ij}. Assuming that these quantities are small enough we can take the relationship to be linear, and T_{ij} can be written as the linear relation

$$T_{ij} = \sum_k \sum_l c_{ijkl}^E S_{kl} - \sum_k e_{kij} E_k. \tag{2.8}$$

Here, the superscript on c_{ijkl}^E identifies this as the stiffness tensor for constant electric field; that is, this tensor relates changes of T_{ij} to changes of S_{kl} when

E is held constant. Similarly, the electric displacement **D** is usually determined by the field **E** and the permittivity tensor ε_{ij}, but in a piezoelectric material it is also related to the strain, such that

$$D_i = \sum_j \varepsilon_{ij}^S E_j + \sum_j \sum_k e_{ijk} S_{jk}, \tag{2.9}$$

where ε_{ij}^S is the permittivity tensor for constant strain. The forms of these equations are justified by thermodynamic arguments which are not considered here. The tensor e_{ijk}, relating elastic to electric fields, is called the piezoelectric tensor. From eq. (2.8) and the symmetry of T_{ij}, this tensor has the symmetry

$$e_{ijk} = e_{ikj}. \tag{2.10}$$

It is equally valid to relate **D** to the stress instead of the strain, and this can be done by eliminating S_{ij} from eqs (2.8) and (2.9). The result is expressed in the form

$$D_i = \sum_j \varepsilon_{ij}^T E_j + \sum_j \sum_k d_{ijk} T_{jk}, \tag{2.11}$$

where the new tensors ε_{ij}^T and d_{ijk} are related in a rather complicated manner to the tensors in eqs (2.8) and (2.9). The tensor ε_{ij}^T is the permittivity tensor for constant stress. We can also eliminate **E** between eqs (2.8) and (2.9) to obtain an equation giving T_{ij} in terms of S_{kl} and **D**; the coefficients of S_{kl} then give a stiffness tensor for constant electric displacement.

Equation of motion

The mechanical equation of motion, eq. (2.7), is valid for a piezoelectric material. It is convenient to express this in terms of the displacements u_i and the electric potential Φ. Because elastic disturbances travel much more slowly than electromagnetic ones, the electric field can be expressed using a quasi-static approximation, so that

$$E_i = -\partial\Phi/\partial x_i \tag{2.12}$$

In this approximation, the electrical variables are governed by electrostatics. Using eq. (2.12) in eq. (2.8), with eq. (2.1) for the strain, the equation of motion becomes

$$\rho\frac{\partial^2 u_i}{\partial t^2} = \sum_j \sum_k \left\{ e_{kij}\frac{\partial^2 \Phi}{\partial x_j \partial x_k} + \sum_l c_{ijkl}^E \frac{\partial^2 u_k}{\partial x_j \partial x_l} \right\}. \tag{2.13a}$$

In addition there are no free charges, since the material is taken to be an insulator. Hence div $\mathbf{D} = 0$, and using eq. (2.9) this gives

$$\sum_i \sum_j \left\{ \varepsilon_{ij}^S \frac{\partial^2 \Phi}{\partial x_i \partial x_j} - \sum_k e_{ijk} \frac{\partial^2 u_j}{\partial x_i \partial x_k} \right\} = 0. \tag{2.13b}$$

Equations (2.13) give four relations between the four quantities u_i and Φ, and hence the motion is determined if appropriate boundary conditions are specified.

Matrix notation

When specifying the stiffness and piezoelectric tensors for a particular material, it is usual to adopt a special notation known as the matrix notation. This is convenient because it reduces the number of elements to be specified. The stiffness tensor c_{ijkl}^E has at most 36 independent components because of its symmetry, eq. (2.5), and it is expressed in terms of a stiffness matrix c_{mn}^E. This is defined by

$$c_{mn}^E = c_{ijkl}^E, \quad m, n = 1, 2, \ldots, 6, \tag{2.14}$$

where m is related to i and j by

$$m = i \quad \text{for } i = j, \quad m = 9 - i - j \quad \text{for } i \neq j, \quad i, j = 1, 2, 3. \tag{2.15}$$

A similar definition relates n to k and l. A simplified piezoelectric matrix is also used, defined by

$$e_{km} = e_{kij}, \quad k = 1, 2, 3, \quad m = 1, 2, \ldots, 6, \tag{2.16}$$

with m related to i and j as above.

2.2 WAVES IN ISOTROPIC MATERIALS

In this book we are concerned mainly with wave motion in anisotropic materials. The complexity of the equations of elasticity, described in the previous section, is such that the properties of the waves can usually be found only by numerical techniques. In contrast, solutions for isotropic materials are much easier to obtain, and since they have many features in common with the solutions for anisotropic materials it is helpful to consider the isotropic case first [3, 4, 7–10]. Numerical examples are given here for fused quartz, which has acoustic properties somewhat similar to crystalline quartz, used in many surface-wave devices.

In an isotropic material the stiffness tensor c_{ijkl} has only two independent components. From symmetry it can be shown [3] that this tensor can be written in the form

$$c_{ijkl} = \lambda\delta_{ij}\delta_{kl} + \mu(\delta_{ik}\delta_{jl} + \delta_{il}\delta_{jk}), \tag{2.17}$$

where the Kronecker delta is defined by $\delta_{ij} = 1$ for $i = j$ and $\delta_{ij} = 0$ for $i \neq j$. The constants λ and μ are known as Lamé constants and in practice these are always positive; μ is also called the rigidity. Substituting into eq. (2.4), the stress can be written in the form

$$T_{ij} = \lambda\delta_{ij}\Delta + 2\mu S_{ij}, \tag{2.18}$$

where

$$\Delta = \sum_i S_{ii} = \sum_i \frac{\partial u_i}{\partial x_i}. \tag{2.19}$$

The equation of motion, eq. (2.7), becomes

$$\rho\frac{\partial^2 u_j}{\partial t^2} = (\lambda + \mu)\frac{\partial\Delta}{\partial x_j} + \mu\nabla^2 u_j, \tag{2.20}$$

where eq. (2.18) has been used and

$$\nabla^2 = \sum_i \frac{\partial^2}{\partial x_i^2}.$$

2.2.1 Plane waves

We first consider an infinite medium supporting plane waves, with frequency ω, in which the displacement $\mathbf{u}$ takes the form

$$\mathbf{u} = \mathbf{u}_0 \exp[j(\omega t - \mathbf{k} \cdot \mathbf{x})], \tag{2.21}$$

where $\mathbf{u}_0$ is a constant vector, independent of $\mathbf{x}$ and t. The actual displacement is the real part of eq. (2.21), but the complex form can be used throughout the analysis because the equations are linear. The wave vector is $\mathbf{k} = (k_1, k_2, k_3)$, which gives the direction of propagation. The wavefronts are solutions of $\mathbf{k} \cdot \mathbf{x} = $ constant, and they are perpendicular to $\mathbf{k}$. The phase velocity of the wave is $V = \omega/|\mathbf{k}|$. With this form for $\mathbf{u}$, we have $\partial\mathbf{u}/\partial x_j = -jk_j\mathbf{u}$, and on substituting into eq. (2.20) we obtain

$$\omega^2\rho u_j = (\lambda + \mu)(\mathbf{k} \cdot \mathbf{u})k_j + \mu|\mathbf{k}|^2 u_j, \quad j = 1, 2, 3,$$

where $|\mathbf{k}|^2 = k_1^2 + k_2^2 + k_3^2$. Substituting for u_j using eq. (2.21), and writing the result in vector form, gives

$$\omega^2\rho\mathbf{u}_0 = (\lambda + \mu)(\mathbf{k} \cdot \mathbf{u}_0)\mathbf{k} + \mu|\mathbf{k}|^2\mathbf{u}_0 \tag{2.22}$$

Here, there are two terms parallel to $\mathbf{u}_0$ and one term parallel to $\mathbf{k}$, with the latter including the scalar product $\mathbf{k} \cdot \mathbf{u}_0$. There are therefore two cases to consider. Firstly, if $\mathbf{u}_0$ is perpendicular to $\mathbf{k}$ the scalar product $\mathbf{k} \cdot \mathbf{u}_0$ is zero, and the remaining terms in the equation are parallel. Secondly, if $\mathbf{u}_0$ is not perpendicular to $\mathbf{k}$ the product $\mathbf{k} \cdot \mathbf{u}_0$ is non-zero, so that for non-trivial solutions we must have $\mathbf{u}_0$ parallel to $\mathbf{k}$. These two cases give shear-wave solutions and longitudinal-wave solutions, respectively.

Taking $\mathbf{u}_0$ to be perpendicular to $\mathbf{k}$ gives *shear*, or transverse, waves. For these the wave vector is denoted by $\mathbf{k}_t$, and eq. (2.22) gives

$$|\mathbf{k}_t|^2 = \omega^2 \rho / \mu.$$

The phase velocity for shear waves is denoted by V_t, equal to $\omega / |\mathbf{k}_t|$, so that

$$V_t = \sqrt{\mu / \rho}, \tag{2.23}$$

taking V_t to be positive. Since this is independent of frequency ω, the wave is non-dispersive. The displacement $\mathbf{u}_0$ can be in any direction in the plane of the wavefront, perpendicular to $\mathbf{k}_t$.

For *longitudinal* waves we consider solutions of eq. (2.22) in which $\mathbf{u}_0$ is parallel or antiparallel to $\mathbf{k}$. Thus, $\mathbf{k}$ is given by

$$\mathbf{k} = \pm \mathbf{u}_0 \frac{|\mathbf{k}|}{|\mathbf{u}_0|}. \tag{2.24}$$

With this relation we find

$$(\mathbf{k} \cdot \mathbf{u}_0)\mathbf{k} = \mathbf{u}_0 |\mathbf{k}|^2,$$

irrespective of the sign in eq. (2.24). In this case the wave vector is denoted by $\mathbf{k}_l$, and substitution into eq. (2.22) gives

$$|\mathbf{k}_l|^2 = \omega^2 \rho / (\lambda + 2\mu).$$

The velocity for this case is denoted by V_l, and is thus given by

$$V_l = \sqrt{\frac{\lambda + 2\mu}{\rho}}, \tag{2.25}$$

and hence the wave is non-dispersive. Since λ and μ are always positive, the velocity of longitudinal waves is always greater than the velocity of shear waves.

The velocities are typically in the region of 3000 m/s for shear waves and 6000 m/s for longitudinal waves. For example, in fused quartz [11], $V_t = 4100$ m/s and $V_l = 6050$ m/s. The displacements involved are usually very small. In fused quartz, a shear wave with a power density of 1 mW/mm^2 and a

frequency of 10 MHz has a peak displacement of about 0.3 nm, and a similar figure applies for longitudinal waves. This displacement is some six orders
of magnitude smaller than the wavelengths.

In Sections 2.2.2 to 2.2.4 below, we consider several configurations involving
plane boundaries. The solutions are obtained by summing plane-wave solutions
of the above types. We consider solutions in which the displacements are proportional to $\exp(-j\beta x_1)$, with the x_1-axis parallel to the boundaries. In each
case this gives one or more characteristic solutions, and these are often called
modes. It should however be noted that these modes are not the only solutions
that can exist.

In most cases we are concerned with waves propagating on a half-space of
some material, which may have a layer of another material on the surface. The
solutions of most interest are surface acoustic waves (SAW) or, more simply,
surface waves. These are solutions for which the displacements in the half-space
decay rapidly in the direction normal to the surface, and the energy of the wave
is transported parallel to the surface. The Rayleigh wave considered next is one
type of surface wave.

2.2.2 Rayleigh waves in a half-space

Consider an isotropic medium with infinite extent in the x_1 and x_2 directions
but with a boundary at $x_3 = 0$, so that the medium occupies the space $x_3 < 0$.
The space $x_3 > 0$ is a vacuum. The surface wave solution for this case is named
after Lord Rayleigh, who first discovered it [10]. We assume propagation in the
x_1 direction, so that the wavefronts are parallel to x_2, as shown in Fig. 2.1. The
(x_1, x_3) plane, which contains the surface normal and the propagation direction,
is known as the *sagittal plane*. The solution must satisfy the equation of motion,
eq. (2.20), and the boundary conditions that there should be no forces on the
free surface at $x_3 = 0$.

The solution is found by adding two components corresponding to plane shear
and longitudinal waves, and these components are described as *partial* waves.
The wave vectors, $\mathbf{k}_t$ and $\mathbf{k}_l$, have magnitudes given by

$$|\mathbf{k}_t|^2 = \omega^2/V_t^2, \quad |\mathbf{k}_l|^2 = \omega^2/V_l^2. \tag{2.26}$$

Since the Rayleigh wave has no variation in the x_2 direction, these vectors must
have no x_2 components and they are therefore in the sagittal plane. In the x_1
direction the displacements are assumed to vary as $\exp(-j\beta x_1)$, where β is
the wavenumber of the Rayleigh wave. The x_1 components of $\mathbf{k}_t$ and $\mathbf{k}_l$ must

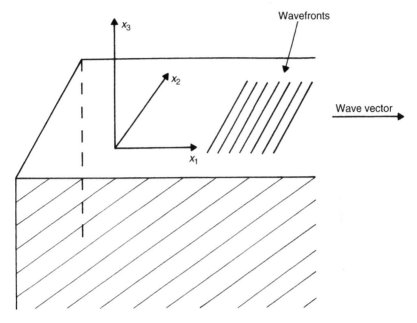

Figure 2.1 Axes for surface-wave analysis.

therefore be equal to β. The x_3 components are, respectively, denoted by T and L, and we thus have

$$T^2 = \omega^2/V_t^2 - \beta^2, \quad L^2 = \omega^2/V_l^2 - \beta^2. \tag{2.27}$$

The displacement of the longitudinal wave, $\mathbf{u}_l$, must be parallel to the wave vector $\mathbf{k}_l = (\beta, 0, L)$. Thus, omitting a factor $\exp(\mathrm{j}\omega t)$, we can write

$$\mathbf{u}_l = A(1, 0, L/\beta)\exp[-\mathrm{j}(\beta x_1 + Lx_3)], \tag{2.28}$$

where A is a constant. The displacement of the shear wave, $\mathbf{u}_t$, is perpendicular to the wave vector $\mathbf{k}_t = (\beta, 0, T)$. This does not determine the direction of $\mathbf{u}_t$, but we assume for the present that $\mathbf{u}_t$ is in the sagittal plane (as is $\mathbf{u}_l$). Thus $\mathbf{u}_t$ is given by

$$\mathbf{u}_t = B(1, 0, -\beta/T)\exp[-\mathrm{j}(\beta x_1 + Tx_3)], \tag{2.29}$$

where B is a constant. The total displacement $\mathbf{u}$ is the sum

$$\mathbf{u} = \mathbf{u}_t + \mathbf{u}_l \tag{2.30}$$

Now, in the x_3 direction the shear-wave displacement $\mathbf{u}_t$ varies as $\exp(-\mathrm{j}Tx_3)$. For a surface-wave solution the displacement must decay for negative x_3, and hence the value of T must be positive imaginary. Thus β must be large enough to ensure this, and since the Rayleigh-wave velocity V_R will be given by $V_R = \omega/\beta$ we must have

$$V_R < V_t. \tag{2.31}$$

Similarly, L must be positive imaginary and this implies $V_R < V_l$, but this is already implied by eq. (2.31) because V_l is greater than V_t. With T and L imaginary, the partial-wave solutions, $\mathbf{u}_l$ of eq. (2.28) and $\mathbf{u}_t$ of eq. (2.29), are no longer plane waves. However, they are still valid as solutions for an infinite medium, satisfying the equation of motion, eq. (2.20). If the total displacement $\mathbf{u}$ of eq. (2.30) is to be a valid solution for a half-space, it must also satisfy the boundary conditions $T_{13} = T_{23} = T_{33} = 0$ at the surface $x_3 = 0$, where the stresses T_{i3} are given by eq. (2.18). This gives two linear homogeneous equations relating the constants A and B. For non-trivial solutions the determinant of the coefficients must be zero, and this gives

$$(T^2 - \beta^2)^2 + 4\beta^2 LT = 0. \tag{2.32}$$

Using eq. (2.27) for T and L, and defining the phase velocity $V = \omega/\beta$, this gives

$$(2 - V^2/V_t^2)^2 = 4\sqrt{1 - V^2/V_t^2} \sqrt{1 - V^2/V_l^2}. \tag{2.33}$$

For Rayleigh waves we need a solution giving V^2 a real positive value, less than V_t^2. The equation has only one such solution, which is denoted by V_R^2, and we take V_R to be positive. The ratio V_R/V_t is determined by the ratio V_l/V_t of the plane-wave velocities and it is shown in Fig. 2.2, which thus gives the Rayleigh-wave velocity for any isotropic material. V_R is usually close to V_t, and it is independent of frequency.

The above analysis also gives the displacements. Omitting a factor $\exp[j(\omega t - \beta x_1)]$ and an arbitrary multiplier, the displacements are

$$u_1 = \gamma \exp(a\beta x_3) - \exp(b\beta x_3),$$
$$u_3 = j[\gamma a \exp(a\beta x_3) - b^{-1} \exp(b\beta x_3)], \tag{2.34}$$

where a, b and γ are real positive quantities given by $a = -jL/\beta, b = -jT/\beta$ and $\gamma = (2 - V_R^2/V_t^2)/(2ab)$ and $\beta = \omega/V_R$. These displacements are shown in Fig. 2.3, as functions of the depth normalized to the Rayleigh wavelength $\lambda_R = 2\pi V_R/\omega$. Since u_3 is in phase quadrature with u_1, the motion of each particle is an ellipse. Because of the change of sign of u_1 at a depth of about 0.2 wavelengths, the ellipse is described in different directions above and below this point; at the surface the motion is retrograde, while lower down it is prograde. The distortion of the material at one instant is shown in Fig. 2.4, with

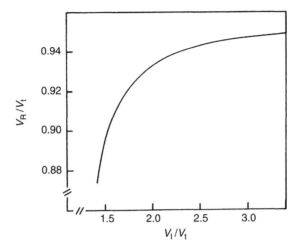

Figure 2.2 Normalized Rayleigh-wave velocity for isotropic materials.

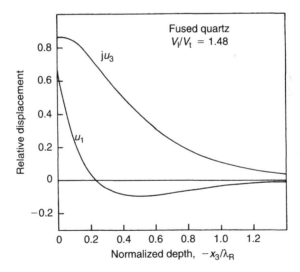

Figure 2.3 Rayleigh-wave displacements for isotropic material.

the displacements exaggerated. The dots in this figure represent the equilibrium positions of the particles within the material, while the lines show the displacements when a Rayleigh wave is present. Note that there is little motion at a depth greater than one wavelength.

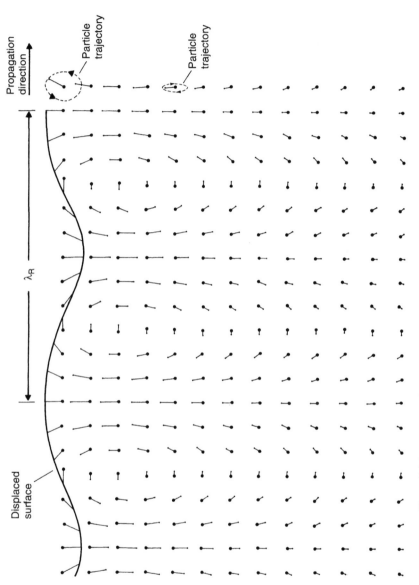

Figure 2.4 Instantaneous displacements for Rayleigh-wave propagation in isotropic material.

2.2.3 Shear-horizontal waves in a half-space

It was assumed above that for surface-wave solutions the displacements would be confined to the sagittal plane (x_1, x_3). We now consider whether a valid solution can be obtained with a perpendicular component, in the x_2 direction. As before, the wave vector of any partial wave must be confined to the sagittal plane, and hence a displacement normal to this plane can only be produced by shear waves. The wave vector must therefore be $\mathbf{k}_t = (\beta, 0, T)$ as before, and the displacement must be

$$\mathbf{u} = A(0, 1, 0) \exp[-j(\beta x_1 + Tx_3)],$$

where A is a constant. For this wave the stresses T_{13} and T_{33} are zero, while $T_{23} = -j\mu T u_2$. The boundary conditions, $T_{i3} = 0$ at $x_3 = 0$, cannot therefore be satisfied if T is finite, and hence there is no solution representing a wave bound to the surface. However, if $T = 0$ the stress components T_{i3} are zero everywhere, and this satisfies the boundary conditions. This solution is simply a plane shear wave propagating parallel to the surface, with its amplitude independent of x_3 within the material. It is called a shear-horizontal, or SH, wave, since its displacements are parallel to the surface. The phase velocity is equal to V_t.

2.2.4 Waves in a layered half-space

Now consider a half-space of material covered by a layer of another material, with layer thickness d as shown in Fig. 2.5. Structures of this type are common in surface-wave devices, where the layer may for example be a metal film. Often, one is concerned with minimizing the perturbing effect of the film, for example to minimize the dispersion. In seismology this structure is of considerable interest because the layering of rocks influences the propagation of surface waves, and consequently the solutions have been studied in detail [12–16].

We first consider waves with their displacements confined to the sagittal plane. If the layer thickness is small, the solutions will be similar to the Rayleigh wave for a half-space, described in Section 2.2.2 above, so the solutions here are described as 'layered Rayleigh waves'. The solutions may be found by summing partial waves, as before, though the calculation is more complex. The layer material is taken to have plane-wave velocities V_l and V_t. Assuming the displacements are proportional to $\exp(-j\beta x_1)$, a partial shear wave in the layer gives displacements with the form of eq. (2.29), with T given by eq. (2.26). However, there are two solutions for T. To account for this we define T as the positive solution of eq. (2.26), so that the partial shear waves have wavenumbers $\mathbf{k}_t = (\beta, 0, \pm T)$. The displacements of these two waves then give

$$\mathbf{u}_t = A(1, 0, -\beta/T) \exp(-jTx_3) + B(1, 0, \beta/T) \exp(jTx_3), \tag{2.35}$$

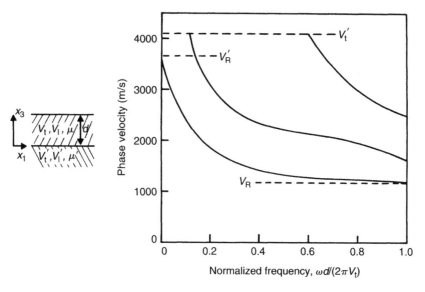

Figure 2.5 Velocities for layered Rayleigh waves, gold layer on fused quartz substrate. From Farnell and Adler [11]. Copyright 1972, with permission from Elsevier.

where A and B are constants, and the variations with x_1 and t have been omitted. Similarly, there are two longitudinal partial waves in the layer, with displacements

$$\mathbf{u}_l = C(1,0,L/\beta)\exp(-jLx_3) + D(1,0,-L/\beta)\exp(jLx_3), \qquad (2.36)$$

where L is the positive solution of eq. (2.27), and C and D are constants. The total displacement in the layer is $\mathbf{u} = \mathbf{u}_t + \mathbf{u}_l$.

We define V_t' and V_l' as the velocities of plane shear and longitudinal waves in the half-space material. The displacement $\mathbf{u}'$ in the half-space is obtained as before; comparing with eqs (2.28) and (2.29) we can write

$$\mathbf{u}' = E(1,0,-\beta/T')\exp(-jT'x_3) + F(1,0,L'/\beta)\exp(-jL'x_3), \qquad (2.37)$$

where E and F are constants. T' and L' are the x_3 components of the wavenumbers, given by eq. (2.27) but with V_t' and V_l' replacing V_t and V_l. For a surface-wave solution, with displacements decaying in the half-space, T' and L' must be positive imaginary, and hence the phase velocity must be less than V_t'.

The boundary conditions require that the stresses T_{i3} should be zero on the free upper surface, while on the lower surface the stresses T_{i3} and the displacements

u_i should be continuous. This gives six homogeneous equations relating the constants $A, B, \ldots, F$, and the determinant of coefficients must be zero for a valid solution. A search is done, using trial values of β until the determinant becomes zero. The velocity is then $V = \omega/\beta$, and the relative displacements are given by the above equations.

Figure 2.5 shows the calculated velocities for a gold layer on fused quartz, after Farnell and Adler [11]. The horizontal axis here gives normalized frequency, but it may also be read as the thickness of the layer divided by the wavelength of plane shear waves in the layer material. The result is typical of cases in which the layer material has acoustic velocities much less than those of the half-space material. The fundamental mode, that is, the solution with lowest velocity, is of primary importance. At low frequencies the layer thickness is much less than the wavelength, so the velocity approaches V_R', the Rayleigh-wave velocity for the half-space material. At high frequencies the structure can support a Rayleigh wave with its energy concentrated near the upper surface, so the velocity approaches the Rayleigh-wave velocity for the layer material, denoted V_R. The velocity thus varies from V_R' to V_R. Clearly, if we wish to minimize dispersion, the materials should be chosen such that V_R is not substantially less than V_R'. For this reason the metal film used for the electrodes in surface-wave devices is usually aluminum, which has a Rayleigh-wave velocity similar to those of the common substrate materials, quartz and lithium niobate. In addition to the fundamental there is a series of higher modes, which are named after Sezawa [14].

If the layer material has acoustic velocities greater than those of the half-space, the velocity of the fundamental is V_R' at zero frequency, rising to a value V_t', at which point there is a cut-off. There is therefore little dispersion in this case.

In addition to layered Rayleigh waves, the layered system can also support surface waves with the displacements normal to the sagittal plane. These are known as Love waves [16]. In this case the partial waves are shear waves, with wave vectors $\mathbf{k}_t = (\beta, 0, \pm T)$ in the layer and $\mathbf{k}_t' = (\beta, 0, \pm T')$ in the half-space. Thus the displacements in the layer can be written

$$\mathbf{u}_t = A(0, 1, 0) \exp(-jTx_3) + B(0, 1, 0) \exp(jTx_3), \qquad (2.38)$$

and the displacement in the half-space is

$$\mathbf{u}_t' = C(0, 1, 0) \exp(-jT'x_3),$$

where A, B and C are constants. The boundary conditions are the same as for the Rayleigh-wave case, and applying these gives the dispersion relation

$$\tan(Td) = j\mu'T'/(\mu T), \qquad (2.39)$$

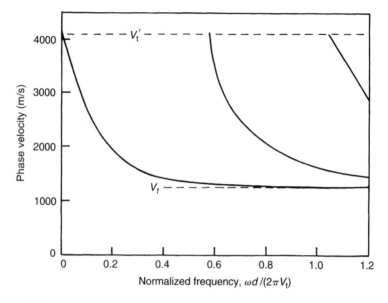

Figure 2.6 Velocities for Love waves, gold layer on fused quartz substrate. After Farnell and Adler [11]. Copyright 1972, with permission from Elsevier.

where μ and μ' are respectively the rigidities of the layer material and the half-space material. Solving for β gives in general a number of modes, and the velocities for gold on fused quartz are shown in Fig. 2.6. Solutions are obtainable only for $V_t < V_t'$, and the solutions must have velocities less than V_t', so that T' is imaginary and the displacement decays in the half-space. At zero frequency the Love-wave solution becomes identical to the SH plane-wave solution for a half-space. Thus Love waves can be regarded as modified forms of the SH plane wave, where the presence of a layer with low acoustic velocity converts the plane wave into a surface wave and causes dispersion.

If we take $d = \infty$ in Fig. 2.5, the configuration becomes two semi-infinite materials with a common plane boundary. In this case, there can be a non-dispersive wave propagating along the boundary, with amplitude decaying exponentially into each material. This wave, with displacements in the sagittal plane, is a Stoneley wave. Solutions exist only if the acoustic velocities in the two materials are fairly similar [12, p. 106]. For isotropic materials, there is no guided SH-wave solution. However, a guided SH-wave solution can exist if the materials are anisotropic and piezoelectric. This solution, called a Maerfeld–Tournois wave, is related to the Bleustein–Gulyaev wave considered in Section 2.3.3 below [12, p. 146].

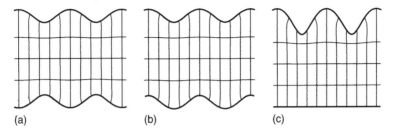

Figure 2.7 Displacement of a rectangular grid by a Lamb wave. (a) Lowest-velocity symmetric mode. (b) Lowest-velocity antisymmetric mode. (c) The sum of (a) and (b), which approximates a Rayleigh surface wave.

2.2.5 Waves in a parallel-sided plate

We now consider a plate of material with boundaries at $x_3 = \pm d/2$ and with infinite extent in the x_1 and x_2 directions. As for the layered substrate, there are two types of solution [3, 12]. For displacements confined to the sagittal plane (x_1, x_3) the solutions are called Lamb waves, while there are also SH-wave solutions with displacements perpendicular to the sagittal plane. Both types of wave have some relevance to surface-wave devices.

The parallel-sided plate is rather similar to the layered half-space considered above, with the half-space omitted. Thus, for Lamb waves, the partial waves have the same form as those for layered Rayleigh waves, given by eqs (2.35) and (2.36). The allowed values for β, and the relative values for the constants A, B, C and D, are obtained by applying the boundary conditions $T_{i3} = 0$ at the two surfaces $x_3 = \pm d/2$. This gives two families of dispersive solutions, known as symmetric and antisymmetric modes. The dispersion relation is

$$\left[\frac{\tan(Ld/2)}{\tan(Td/2)} \right]^{\pm 1} = -\frac{(T^2 - \beta^2)^2}{4LT\beta^2}, \tag{2.40}$$

taking the upper sign for symmetric modes and the lower sign for antisymmetric modes. The displacement of the material is illustrated in Fig. 2.7.

At high frequencies Td and Ld are large, and for the lowest-velocity solutions they are found to be imaginary. In this case, the left side of eq. (2.40) approaches unity. Comparison with eq.(2.32) shows that the velocity approaches the Rayleigh-wave velocity, V_R. Thus the two Lamb modes, one symmetric and one antisymmetric, each become equivalent to a Rayleigh wave on the upper surface plus a Rayleigh wave on the lower surface. This gives some insight into the behavior of a wave generated on one surface of a parallel-sided plate. The wave is equivalent to the sum of the two Lamb waves, with equal

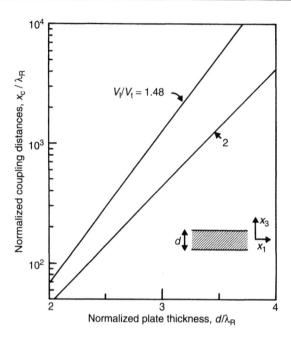

Figure 2.8 Coupling distance for transfer of Rayleigh waves between the two surfaces of a parallel-sided plate.

amplitudes and phases. Owing to the symmetry the Lamb wave displacements are additive near the upper surface but cancel near the lower surface, so that if d is large enough the wave is essentially a Rayleigh wave on the upper surface, as in Fig. 2.7c. However, for finite d the Lamb waves have different wavenumbers, β_s and β_a say, so that the disturbance on the upper surface has the form $[\exp(-j\beta_a x_1) + \exp(-j\beta_s x_1)]$, with an amplitude proportional to $\cos[(\beta_a - \beta_s)x_1/2]$. Thus, after traveling a distance

$$x_c = \pi/|\beta_a - \beta_s|$$

the amplitude at the top surface is zero. At this point, the two Lamb waves are in antiphase, so that they reinforce at the lower surface. In effect, the Rayleigh wave has been transferred from one surface to the other. The distance x_c is therefore called the coupling distance. At a distance $2x_c$, the amplitude on the upper surface is again maximized, so that the Rayleigh wave has been transferred back again.

The coupling distance may be evaluated by solving eq. (2.40) for β_a and β_s, giving the result shown in Fig. 2.8, where x_c is normalized to the Rayleigh wavelength. The coupling distance is many thousands of wavelengths, even when the plate is only a few wavelengths thick. In surface-wave devices the wave is generated on a rectangular substrate, and it can be concluded that a substrate thickness of a few wavelengths should be sufficient to prevent the rear surface from having any significant effect. This conclusion remains valid if the substrate is mounted on a carrier using an adhesive, as is usually the case.

Setting $\beta = 0$ in eq. (2.40) gives the resonant frequencies of the plate. These are the frequencies at which the thickness $2d$ is a multiple of the shear-wave wavelength $2\pi V_t/\omega$, or of the longitudinal-wave wavelength $2\pi V_l/\omega$. Resonances of this type are used in the parallel-plate crystal resonator. They are sometimes observed in surface-wave devices, though usually most of them are damped out because of the mounting of the substrate.

The parallel-sided plate also supports SH modes, with the displacements in the x_2 direction. These solutions can be obtained by assuming partial waves of the form given by eq. (2.38). This gives a series of modes, most of which are dispersive. However, at low frequencies there is only one solution, a non-dispersive mode with velocity V_t'; this is simply a plane shear wave propagating in the x_1 direction. The same solution was found in Section 2.2.3 for propagation in a half-space, and it was shown that this wave gives no stress on planes perpendicular to x_3.

The non-dispersive SH mode of the parallel-sided plate is used in a type of dispersive delay line called the IMCON, in which the wave is reflected by arrays of grooves. Dispersive modes (Lamb waves and SH waves) have also been used in dispersive delay lines.

2.3 WAVES IN ANISOTROPIC MATERIALS

This section is concerned with acoustic waves in piezoelectric materials, which must of necessity be anisotropic. Because of the complexity of the equations for this case, the solutions can usually be found only by numerical methods [2, 17–19]. The account here is therefore mainly descriptive.

2.3.1 Plane waves in an infinite medium

For a piezoelectric material, the equation of motion takes the form of eqs (2.13), in terms of the displacements $\mathbf{u}$ and the potential Φ. We consider plane-wave solutions with frequency ω and wave vector $\mathbf{k}$, with $\mathbf{u}$ and Φ having the forms

$$\mathbf{u} = \mathbf{u}_0 \exp[j(\omega t - \mathbf{k} \cdot \mathbf{x})],$$

$$\Phi = \Phi_0 \exp[j(\omega t - \mathbf{k} \cdot \mathbf{x})],$$

where $\mathbf{u}_0$ and Φ_0 are constants, independent of $\mathbf{x}$ and t, and $\mathbf{k}$ is real. For isotropic materials, Section 2.2.1, there are two solutions – the shear wave and the longitudinal wave, with $\mathbf{u}_0$ respectively perpendicular and parallel to $\mathbf{k}$.

To find the solutions for anisotropic materials the above functions $\mathbf{u}$ and Φ are substituted into eqs (2.13), giving four equations in the four variables $\mathbf{u}_0$, Φ_0. Setting the determinant of coefficients to zero generally gives four solutions, with different $\mathbf{k}$ values. One of these solutions is essentially electrostatic in nature – it corresponds to the electrostatic solution for an isotropic material, for which $e_{ijk} = 0$ and $\varepsilon_{ij}^S = \varepsilon$ so that eq. (2.13b) reduces to Laplace's equation. This solution is of little interest here. The other three solutions are non-dispersive acoustic waves. Usually, one solution has the displacement $\mathbf{u}_0$ almost parallel to $\mathbf{k}$, and this is called the 'quasi-longitudinal', or longitudinal, wave. The other two solutions, generally with different velocities, usually have $\mathbf{u}_0$ almost perpendicular to $\mathbf{k}$, and they are called 'quasi-shear', or shear, waves. Relative to the crystal lattice, there can be propagation directions such that the longitudinal wave has $\mathbf{u}_0$ parallel to $\mathbf{k}$, and it is then called a 'pure longitudinal' wave. Similarly, shear waves may have $\mathbf{u}_0$ perpendicular to $\mathbf{k}$, and they are then called 'pure shear' waves. Owing to anisotropy, each of the three waves has a phase velocity (equal to $\omega/|\mathbf{k}|$) dependent on the propagation direction. In addition, all three solutions may have associated electric potentials, though for particular propagation directions the potential may disappear.

2.3.2 Theory for a piezoelectric half-space

In Section 2.2.2 we saw that the Rayleigh-wave solution for an isotropic half-space can be obtained by adding two partial waves, corresponding to plane shear and longitudinal waves, with the x_1 components of their wave vectors equal. For anisotropic materials a related method is used [1–3, 18, 19], though a numerical procedure must be adopted to obtain the solutions.

Care is needed in specifying the orientation of the material. For crystalline materials the internal structure can be referenced to a set of axes denoted by the uppercase symbols X, Y, Z, with directions defined by convention in relation to the crystal lattice [20]. The surface orientation and the wave propagation direction must be defined in relation to these axes. The usual convention is to specify the outwardly directed surface normal x_3, followed by the propagation direction x_1. For example, 'Y–Z lithium niobate' indicates that x_3 is parallel to the crystal Y axis and x_1 is parallel to the Z axis. The orientation of x_3 is also referred to as the cut, so that for Y–Z lithium niobate the crystal is Y-cut. The material tensors – the stiffness, permittivity and piezoelectric tensors – are specified in relation to the X, Y and Z axes, so for the analysis they must first be rotated into the frame defined by x_1, x_2 and x_3 using tensor transformations [3, 4]. More general orientations can be specified using Euler angles (Chapter 4).

For a piezoelectric material it is necessary to use an electrical boundary condition at the surface, in addition to the force-free condition which applies for isotropic materials. Two cases are usually considered. In the first case the space above the surface is a vacuum and conductors are excluded, so that there are no free charges. This is called the *free-surface* case, with wave velocity v_f. In general there will be a potential in the vacuum above the surface. In the second case the surface is assumed to be covered with a thin metal layer with infinite conductivity, which shorts out the horizontal component of **E** at the surface but does not affect the mechanical boundary conditions. This is called the *metallized* case, with velocity v_m. These two cases generally give different velocities. The velocity difference is a measure of the coupling between the wave and electrical perturbations at the surface. It will be seen in Chapter 3 that this is of crucial importance to the performance of surface-wave transducers.

For the free-surface case the potential in the vacuum satisfies Laplace's equation $\nabla^2 \Phi = 0$. If the wavenumber of the surface wave is β, the potential Φ in the vacuum can be written as

$$\Phi = f(x_3) \exp[j(\omega t - \beta x_1)].$$

Using Laplace's equation shows that the function $f(x_3)$ has the form $\exp(\pm \beta x_3)$, and since Φ must vanish at $x_3 = \infty$ the potential for $x_3 \geq 0$ is given by

$$\Phi = \Phi_0 \exp(-|\beta| x_3) \exp[j(\omega t - \beta x_1)], \tag{2.41}$$

where Φ_0 is a constant. Since there are no free charges, D_3 must be continuous, so that in both the piezoelectric and the vacuum we have

$$D_3 = \varepsilon_0 |\beta| \Phi, \quad \text{at } x_3 = 0. \tag{2.42}$$

For the metallized case the potential at the surface is zero, so

$$\Phi = 0, \quad \text{at } x_3 = 0. \tag{2.43}$$

In addition, for either case there are no forces on the surface, so

$$T_{13} = T_{23} = T_{33} = 0, \quad \text{at } x_3 = 0. \tag{2.44}$$

To find the surface-wave solutions, we first consider partial waves in which the displacements and potential, denoted by $\mathbf{u}'$ and Φ', take the form

$$\mathbf{u}' = \mathbf{u}_0' \exp(j\gamma x_3) \exp[j(\omega t - \beta x_1)],$$
$$\Phi' = \Phi_0' \exp(j\gamma x_3) \exp[j(\omega t - \beta x_1)], \tag{2.45}$$

where β is the wavenumber of the surface wave, assumed to be real. These expressions are to satisfy the equations of motion, eqs (2.13), for an infinite material. As in the isotropic case, if β is fixed there are a number of specific solutions for γ, the x_3 component of the wave vector. We assume a particular real value of β and substitute eq. (2.45) into eqs (2.13). These can then be solved numerically, giving eight solutions for γ, and for each solution the relative values of $\mathbf{u}'_0$ and Φ'_0 are obtained. The values of γ are generally complex, and we can only allow values whose imaginary parts are negative, so that $\mathbf{u}'$ and Φ' vanish at $x_3 = -\infty$. There are four such values of γ in general, and these are denoted by $\gamma_1, \ldots, \gamma_4$. The partial waves are therefore

$$\mathbf{u}'_m = \mathbf{u}'_{0m} \exp(j\gamma_m x_3) \exp[j(\omega t - \beta x_1)],$$
$$\Phi'_m = \Phi'_{0m} \exp(j\gamma_m x_3) \exp[j(\omega t - \beta x_1)], \quad m = 1, 2, 3, 4, \qquad (2.46)$$

where $\mathbf{u}'_{0m}$ and Φ'_{0m} are the displacement and potential corresponding to γ_m.

In the half-space it is assumed that the solution is a linear sum of these partial waves, so that

$$\mathbf{u} = \sum_{m=1}^{4} A_m \mathbf{u}'_m,$$

$$\Phi = \sum_{m=1}^{4} A_m \Phi'_m. \qquad (2.47)$$

The coefficients A_m are to be such that the solution satisfies the boundary conditions, given by eq. (2.44) and either eq. (2.42) (for the free-surface case) or eq. (2.43) (for the metallized case). These conditions give a determinant which must be zero for a valid solution. The determinant will only be zero if the correct value of β has been chosen. Thus, to find the solution the entire procedure is iterated using different values for β until the determinant vanishes. The velocity of the wave is then ω/β, and the displacements and potential are given by eq. (2.47).

An alternative approach is to evaluate the effective permittivity $\varepsilon_s(\beta)$, described in Chapter 3. This function is zero for $\omega/\beta = v_f$, and infinite for $\omega/\beta = v_m$.

2.3.3 Surface-wave solutions

The solutions obtained by the above procedure are of course significantly affected by the anisotropy of the material and its orientation. The determination of surface-wave characteristics in general is a considerable task because of the variety of crystal symmetries, and because for any one symmetry the orientation

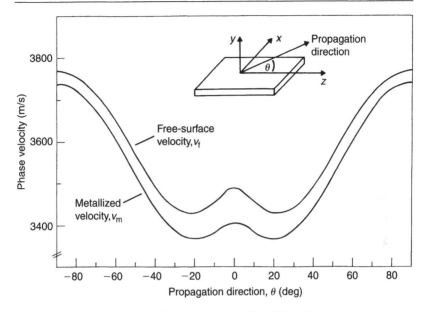

Figure 2.9 Rayleigh-wave velocities for *Y*-cut lithium niobate.

depends on three angular variables. However, a very extensive range of cases has been studied, as shown for example by Refs. [1, 2, 18, 19]. Except in very rare cases [21], there is always one or more surface-wave solution, whatever the symmetry and orientation. In general the solution involves all the three components of the displacement, so that the motion is not confined to the sagittal plane. However, since there is no variation in the x_2 direction the electric field **E**, given by the negative gradient of Φ, is always in the sagittal plane. For a metallized surface the parallel component, E_1, is zero at the surface though the normal component, E_3, is not generally zero. In some orientations the solution has no associated electric field, and can thus be described as non-piezoelectric. In this case the solution is not affected by metallization of the surface.

In general, the velocity of a surface wave must be less than the velocities of plane waves propagating in the x_1 direction in an infinite material. The reason for this is that, as for isotropic materials, the wave vectors of the partial waves must not have real x_3 components. Of the three plane waves (longitudinal, fast shear, slow shear) the slow shear wave has the lowest velocity, so the surface-wave velocity must be less than this. Usually the surface-wave velocity is quite close to the slow shear-wave velocity.

The solution encountered most frequently has its displacement **u** directed parallel, or almost parallel, to the sagittal plane, and it has an associated electric

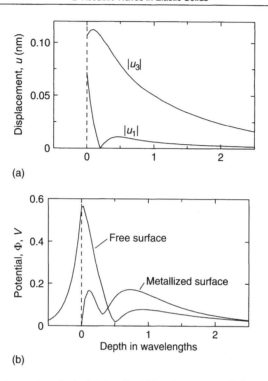

(a)

(b)

Figure 2.10 Displacements and potentials for a Rayleigh surface wave on $Y-Z$ lithium niobate. (a) Displacements for free-surface case. (b) Potentials for free-surface and metallized cases. For the metallized case, the displacements are almost identical to those for the free-surface case.

field. This solution is called the *piezoelectric Rayleigh wave*. It is similar to the Rayleigh wave for an isotropic material, with its behavior modified somewhat by anisotropy and piezoelectricity. The penetration depth is typically about one wavelength. This type of solution is found in, for example, Y-cut lithium niobate. For this material the velocities are shown in Fig. 2.9, as functions of propagation direction. Note that the metallized velocity v_m is less than the free-surface velocity v_f, which is always the case for a piezoelectrically coupled wave. For Y-cut lithium niobate the marked difference between the two velocities shows that the piezoelectric coupling is strong. The coupling is strongest for propagation in the Z direction and this orientation, described as Y-Z, is often used for surface-wave devices. The displacements and potential are shown in Fig. 2.10, where the scales are appropriate for a power density of 1 mW/mm and a frequency of 100 MHz.

A quite different solution, called the Bleustein–Gulyaev wave [22–25], occurs if the sagittal plane is normal to an even-order axis (2-, 4- or 6-fold) of the crystal. This wave has its displacement normal to the sagittal plane, and it has an associated electric field. It is closely related to the plane SH wave that can propagate in an isotropic half-space; in effect, piezoelectricity has caused the wave to become bound to the surface. However, the wave is not very strongly bound, even in a strongly piezoelectric material. For example, in cadmium sulfide the penetration depth is typically four wavelengths for a metallized surface, and 44 wavelengths for a free surface. The velocity is very close to that of slow shear plane waves. For the same orientation there is also a separate non-piezoelectric Rayleigh-wave solution, with its displacements in the sagittal plane. For a given surface orientation, the Bleustein–Gulyaev solution is generally found only for a limited range of propagation directions [25].

The effects associated with crystal symmetry can be summarized as follows [2, 3]:

(a) If the sagittal plane (x_1, x_3) is a plane of mirror symmetry of the crystal, then the transverse displacement u_2 is decoupled from u_1, u_3 and Φ. In this case, there may be a piezoelectric Rayleigh-wave solution with **u** in the sagittal plane.

(b) If the sagittal plane is normal to an even-order axis of the crystal, then the components u_1 and u_3 are decoupled from u_2 and Φ. In this case, there may be a non-piezoelectric Rayleigh wave with **u** confined to the sagittal plane, and possibly also a Bleustein–Gulyaev-wave solution.

These solutions are called 'pure' modes. They have the property that the wave velocity in the $x_1 - x_2$ plane is symmetric about the x_1 direction, and therefore it has a maximum or minimum in this direction.

The Rayleigh-wave and Bleustein–Gulyaev-wave solutions are strongly related to the Rayleigh- and SH-wave solutions for an isotropic half-space, and these are in turn related to the solutions for a layered half-space. These relationships are summarized in Table 2.1, which includes other solutions described below.

2.3.4 Other solutions

In some cases there can be another type of solution known as a *pseudo-surface acoustic wave* (PSAW). Rather surprisingly, the PSAW has a phase velocity higher than that of the slow shear wave, though it is less than that of the fast shear wave. Despite this the PSAW is a true surface wave, since its displacements decay exponentially with depth and it propagates without attenuation. This behavior is possible because the displacement of the PSAW is perpendicular

Table 2.1 Summary of wave types.

Isotropic half-space		Anisotropic half-space
Non-layered	Layered	
Rayleigh wave	Rayleigh wave (dispersive)	Rayleigh wave Pseudo-surface wave Leaky surface wave
SH plane wave	Love wave SH (dispersive)	Bleustein–Gulyaev wave

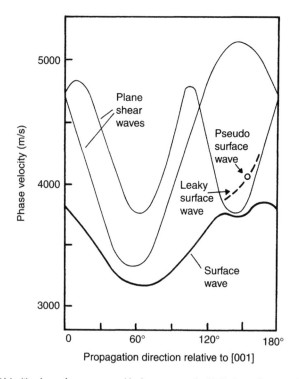

Figure 2.11 Velocities for surface waves and leaky waves on the Y–Z plane of quartz. After Farnell [1]. Copyright 1970, with permission from Elsevier.

to the displacement of the slow shear wave, so that the PSAW does not have a partial wave component corresponding to the slow shear wave. An example is on the Y–Z plane of quartz [1, 26], which gives velocities shown in Fig. 2.11. For all directions in this plane there is a Rayleigh-wave solution, but in addition a PSAW exists for propagation at an angle of 153° from the Z-axis. The Rayleigh wave and the PSAW are both piezoelectric. The plane shear-wave velocities for an unbounded medium are also shown.

For propagation in a slightly different direction the boundary conditions cannot be satisfied without including a partial wave corresponding to the slow shear bulk wave, and the latter carries some energy away from the surface and thus causes attenuation. However, the attenuation can be small, so that the wave behaves much like a true surface wave. This is a *leaky surface acoustic wave* (LSAW). The solution may be found by the procedure outlined above for surface waves (Section 2.3.2), allowing the wavenumber β to become complex. For the $Y-Z$ plane of quartz the attenuation is less than 0.006 dB per wavelength for propagation directions covering a range of about 30°. Experimentally, an attenuation as small as this is difficult to detect, and the wave appears to behave as a true surface wave. The term 'leaky wave' is often taken to include the PSAW case. Other leaky-wave solutions are found in lithium niobate and lithium tantalate, and the important role of these waves in practical devices will be described later in Chapter 11, Section 11.4.

2.3.5 Surface waves in layered substrates: perturbation theory

The effect of a layer on surface waves was considered for isotropic materials in Section 2.2.4, where it was shown that the layer causes Rayleigh waves to become dispersive, and a series of modes can exist. In the piezoelectric case the analysis becomes much more complex. Practical interest in layers arises for several reasons. For example, metal films are used for transducer electrodes, dielectric films can be used to improve the temperature stability of some devices, and piezoelectric films can be deposited on non-piezoelectric substrates to enable transducers to function. These topics are considered in more detail in Chapter 4.

The analysis of layered systems including piezoelectricity employs principles similar to the analysis of a half-space, described above. However, it is substantially more complex [27–29], and details are not given here.

A case of common interest concerns a metal layer on a piezoelectric substrate, since a metal layer is used for transducers and other structures. Consider a piezoelectric half-space with a uniform metal layer, as in Fig. 2.5. The layer is taken to be isotropic, and the wave is assumed to be a piezoelectric Rayleigh wave. A very thin layer reduces the velocity from v_f to v_m because it shorts out the parallel electric field at the surface, as noted before. This effect is called *electrical loading*. If the layer thickness is finite its mechanical properties become relevant, causing an additional change of velocity. This effect is called *mass loading*. This is change is frequency dependent, that is, it is dispersive. A series of higher modes can exist, but in pratical devices this is not usually relevant.

In practical cases the loading effects are small – electrical loading causes a velocity change of typically a few percent, and mass loading must be controlled to avoid undue dispersion. For such cases a perturbation theory gives a good estimate of the velocity change involved. For electrical loading, the theory gives [2, 12]

$$\frac{\Delta v}{v} \equiv \frac{v_f - v_m}{v_f} \approx \frac{(\varepsilon_0 + \varepsilon_p^T)|\phi_s|^2 \omega}{4 P_s}, \tag{2.48}$$

where $\varepsilon_p^T = [\varepsilon_{33}^T \varepsilon_{11}^T - (\varepsilon_{13}^T)^2]^{1/2}$ and the permittivities ε_{ij}^T are measured at constant stress. In addition, ϕ_s is the surface potential associated with the wave and P_s is the wave power per unit width in the x_2 direction. Equation (2.48) will be derived from a Green's function theory in Chapter 3. The perturbation theory can also be used for other problems, such as the velocity change due to mass loading [2, 12].

REFERENCES

1. G.W. Farnell. 'Properties of elastic surface waves', in W.P. Mason and R.N. Thurston (eds.), *Physical Acoustics*, Vol. 6, Academic Press, 1970, pp. 109–166.
2. G.W. Farnell. 'Elastic surface waves', in H. Matthews (ed.), *Surface Wave Filters*, Wiley, 1977, pp. 1–53.
3. D. Royer and E. Dieulesaint. *Elastic Waves in Solids*, Vol. 1, Springer, 2000.
4. B.A. Auld. *Acoustic Fields and Waves in Solids*, Vol. 1, Krieger, 1990.
5. J.F. Nye. *Physical Properties of Crystals – Their Representation by Tensors and Matrices*, Oxford University Press, 1957.
6. H.F. Tiersten. *Linear Piezoelectric Plate Vibrations*, Plenum Press, 1969.
7. I.A. Viktorov. *Rayleigh and Lamb Waves*, Plenum Press, 1967.
8. A.A. Oliner. 'Microwave network methods for guided elastic waves', *IEEE Trans.*, **MTT-17**, 812–826 (1969).
9. M. Redwood. *Mechanical Waveguides*, Pergamon, 1960.
10. Lord Rayleigh. 'On waves propagated along the plane surface of an elastic solid', *Proc. Lond. Math. Soc.*, **17**, 4–11 (1885).
11. G.W. Farnell and E.L. Adler. 'Elastic wave propagation in thin layers', in W.P. Mason and R.N. Thurston (eds.), *Physical Acoustics*, Vol. 9, Academic Press, 1972, Chapter 2.
12. B.A. Auld. *Acoustic Fields and Waves in Solids*, Vol. 2, Krieger, 1990.
13. L.M. Brekhovskikh and O.A. Godin. *Acoustics of Layered Media*, Springer, 1990 (Vol. 1), 1992 (Vol. 2).
14. K. Sezawa. 'Dispersion of elastic waves propagated on the surface of stratified bodies and on curved surfaces', *Bull. Earthquake Res. Inst. Tokyo*, **3**, 1–18 (1927).
15. H.M. Mooney and B.A. Bolt. 'Dispersion characteristics of the first three Rayleigh modes for a single surface layer', *Bull. Seism. Soc. Am.*, **56**, 43–67 (1966).
16. A.E.H. Love. *Some Problems of Geodynamics*, Cambridge University Press, 1911; Dover, 1967.
17. M.J.P. Musgrave. *Crystal Acoustics*, Acoustical Society of America, 2002.

18. J.J. Campbell and W.R. Jones. 'A method for estimating optimal crystal cuts and propagation directions for excitation of piezoelectric surface waves', *IEEE Trans.*, **SU-15**, 209–217 (1968).

19. A.J. Slobodnik, E.D. Conway and R.T. Delmonico. *Microwave Acoustics Handbook*, Vol. 1A, Surface wave velocities, Air Force Cambridge Research Laboratories, AFCRL-TR-73-0597, 1973.

20. IEEE Standard 176 on Piezoelectricity. *IEEE Trans,* **SU-31**, March 1984, Part II. Also T.R. Meitzler, *IEEE Trans,* **SU-31**, 135–136 (1984).

21. R.C. Peach. 'On the existence of surface acoustic waves on piezoelectric substrates', *IEEE Trans. Ultrason. Ferroelect. Freq. Contr.*, **48**, 1308–1320 (2001).

22. J.L. Bleustein. 'A new surface wave in piezoelectric crystals', *Appl. Phys. Lett.*, **13**, 412–413 (1968).

23. Y.V. Gulyaev. 'Electroacoustic surface waves in solids', *Soviet Phys. JETP Lett.*, **9**, 37–38 (1969).

24. Y. Ohta, K. Nakamura and H. Shimizu. 'Surface concentration of shear wave on piezoelectric materials with a conductor', *Technical Report of IECE, Japan*, **US69-3**, 1969 (in Japanese).

25. G. Koerber and R.F. Vogel. 'SH-mode piezoelectric surface waves on rotated cuts', *IEEE Trans.*, **SU-20**, 9–12 (1973).

26. H. Engan, K.A. Ingebrigtsen and A. Tonning. 'Elastic surface waves in alpha quartz: observation of leaky surface waves', *Appl. Phys. Lett.*, **10**, 311–313 (1967).

27. E.L. Adler. 'SAW and pseudo-SAW properties using matrix methods', *IEEE Trans. Ultrason., Ferroelect., Freq. Contr.*, **41**, 876–882 (1994).

28. N. Nakahata, A. Hachigo, K. Higaki, S. Fujii, S. Shikata and N. Fujimori. 'Theoretical study on SAW characteristics of layered structures including a diamond layer', *IEEE Trans. Ultrason. Ferroelect. Freq. Contr.*, **42**, 362–375 (1995).

29. L. Wang and S.I. Rohklin. 'Compliance/stiffness matrix formulation of general Green's function and effective permittivity for piezoelectric multilayers', *IEEE Trans. Ultrason. Ferroelect. Freq. Contr.*, **51**, 453–463 (2004).

3

ELECTRICAL EXCITATION AT A
PLANE SURFACE

For a piezoelectric material, the surface waves described in Chapter 2 can be generated and detected electrically by means of metal electrodes on the surface. This principle is used in interdigital transducers and multistrip couplers. A basic concept used in the analysis of these components is the effective permittivity, which gives a description of the electrical behavior of the surface taking account of the acoustic behavior of the material. This chapter gives the basic theory. Sections 3.1 to 3.4 describe the effective permittivity and the Green's function for excitation of a half-space, including some approximations that will be used for analysis of transducers in later chapters. In Section 3.5 some other applications of the basic concepts are described briefly.

For practical situations, the electric fields are adequately approximated by taking the electrode thickness to be zero, so this is assumed in this chapter. This also implies that mechanical perturbations are ignored here.

3.1 ELECTROSTATIC CASE

We first consider electrical excitation for the electrostatic case, that is, for a non-piezoelectric material [1–3]. In practical situations the material is always piezoelectric, of course. However, the electrostatic solution is of considerable importance because it can be used as a first approximation in the piezoelectric case. It will be used extensively in later chapters.

Coordinate axes are defined as in Fig. 3.1. A homogeneous anisotropic dielectric occupies the space $x_3 < 0$ with a vacuum above. We shall be concerned with cases where electrodes are present at the surface, so there will in general be free charges in the plane $x_3 = 0$. However, the electrodes are not considered

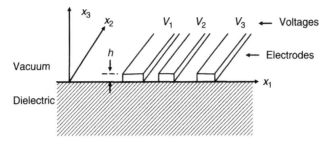

Figure 3.1 General configuration of electrodes on a dielectric substrate. Fields are independent of x_2, and the waves propagate in $\pm x_1$ directions.

explicitly at this stage. They will be allowed for later by applying appropriate boundary conditions. The potential is assumed to be invariant in the x_2 direction, so the electrode edges must be parallel to the x_2 axis. The film thickness h is assumed to be negligible.

We first consider the field in the dielectric region $x_3 < 0$. The fields in the vacuum are strongly related, and they will be considered later. In the dielectric region we have div $\mathbf{D} = 0$ and $\mathbf{E} = -\text{grad } \Phi$, assuming the electric field to be static. Using $\partial/\partial x_2 = 0$ and $\mathbf{D} = \boldsymbol{\varepsilon} \cdot \mathbf{E}$, we find the potential Φ is given by

$$\varepsilon_{11}\frac{\partial^2 \Phi}{\partial x_1^2} + 2\varepsilon_{13}\frac{\partial^2 \Phi}{\partial x_1 \partial x_3} + \varepsilon_{33}\frac{\partial^2 \Phi}{\partial x_3^2} = 0, \quad \text{for } x_3 < 0. \tag{3.1}$$

Consider first a harmonic solution in which the potential varies as $\exp(j\beta x_1)$. The potential is required to vanish at $x_3 = -\infty$. Taking β to be real, a solution that satisfies this requirement and also satisfies eq. (3.1) is given by

$$\tilde{\Phi}(x_1, x_3) = \exp[j\beta(x_1 - x_3\varepsilon_{13}/\varepsilon_{33}) + |\beta|x_3\varepsilon_p/\varepsilon_{33}], \quad \text{for } x_3 < 0, \tag{3.2}$$

where

$$\varepsilon_p = \sqrt{\varepsilon_{11}\varepsilon_{33} - \varepsilon_{13}^2} \tag{3.3}$$

and the tilde is used to indicate the harmonic solution. With this potential, the normal component of the electric displacement is

$$\tilde{D}_3(x_1, x_3) = -\varepsilon_p|\beta|\tilde{\Phi}(x_1, x_3), \quad \text{for } x_3 < 0. \tag{3.4}$$

At the surface $x_3 = 0$, the value of $\tilde{D}_3$ gives $\tilde{\sigma}_-(x_1)$, the density of free charges on the dielectric side of the electrodes. Define $\tilde{D}_{3-}$ as the value of $\tilde{D}_3(x_1, 0)$ in the dielectric, and $\tilde{\phi}(x_1) \equiv \tilde{\Phi}(x_1, 0)$ as the surface potential. We then have

$$\tilde{\sigma}_-(x_1) = -\tilde{D}_{3-} = \varepsilon_p |\beta| \tilde{\phi}(x_1). \tag{3.5}$$

A similar development gives the fields in the vacuum region $x_3 > 0$. The equations are the same, except that we set $\varepsilon_{11} = \varepsilon_{33} = \varepsilon_0$ and $\varepsilon_{13} = 0$. Equation (3.1) reduces to $\nabla^2 \Phi = 0$. In addition, $|\beta|$ is replaced by $-|\beta|$ in eq. (3.2). At $x_3 = 0$ the value of $\tilde{D}_3$ is denoted by $\tilde{D}_{3+}$, and the charge density on the vacuum side of the electrodes is $\tilde{\sigma}_+(x_1) = \tilde{D}_{3+} = \varepsilon_0 |\beta| \tilde{\phi}(x_1)$. The total charge density is therefore

$$\tilde{\sigma}(x_1) = \tilde{\sigma}_+(x_1) + \tilde{\sigma}_-(x_1) = (\varepsilon_0 + \varepsilon_p) |\beta| \tilde{\phi}(x_1). \tag{3.6}$$

The term $(\varepsilon_0 + \varepsilon_p)$ is an effective permittivity. From the above equations, we find that the discontinuity in $\tilde{D}_3$ is related to $\tilde{E}_1$ by $\tilde{D}_{3+} - \tilde{D}_{3-} =$ j $\mathrm{sgn}(\beta)(\varepsilon_0 + \varepsilon_p)\tilde{E}_1(x_1, 0)$, where $\mathrm{sgn}(\beta) = 1$ for $\beta > 0$ and $\mathrm{sgn}(\beta) = -1$ for $\beta < 0$.

We now consider a more general solution. In eq. (3.6), $\tilde{\sigma}(x_1)$ and $\tilde{\phi}(x_1)$ are both proportional to $\exp(j\beta x_1)$. The general solution, with charge density $\sigma(x_1)$ and surface potential $\phi(x_1)$, is obtained by Fourier synthesis. We have

$$\sigma(x_1) = \frac{1}{2\pi} \int_{-\infty}^{\infty} \overline{\sigma}(\beta) \exp(j\beta x_1) d\beta, \tag{3.7}$$

and a similar equation relating $\phi(x_1)$ to $\overline{\phi}(\beta)$, where $\overline{\sigma}(\beta)$ and $\overline{\phi}(\beta)$ are the Fourier transforms of $\sigma(x_1)$ and $\phi(x_1)$. The general solution is an infinite sum of harmonic solutions with different values of β. For each β, eq. (3.6) applies, and we thus have

$$\overline{\sigma}(\beta) = (\varepsilon_0 + \varepsilon_p) |\beta| \overline{\phi}(\beta). \tag{3.8}$$

Thus the charge density $\sigma(x_1)$ may be obtained for any surface potential $\phi(x_1)$.

It should be noted that $\overline{\phi}(\beta)$ must be zero at $\beta = 0$, because if this were not so the potential would be finite at $x_3 = \pm\infty$, as can be seen from eq. (3.2) for $\beta = 0$. It follows from eq. (3.8) that $\overline{\sigma}(\beta)$ is also zero at $\beta = 0$, and from the definition of the Fourier transform this gives

$$\int_{-\infty}^{\infty} \sigma(x_1) dx_1 = 0, \tag{3.9}$$

that is, the sum total of all the charges at the surface $x_3 = 0$ is zero.

Corresponding relationships in the spatial domain can be obtained using Fourier analysis. At the surface, we have $E_1(x_1, 0) = -\partial\phi/\partial x_1$, and hence its Fourier transform is $\overline{E}_1(\beta) = -j\beta\overline{\phi}(\beta)$. Equation (3.8) thus gives

$$(\varepsilon_0 + \varepsilon_p)\overline{E}_1(\beta) = -j\text{sgn}(\beta)\overline{\sigma}(\beta).$$

This can be transformed to the x_1 domain by using the convolution theorem, eq. (A.13), noting that the inverse transform of $\text{sgn}(\beta)$ is $j/(\pi x_1)$ as shown by eq. (A.27). This gives

$$(\varepsilon_0 + \varepsilon_p)E_1(x_1, 0) = \sigma(x_1) * \frac{1}{\pi x_1}, \tag{3.10}$$

where the asterisk indicates convolution. The surface potential $\phi(x_1)$ can be obtained by integrating $E_1(x_1, 0)$, and with the aid of eq. (3.9) this gives

$$\phi(x_1) = G_e(x_1) * \sigma(x_1) = \frac{-\ln|x_1|}{\pi(\varepsilon_0 + \varepsilon_p)} * \sigma(x_1). \tag{3.11}$$

Here $G_e(x_1)$ is the electrostatic Green's function, which can be interpreted as the surface potential associated with a line of charge at $x_1 = 0$. To obtain solutions for $\phi(x_1)$ and $\sigma(x_1)$, it is also necessary to use the boundary conditions that $\phi(x_1)$ must be constant over each electrode and $\sigma(x_1)$ must be zero in any unmetallized region. Methods for obtaining solutions are considered later in Chapter 5, Section 5.1.2.

Relation to free-space solution

Given some set of electrodes with specified voltages, the field in the vacuum region $x_3 \geq 0$ is determined, irrespective of the substrate material. This can be seen from the above equations. The field in the substrate region $x_3 \leq 0$ is also determined. Furthermore, these fields are strongly related [3, 4]. Consequently, it is sufficient to solve for the fields assuming a vacuum everywhere, and then apply simple scaling. Suppose the vacuum case gives a charge density $\sigma_v(x_1)$ on each side of the electrodes. Then, if the vacuum in the region $x_3 < 0$ is replaced by a dielectric substrate, the charge density on this side of the electrodes is $(\varepsilon_p/\varepsilon_0)\sigma_v(x_1)$. Hence, the total charge density is $\sigma(x_1) = (1 + \varepsilon_p/\varepsilon_0)\sigma_v(x_1)$.

There is also a simple derivation of the Green's function. The radial field at a distance r from a line of charge in a vacuum, with charge q per unit length, is $E = q/(2\pi\varepsilon_0 r)$. Integrating from some point r_0 to r, the potential is $\Phi = -q\ln(r/r_0)/(2\pi\varepsilon_0) + \text{constant}$. This gives the Green's function of eq. (3.11).

3.2 PIEZOELECTRIC HALF-SPACE

In Section 3.1 it was shown that the potential and charge density at the surface of a non-piezoelectric half-space are related by an effective permittivity $(\varepsilon_0 + \varepsilon_p)$. Here we consider the effective permittivity for a piezoelectric half-space. The method was first given by Ingebrigtsen [1] and developed later by Greebe *et al.* [5] and Milsom *et al.* [6, 7]. The theory here follows most closely that of Milsom. Other approaches are the perturbation theory [8, 9] and normal mode theory [10, 11]. These give results that are essentially the same as the method used here.

It is assumed that the variables are proportional to $\exp(j\omega t)$, with the frequency ω positive. As in Section 3.1 we consider initially a harmonic solution with variables proportional to $\exp(j\beta x_1)$, with β real, and generalize later using Fourier synthesis. The procedure is similar to that used to obtain surface-wave velocities, Section 2.3.2 in Chapter 2, but here the electrical boundary condition at the surface is not specified. This enables a solution to be obtained for any value of β. Another difference is that positive β values refer to propagation in the $-x_1$ direction instead of the $+x_1$ direction. This is done for convenience when using Fourier synthesis.

As in Chapter 2, Section 2.3.2, we consider partial waves in which the displacements $\mathbf{u}'$ and potential Φ' have the form

$$\mathbf{u}' = \mathbf{u}'_0 \exp(j\gamma x_3) \exp[j(\omega t + \beta x_1)],$$
$$\Phi' = \Phi'_0 \exp(j\gamma x_3) \exp[j(\omega t + \beta x_1)], \tag{3.12}$$

where $\mathbf{u}'_0$ and Φ'_0 are constants and γ is the x_3 component of the wave vector, which by definition has no x_2 component. These expressions are required to satisfy the equation of motion, eqs (2.13), for an infinite medium, with the material tensors rotated into the frame of the axes x_1, x_2, x_3. Substitution into eqs (2.13) gives four linear homogeneous equations in the four variables $\mathbf{u}'_0$ and Φ'_0, and for non-trivial solutions the determinant of coefficients is set to zero. The determinant is an eighth-order polynomial in γ and therefore has eight roots, and for each root the equations give the relative values of $\mathbf{u}'_0$ and Φ'_0. Four of the roots are unacceptable, however, because they do not correspond to excitation at the surface. Care is needed in choosing acceptable roots, because the solution is not necessarily a surface-wave solution. Complex or imaginary values of γ are acceptable if the imaginary part is negative, so that $\mathbf{u}'_0$ and Φ'_0 decay away from the surface. Real values of γ give plane waves, and these are acceptable only if they carry energy away from the surface. Usually this requires γ to have its sign opposite to β.

The four acceptable partial-wave solutions are written as

$$\mathbf{u}'_m = \mathbf{u}'_{0m} \exp(j\gamma_m x_3) \exp[j(\omega t + \beta x_1)],$$

$$\Phi'_m = \Phi'_{0m} \exp(j\gamma_m x_3) \exp[j(\omega t + \beta x_1)], \quad m = 1, 2, 3, 4. \tag{3.13}$$

The total solution in the half-space has displacements $\tilde{\mathbf{u}}$ and $\tilde{\Phi}$, where the tilde indicates a harmonic solution with variables proportional to $\exp(j\beta x_1)$. The total solution is taken as a linear combination of the partial waves, so that

$$\tilde{\mathbf{u}} = \sum_{m=1}^{4} A_m \mathbf{u}'_m, \quad \tilde{\Phi} = \sum_{m=1}^{4} A_m \Phi'_m. \tag{3.14}$$

The mechanical boundary conditions require that there are no forces on the surface, so that $T_{13} = T_{23} = T_{33} = 0$ at $x_3 = 0$, with the stresses given by eq. (2.8). The electrical boundary conditions are not specified here. We thus have three homogeneous equations relating the four constants A_m, and the relative values of these constants can be found. The relative values of $\tilde{\mathbf{u}}$ and $\tilde{\Phi}$ can be obtained from eq. (3.14), giving a solution for any value of β.

The surface boundary conditions concern the potential and the displacement $\tilde{\mathbf{D}}$. The latter can be calculated from $\tilde{\Phi}$ and $\tilde{\mathbf{u}}$ using eq. (2.9). At the surface, the value of $\tilde{D}_3$ is denoted by $\tilde{D}_{3-}$ on the piezoelectric side, and $\tilde{D}_{3+}$ on the vacuum side. The surface potential is $\tilde{\phi}(x_1) \equiv \tilde{\Phi}(x_1, 0)$. The ratio $\tilde{D}_{3-}/\tilde{\phi}(x_1)$ is determined by the above solution, and in general it is a function of β.

In the vacuum $x_3 > 0$, the potential satisfies Laplace's equation $\nabla^2 \tilde{\Phi} = 0$. Since $\tilde{\Phi}$ is proportional to $\exp(j\beta x_1)$ and it must vanish at $x_3 = \infty$, the x_3 dependence is $\exp(-|\beta|x_3)$, and so $\tilde{\Phi}(x_1, x_3) = \tilde{\phi}(x_1) \exp(-|\beta|x_3)$. It follows that

$$\tilde{D}_{3+} = \varepsilon_0 |\beta| \tilde{\phi}(x_1). \tag{3.15}$$

At the surface, the discontinuity in $\tilde{D}_3$ is related to the potential by the *effective permittivity* $\varepsilon_s(\beta)$, defined by

$$\varepsilon_s(\beta) = \frac{\tilde{D}_{3+} - \tilde{D}_{3-}}{|\beta| \tilde{\phi}(x_1)}. \tag{3.16}$$

The effective permittivity thus gives the electrical behavior of the interface between the vacuum and the piezoelectric half-space.

If $\tilde{D}_{3+}$ and $\tilde{D}_{3-}$ differ, there must be free charges present at the surface, implying the presence of electrodes. The charge densities on the upper and lower sides are

$\tilde{D}_{3+}$ and $-\tilde{D}_{3-}$, respectively. Thus, if the total charge density is $\tilde{\sigma}(x_1)$, we have

$$\varepsilon_s(\beta) = \frac{\tilde{\sigma}(x_1)}{|\beta|\tilde{\phi}(x_1)}, \qquad (3.17)$$

where $\tilde{\sigma}(x_1)$ and $\tilde{\phi}(x_1)$ are both proportional to $\exp[j(\omega t + \beta x_1)]$. This corresponds to the electrostatic case, eq. (3.6). Some authors exclude the charges on the vacuum side when defining $\varepsilon_s(\beta)$, that is, they omit the term $\tilde{D}_{3+}$ in eq. (3.16). This reduces the value of $\varepsilon_s(\beta)$ by an amount ε_0.

In the above equations the potential $\tilde{\phi}(x_1)$ and charge density $\tilde{\sigma}(x_1)$ are proportional to $\exp(j\omega t)$, and the frequency ω was taken to be constant throughout. If ω is changed the value of $\varepsilon_s(\beta)$ changes, so $\varepsilon_s(\beta)$ is a function of ω as well as β. However, $\varepsilon_s(\beta)$ is essentially the ratio of $\tilde{D}_3$ to $\tilde{E}_1$, as shown by eq. (3.16), and it can be seen that it remains unchanged if ω and β are changed in proportion. Thus, $\varepsilon_s(\beta)$ is a function of the normalized quantity β/ω. This has dimensions the same as the reciprocal of velocity, and it is referred to as the 'slowness'. In this chapter the analysis applies for constant frequency, and the effective permittivity is written as $\varepsilon_s(\beta)$ without showing the frequency dependence explicitly.

A more general solution, with surface potential $\phi(x_1)$ and charge density $\sigma(x_1)$, is readily obtained by Fourier synthesis. The method is the same as in the electrostatic case of Section 3.1, so the result follows directly. Thus, comparing with eq. (3.6), the general solution obtained from eq. (3.17) gives

$$\varepsilon_s(\beta) = \frac{\overline{\sigma}(\beta)}{|\beta|\overline{\phi}(\beta)}, \qquad (3.18)$$

where $\overline{\sigma}(\beta)$ and $\overline{\phi}(\beta)$ are respectively the Fourier transforms of $\sigma(x_1)$ and $\phi(x_1)$. Thus, given some general potential function $\phi(x_1)$, the corresponding charge density may be obtained by transforming to obtain $\overline{\phi}(\beta)$, using $\varepsilon_s(\beta)$ to obtain $\overline{\sigma}(\beta)$, and then transforming back to the x_1 domain.

For the general solution the potential $\phi(x_1)$ and charge density $\sigma(x_1)$ are proportional to $\exp(j\omega t)$, with the frequency ω regarded as a constant in the above equations. In solving a particular problem it is usually found that the potential and charge density are functions of frequency, so their transforms $\overline{\phi}(\beta)$ and $\overline{\sigma}(\beta)$ will also be functions of frequency. In the Fourier transform, the frequency ω is held constant during the integration. The relationship given by the effective permittivity, eq. (3.18), applies for all values of ω.

3.3 SOME PROPERTIES OF THE EFFECTIVE PERMITTIVITY

The effective permittivity described above is a powerful tool for solving problems concerning a one-dimensional set of electrodes on the surface of a piezoelectric half-space. For variables proportional to $\exp(j\omega t)$, the surface potential $\phi(x_1)$ and charge density $\sigma(x_1)$ are related by the effective permittivity, and the solution is then determined if appropriate boundary conditions are applied. Usually, $\phi(x_1)$ is specified at the electrode locations, while $\sigma(x_1)$ must be zero on all unmetallized regions. Acoustic-wave excitation is allowed for implicitly by the definition of the effective permittivity. This includes all forms of acoustic wave that can be excited; thus, in addition to the usual excitation of piezoelectric Rayleigh waves, the effective permittivity will when appropriate include the effects of pseudo-surface waves, bulk waves and Bleustein–Gulyaev waves. In fact, many of the properties of these waves in the material under consideration may be deduced by examining the effective permittivity.

However, it should be noted that the permittivity does not show the effect of any acoustic waves that are not piezoelectrically coupled at the surface. Such waves, which may occur in a piezoelectric material, cannot of course be excited by electrodes on the surface; nevertheless, they may be present in a practical device owing to mode conversion at a discontinuity, for example the edge of a substrate.

An important limitation of the method follows from the assumption that there are no mechanical forces on the surface. This implies that any electrodes must be sufficiently thin that they cause negligible mechanical loading. In transducers, a common practical effect of mechanical loading is internal reflectivity due to electrode reflections. This is often avoided by using double-electrode transducers. Another cause of reflectivity is electrical loading, and this can be analyzed using the effective permittivity. These effects are discussed further in Chapter 8, Section 8.1.4.

Generally, the effective permittivity is a complicated function of β, and it must be found numerically using the method of Section 3.2 above. There are, however, some important properties that are readily deduced. Firstly, the function is symmetrical, so that $\varepsilon_s(-\beta) = \varepsilon_s(\beta)$. This follows from the general reciprocity relation, as shown in Appendix B, Section B.4. Secondly, if $\varepsilon_s(\beta)$ has an imaginary part this indicates that energy is being radiated away from the surface, in the form of bulk waves. This can be seen from the definition involving the harmonic solution, eq. (3.17). The quantity $\tilde{\sigma}(x_1)$ is the charge density on some set of electrodes with edges parallel to the x_2 axis. Since the charge density is time variant there must in general be currents entering these electrodes from outside. For unit length in the x_2 direction, the current in an interval dx_1 is $J(x_1)$ dx_1, where $J(x_1) = j\omega\tilde{\sigma}(x_1)$ is the current density. Now, if there is no power transferred into the system from outside, $J(x_1)$ must be in phase quadrature with the potential $\tilde{\phi}(x_1)$, and eq. (3.17) shows that $\varepsilon_s(\beta)$ must then be real. In the steady-state condition a net transfer of power can only occur if bulk acoustic waves are radiated away from the surface. Hence a complex value of $\varepsilon_s(\beta)$ indicates that bulk waves are being excited. It is usually found that $\varepsilon_s(\beta)$ is complex for some values of β, and real for other values.

The form of $\varepsilon_s(\beta)$ depends markedly on the type of acoustic wave involved. Usually we are concerned with piezoelectric Rayleigh waves, though bulk-wave excitation has been investigated extensively because it occurs in most surface-wave devices to some extent. For excitation of Bleustein–Gulyaev waves, $\varepsilon_s(\beta)$ can be expressed as an analytic formula [5, 12].

In the following description the wave is taken to be a piezoelectric Rayleigh wave. For example, the permittivity for a Y–Z lithium niobate half-space is shown in Fig. 3.2, as a function of slowness $s = \beta/\omega$. The permittivity has a pole and a zero at two slowness values close together. From eq. (3.17), the zero corresponds to a surface-wave solution for a free surface, since the charge density is zero. Here the slowness is s_f and the wavenumber is $k_f > 0$, so that $\varepsilon_s(\pm k_f) = 0$. The pole of $\varepsilon_s(\beta)$ indicates a surface-wave solution for a metallized surface, since it gives finite charge density and zero potential. Here the slowness is s_m and the wavenumber is $k_m > 0$, and $\varepsilon_s(\pm k_m) = \infty$. The surface-wave velocities for these two cases are $v_f = \omega/k_f$ and $v_m = \omega/k_m$. At a lower value of β the permittivity becomes complex, and it remains complex for all smaller

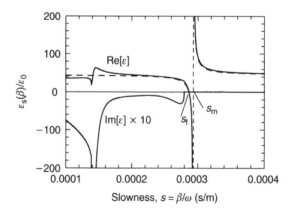

Figure 3.2 Effective permittivity for Y–Z lithium niobate. Solid line: calculated using program EPS [13]. Broken line: calculated using Ingebrigtsen's approximation.

values of β. In this region, bulk-wave radiation is occurring. For many practical purposes it is not necessary to consider bulk waves in any detail.

An important quantity is the differential of $\varepsilon_s(\beta)$ at the free-surface wave number k_f. This is related directly to the amplitude of a wave generated by a transducer, as shown in Appendix B, Section B.6. The same quantity relates the surface-wave power flow to the surface potential, as we now show.

Surface-wave power flow

Following Ingebrigtsen's approach [1], we consider a free-surface wave propagating with no attenuation. The surface potential $\tilde{\phi}(x_1)$ and charge density $\tilde{\sigma}(x_1)$ are both proportional to $\exp(j\beta x_1)$. Applying electrical power via some system of electrodes, the wave amplitude can be made to rise with x_1, and β will then have a small imaginary part. The surface-wave power density may be obtained by considering the electrical power applied to the system from outside.

Consider the power transferred in a width W in the x_2 direction. The current density at the surface is $J(x_1) = j\omega\tilde{\sigma}(x_1)$. The power flowing into the system, in a small interval Δx_1, is

$$\Delta P(x_1) = \frac{1}{2} W \mathrm{Re}[\tilde{\phi}^*(x_1)J(x_1)]\Delta x_1, \qquad (3.19)$$

where the asterisk indicates a complex conjugate. The potential and current density are written as $\tilde{\phi}(x_1) = \tilde{\phi}_0 \exp(j\beta x_1)$ and $\tilde{\sigma}(x_1) = \tilde{\sigma}_0 \exp(j\beta x_1)$, where $\tilde{\phi}_0$ and $\tilde{\sigma}_0$ are constants and the time variation $\exp(j\omega t)$ is omitted. Setting $\beta = \beta_r + j\beta_i$ gives

$$\Delta P(x_1) = -\frac{1}{2}\omega W \exp(-2\beta_i x_1)\,\mathrm{Im}[\tilde{\phi}_0^*\tilde{\sigma}_0]\Delta x_1. \qquad (3.20)$$

Now, for real β we have $\tilde{\sigma}_0 = f(\beta)\tilde{\phi}_0$ with, from eq. (3.17),

$$f(\beta) = |\beta|\varepsilon_s(\beta). \qquad (3.21)$$

It is assumed that $f(\beta)$ can be continued analytically for complex values of β near the real axis. Equation (3.20) thus becomes

$$\Delta P(x_1) = -\frac{1}{2}\omega W |\tilde{\phi}_0|^2 \exp(-2\beta_i x_1)\,\mathrm{Im}[f(\beta)]\Delta x_1. \qquad (3.22)$$

Using the Cauchy–Riemann equations and noting that $f(\beta)$ is real for $\beta_i = 0$, we have, for small β_i,

$$\mathrm{Im}[f(\beta)] = \beta_i \left[\frac{\partial f(\beta)}{\partial \beta}\right]_{\beta_i=0}. \qquad (3.23)$$

Now, since the potential $\tilde{\phi}(x_1)$ is proportional to $\tilde{\phi}_0 \exp(-\beta_i x_1)$, the wave power $P_s(x_1)$ must have the form $P_s(x_1) = C|\tilde{\phi}_0|^2 W \exp(-2\beta_i x_1)$, where C is a constant. The change of $P_s(x_1)$ over a distance Δx_1 is equated with $\Delta P(x_1)$ of eq. (3.22), giving

$$C = -\frac{1}{4}\omega\left[\frac{\partial f(\beta)}{\partial \beta}\right]_{\beta_i=0}. \tag{3.24}$$

where eq. (3.23) has also been used.

In the limit $\beta_i \to 0$, the wave power P_s is independent of x_1. We consider a wave on a free surface, so that $\beta = k_f$. For this case, using eq. (3.21) and noting that $\varepsilon_s(k_f) = 0$, we have

$$P_s = -\frac{1}{4}\omega W|\tilde{\phi}_0|^2 k_f\left[\frac{\partial \varepsilon_s(\beta)}{\partial \beta}\right]_{k_f}. \tag{3.25}$$

We define a real positive constant Γ_s by the equation

$$\frac{1}{\Gamma_s} = -k_f\left[\frac{\partial \varepsilon_s(\beta)}{\partial \beta}\right]_{k_f}. \tag{3.26}$$

This is independent of ω; it depends only on the properties of the half-space and its orientation. The surface-wave power flow is thus

$$P_s = \frac{1}{4}\omega W|\tilde{\phi}_0|^2/\Gamma_s. \tag{3.27}$$

Milsom *et al.* [6] give an alternative derivation.

An analogous formula applies if the surface surrounding the transducer is considered to be metallized, in which case the differential is evaluated at k_m instead of k_f [6]. Either of these expressions can be used to define a surface-wave coupling constant, and for Rayleigh waves the results are very similar [14].

Ingebrigtsen's approximation

A convenient approximate form for $\varepsilon_s(\beta)$, suitable when the main acoustic wave present is a piezoelectric Rayleigh wave, was given by Ingebrigtsen [1, 15]. Since $\varepsilon_s(\beta)$ is zero at $\beta = k_f$, it must be proportional to $(\beta - k_f)$ when β is close to k_f. This assumes that $\varepsilon_s(\beta)$ is well behaved here, so that Taylor's theorem applies. Similarly, $1/\varepsilon_s(\beta)$ is zero when $\beta = k_m$, so it must be proportional to

$(\beta - k_m)$ when β is close to k_m. Thus, $\varepsilon_s(\beta)$ is proportional to $(\beta - k_f)/(\beta - k_m)$. This is modified a little to make it an even function of β, giving the formula

$$\varepsilon_s(\beta) \approx \varepsilon_\infty \frac{\beta^2 - k_f^2}{\beta^2 - k_m^2}, \tag{3.28}$$

where ε_∞ is a constant. By setting $\beta \to \infty$, we have $\varepsilon_\infty \equiv \varepsilon_s(\infty)$. The notation ε_∞ is introduced for brevity. The value can be found by evaluating $\varepsilon_s(\beta)$ numerically. To find an approximate value we can compare eqs (3.8) and (3.18), showing that $\varepsilon_\infty = (\varepsilon_0 + \varepsilon_p)$ for the electrostatic case. For the piezoelectric case this remains approximately true because we are concerned with weakly piezoelectric materials. However, the mechanical conditions affect the dielectric constant ε_{ij}. Since the wave exists near the force-free surface, it seems a reasonable approximation to use the constant-stress components ε_{ij}^T. We therefore define ε_p^T as in eq. (3.3) but using these components, so that

$$\varepsilon_p^T = \sqrt{\varepsilon_{11}^T \varepsilon_{33}^T - (\varepsilon_{13}^T)^2}, \tag{3.29}$$

and the constant in eq. (3.28) is then

$$\varepsilon_\infty \equiv \varepsilon_s(\infty) \approx \varepsilon_0 + \varepsilon_p^T. \tag{3.30}$$

Differentiating eq. (3.28), the constant Γ_s of eq. (3.26) is given by

$$\Gamma_s \approx \frac{1}{\varepsilon_\infty} \frac{v_f - v_m}{v_f}. \tag{3.31}$$

These equations are consistent with the power flow formula obtained from the perturbation theory, eq. (2.48), and with normal mode theory [11].

Ingebrigtsen's approximation is compared with the accurate result on Fig. 3.2, for a Y–Z lithium niobate substrate. The agreement is very good except in the small-slowness region where the actual permittivity becomes complex; here the approximate form remains real. The approximate form of the permittivity is convenient for analysis of practical devices because it depends only on three variables, and numerical data for these are readily available. The main limitation is that bulk-wave excitation is excluded, since the function is real for all β. The approximation is valid only for piezoelectric Rayleigh waves; for example, it is not valid for leaky waves or Bleustein–Gulyaev waves.

3.4 GREEN'S FUNCTION

The effective permittivity $\varepsilon_s(\beta)$ relates the charge density to the surface potential in the β domain. However, we are commonly concerned with problems in which the boundary conditions are expressed in the x_1 domain. The relationship in this domain can be expressed using a Green's function [6, 7].

From eq. (3.18) the potential and charge density are related in the β domain by

$$\overline{\phi}(\beta) = \overline{G}(\beta, \omega)\,\overline{\sigma}(\beta), \qquad (3.32)$$

where

$$\overline{G}(\beta, \omega) = [|\beta|\varepsilon_s(\beta)]^{-1}. \qquad (3.33)$$

In the x_1 domain, the surface potential $\phi(x_1)$ and charge density $\sigma(x_1)$ are the inverse transforms of $\overline{\phi}(\beta)$ and $\overline{\sigma}(\beta)$. Generally, these functions also depend on the frequency ω, but here ω is taken to be constant. Using the convolution theorem of Fourier analysis [Appendix A, eq. (A.13)], we can transform eq. (3.32) to the x_1 domain; the two functions on the right are each transformed to the x_1 domain, and then they are convolved. Thus, if $G(x_1, \omega)$ is the inverse transform of $\overline{G}(\beta, \omega)$, we have

$$\phi(x_1) = G(x_1, \omega) * \sigma(x_1) \equiv \int_{-\infty}^{\infty} G(x_1 - x_1', \omega)\,\sigma(x_1')\mathrm{d}x_1', \qquad (3.34)$$

where the asterisk indicates convolution. The function $G(x_1, \omega)$ is the *Green's function*. It can be interpreted as surface potential produced by a line of charge at $x_1 = 0$, as seen by putting $\sigma(x_1) = \delta(x_1)$ in eq. (3.34). Transforming eq. (3.33), we have

$$G(x_1, \omega) = \frac{1}{2\pi} \int_{-\infty}^{\infty} \frac{\exp(\mathrm{j}\beta x_1)}{|\beta|\varepsilon_s(\beta)}\mathrm{d}\beta. \qquad (3.35)$$

Using reciprocity, it is shown in Appendix B that this is an even function, so that $G(-x_1, \omega) = G(x_1, \omega)$.

Green's function for piezoelectric Rayleigh waves

We consider the form of the Green's function for a substrate supporting piezoelectric Rayleigh waves, for example Y–Z lithium niobate. In addition to surface waves, excitation of bulk waves can occur, as shown by the fact that $\varepsilon_s(\beta)$ is complex for some values of β (Fig. 3.2). Milsom *et al.* [6] have shown that $G(x_1, \omega)$ may be regarded as a sum of three terms corresponding to surface-wave excitation, electrostatic effects and bulk-wave excitation.

The surface-wave term can be deduced by considering the generation of surface waves by a transducer consisting of a set of electrodes occupying a finite region of x_1, with voltages applied to them. The charge density on the electrodes is $\sigma(x_1)$, with Fourier transform $\overline{\sigma}(\beta)$. For points outside the transducer, the potential associated with these waves is derived in Appendix B, Section B.6, and it is given by

$$\phi(x_1) = j\Gamma_s \overline{\sigma}(\mp k_f, \omega) \exp(\mp jk_f x_1), \qquad (3.36)$$

where the upper signs are for waves radiated in the $+x_1$ direction, with $x_1 > 0$. The lower signs are for radiation in the $-x_1$ direction, with $x_1 < 0$. This potential is considered to arise from a surface-wave component $G_s(x_1, \omega)$ of the Green's function. By eq. (3.34), this is the surface-wave potential obtained when $\sigma(x_1) = \delta(x_1)$, that is, when $\overline{\sigma}(\beta) = 1$. Thus, from eq. (3.34) we have

$$G_s(x_1, \omega) = j\Gamma_s \exp(-jk_f|x_1|). \qquad (3.37)$$

The surface-wave component of $G(x_1, \omega)$ arises because $\varepsilon_s(\beta)$ is zero at $\beta = k_f$, so that the integrand of eq. (3.35) has a pole at this point. There is also a pole at $\beta = 0$. If the variation of $\varepsilon_s(\beta)$ is ignored in this region, eq. (3.32) shows that $\overline{\phi}(\beta)$ is proportional to $\overline{\sigma}(\beta)/|\beta|$. A relation of this form applies for the electrostatic case, as shown by eq. (3.8). In the x_1 domain this potential is given by the electrostatic Green's function $G_e(x_1)$, and by comparison with eq. (3.11) we have

$$G_e(x_1) = -\frac{\ln|x_1|}{\pi\varepsilon_\infty}. \qquad (3.38)$$

The use of the constant ε_∞ will be justified later in this section, though eq. (3.11) shows that it is correct if the material is not piezoelectric. Note that $G_e(x_1)$ is independent of ω.

In addition to surface-wave and electrostatic effects, the total Green's function $G(x_1, \omega)$ must include the effects of bulk waves. It is assumed that these can be accounted for by adding a further term $G_b(x_1, \omega)$, so that the total Green's function of eq. (3.35) is given by

$$G(x_1, \omega) = G_e(x_1) + G_s(x_1, \omega) + G_b(x_1, \omega). \qquad (3.39)$$

The bulk-wave term cannot generally be obtained analytically, though it can be deduced numerically from the effective permittivity [6]. In many devices the bulk-wave effects are small, or they occur at frequencies remote from the main surface-wave response. In such cases, the bulk-wave Green's function can often be ignored.

Approximate form for the Green's function

A convenient approximate form for $G(x_1, \omega)$ is obtained by omitting the bulk-wave term from the accurate expression of eq. (3.39). The Green's function thus becomes

$$G(x_1, \omega) \approx G_e(x_1) + G_s(x_1, \omega) = -\frac{\ln|x_1|}{\pi\varepsilon_\infty} + j\Gamma_s \exp(-jk_f|x_1|). \qquad (3.40)$$

To confirm the validity of this, we deduce the corresponding effective permittivity. From eq. (3.33), the permittivity is the reciprocal of $|\beta|\overline{G}(\beta, \omega)$, where $\overline{G}(\beta, \omega)$ is the transform of the right side of eq. (3.40). The transform of the exponential term is given by eq. (A.31), and the transform of $\ln|x_1|$ can be taken as $-\pi/|\beta|$, which follows by comparing eq. (3.11) with eq. (3.8). We thus find

$$\overline{G}(\beta, \omega) \approx [\varepsilon_\infty|\beta|]^{-1} + j\Gamma_s[\pi\delta(\beta - k_f) + \pi\delta(\beta + k_f) + 2jk_f/(\beta^2 - k_f^2)]. \quad (3.41)$$

This function is infinite at $\beta = \pm k_f$, so that $\varepsilon_s(\beta)$ is zero at these points, as required. It is also necessary that $\overline{G}(\beta, \omega)$ should be zero at $\beta = \pm k_m$, so that $\varepsilon_s(\beta)$ is infinite at these points. This condition determines the value of Γ_s. Setting $\overline{G}(\pm k_m, \omega) = 0$ in eq. (3.41) gives

$$\Gamma_s = \frac{k_m^2 - k_f^2}{2\varepsilon_\infty k_f k_m} \approx \frac{1}{\varepsilon_\infty}\frac{v_f - v_m}{v_f}, \quad (3.42)$$

in good agreement with the previous result of eq. (3.31).

In eq. (3.41), the term $k_f/(\beta^2 - k_f^2)$ makes $\overline{G}(\beta, \omega)$ infinite at $\beta = \pm k_f$, so the reciprocal of $\overline{G}(\beta, \omega)$ is zero at these points. Hence, the reciprocal of $\overline{G}(\beta, \omega)$ is not affected by the delta functions, so these can be omitted when calculating $\varepsilon_s(\beta)$. Using eq. (3.42) for Γ_s, the effective permittivity is

$$\varepsilon_s(\beta) = \frac{1}{|\beta|\overline{G}(\beta, \omega)} = \varepsilon_\infty \frac{\beta^2 - k_f^2}{(|\beta| - k_m)(|\beta| + k_f^2/k_m)}. \quad (3.43)$$

Since $k_f^2/k_m \approx k_m$, this is almost identical to Ingebrigtsen's formula, eq. (3.28). The agreement confirms the validity of the approximate Green's function, eq. (3.40), and it also justifies the use of the constant ε_∞ in the electrostatic part of the Green's function, eq. (3.38).

3.5   OTHER APPLICATIONS OF THE EFFECTIVE PERMITTIVITY

The above development shows how the effective permittivity $\varepsilon_s(\beta)$ relates the charge density and potential at the boundary between a piezoelectric half-space and a vacuum. In Chapter 5, this will be used as a basis for analysis of surface-wave transducers. However, the concept of effective permittivity may be generalized for some other problems, and in this section we digress to discuss these briefly.

We first consider the coupling between a piezoelectric half-space and a plane at a height h above it [1, 8, 9, 11]. As shown in Fig. 3.3a, the piezoelectric occupies

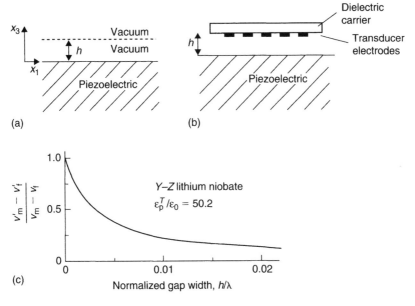

Figure 3.3 Piezoelectric coupling across a gap.

the space $x_3 < 0$, a vacuum occupies the space $x_3 > 0$, and free charges are allowed to exist only at the plane $x_3 = h$. For a harmonic solution, with variables proportional to $\exp(j\beta x_1)$, an effective permittivity $\varepsilon'_s(\beta)$ can be defined by the relation

$$\varepsilon'_s(\beta) = \frac{\tilde{\sigma}}{|\beta|\,\tilde{\Phi}(h)}, \tag{3.44}$$

where $\tilde{\Phi}(h)$ and $\tilde{\sigma}$ are the potential and charge density at $x_3 = h$. This function can be expressed in terms of the permittivity $\varepsilon_s(\beta)$ defined above, eq. (3.17). At $x_3 = 0$, the value of $\tilde{D}_3$ in the piezoelectric is, from eqs (3.15) and (3.16),

$$\tilde{D}_{3-} = \tilde{\phi}|\beta|[\varepsilon_0 - \varepsilon_s(\beta)], \tag{3.45}$$

where $\tilde{\phi}$ is the potential at $x_3 = 0$. In the vacuum region $0 < x_3 < h$, the potential $\tilde{\Phi}$ has terms proportional to $\exp(\pm\beta x_3)$, from Laplace's equation. In the region $x_3 > h$ the potential is proportional to $\exp(-|\beta|x_3)$. The potential must be continuous everywhere, and $\tilde{D}_3$ must be continuous at $x_3 = 0$. The discontinuity of $\tilde{D}_3$ at $x_3 = h$ give the charge density $\tilde{\sigma}$. Using these relations, the effective permittivity at $x_3 = h$ is found to be

$$\varepsilon'_s(\beta) = \frac{\varepsilon_s(\beta)[1 + \tanh(|\beta|h)]}{1 - [1 - \varepsilon_s(\beta)/\varepsilon_0]\tanh(|\beta|h)}. \tag{3.46}$$

This function may be used to analyze excitation by a transducer at the plane $x_3 = h$, as in Fig. 3.3b. In practice the transducer would need to be supported by a dielectric, but this will not appreciably affect $\varepsilon'_s(\beta)$ if its permittivity is not too large. As for excitation at the surface, $\varepsilon'_s(\beta)$ has a zero at $\beta = k'_f$, say, and a pole at $\beta = k'_m$, say, with corresponding wave velocities $v'_f = \omega/k'_f$ and $v'_m = \omega/k'_m$. The difference between these velocities is a measure of the coupling strength. Equation (3.46) shows that $\varepsilon'_s(\beta)$ is zero when $\varepsilon_s(\beta)$ is zero, so that $v'_f = v_f$. The metallized velocity v'_m depends on h, approaching v'_f for large h. Assuming that the wave is a piezoelectric Rayleigh wave, v'_m may be found by using Ingebrigtsen's approximation, eq. (3.28), for $\varepsilon_s(\beta)$. Noting that the four velocities v'_f, v'_m, v_f and v_m are all numerically similar, this is found to give

$$\frac{v'_m - v'_f}{v_m - v_f} = \frac{1}{1 - (\varepsilon_\infty/\varepsilon_0)/[1 - \coth k_f h]}. \tag{3.47}$$

Here the left side gives the coupling at the plane $x_3 = h$, normalized to the coupling at $x_3 = 0$. The function is shown in Fig. 3.3c for a Y–Z lithium niobate half-space, with the height h normalized to the wavelength $\lambda = 2\pi/k_f$. The coupling strength decreases very rapidly with h. It can be concluded that, to be practically effective, a transducer held above the surface would need to be very close, with a gap width much less than the wavelength. A gap this small is technically difficult to obtain, so transducers normally need to be deposited directly on the substrate. Equation (3.47) gives good agreement with accurate calculations [8, 11], and this confirms Ingebrigtsen's approximation and justifies the use of the coefficient ε_∞.

Another use of the effective permittivity concerns coupling to a semiconductor above the piezoelectric surface [5, 11, 12]. Typically, the semiconductor is separated by a small gap. As before, the ratio $\tilde{D}_3/\tilde{\Phi}$ at the piezoelectric surface is determined by $\varepsilon_s(\beta)$, while the same ratio at the semiconductor surface is determined by the semiconductor equations. Taken together, these relations determine β, which is generally complex. As for a transducer held above the surface, the gap needs to be very small to obtain useful coupling. The coupling has been exploited in several ways, as reviewed by Kino [16]. The surface wave can be amplified by applying a drift field in the x_1 direction, such that the carrier velocity in the semiconductor exceeds the surface-wave velocity. A gain of typically 50 dB/cm can be obtained for a drift field of a few kV/cm. A related mechanism is used in a semiconductor convolver, where a non-linear effect in the semiconductor is used to mix the fields of two contra-directed surface waves (Chapter 7, Section 7.6.3).

The effective permittivity can also be used to analyze transducers using piezo-electric layers. A piezoelectric film deposited on a non-piezoelectric substrate enables an interdigital transducer to be used to generate surface waves. For example, zinc oxide films have been used on glass, silicon, diamond or sapphire. For the layered system, an effective permittivity can be defined in the usual way, relating charge density to potential at the transducer plane. As before, two key parameters are the wave velocity with and without an ideal conducting sheet at this plane. The velocity difference gives a measure of the coupling strength. Ingebrigtsen's approximation, eq. (3.28), can be used for the effective permittivity, and much of the analysis then applies as for a piezoelectric half-space. However, there is the complication that the surface wave becomes dispersive, so that the wavenumbers k_f and k_m are no longer proportional to ω. An analysis based on normal mode theory was given by Kino and Wagers [17].

REFERENCES

1. K.A. Ingebrigtsen. 'Surface waves in piezoelectrics', *J. Appl. Phys.*, **40**, 2681–2686 (1969).
2. C.S. Hartmann and B.G. Secrest. 'End effects in interdigital surface wave transducers', *IEEE Ultrason. Symp.*, 1972, pp. 413–416.
3. R.C. Peach. 'A general approach to the electrostatic problem of the SAW interdigital transducer', *IEEE Trans.*, **SU-28**, 96–105 (1981).
4. S.V. Biryukov and V.G. Polevoi. 'The electrostatic problem for the SAW interdigital transducer in an external electric field – Part I: A general solution for a limited number of electrodes', *IEEE Trans. Ultrason. Ferroelect. Freq. Contr.*, **43**, 1150–1159 (1996).
5. C.A.A.J. Greebe, P.A. van Dalen, T.J.N. Swanenberg and J. Wolter. 'Electric coupling properties of acoustic and electric surface waves', *Phys. Rep. Phys. Lett. C*, **1C**, 235–268 (1971).
6. R.F. Milsom, N.H.C. Reilly and M. Redwood. 'Analysis of generation and detection of surface and bulk acoustic waves by interdigital transducers', *IEEE Trans.*, **SU-24**, 147–166 (1977).
7. R.F. Milsom, M. Redwood and N.H.C. Reilly. 'The interdigital transducer', in H. Matthews (ed.), *Surface Wave Filters*, Wiley, 1977, pp. 55–108.
8. K.M. Lakin. 'Perturbation theory for electromagnetic coupling to elastic surface waves on piezoelectric substrates', *J. Appl. Phys.*, **42**, 899–906 (1971).
9. B.A. Auld. *Acoustic Fields and Waves in Solids*, Vol. 2, Krieger, 1990.
10. B.A. Auld and G.S. Kino. 'Normal mode theory for acoustic waves and its application to the interdigital transducer', *IEEE Trans.*, **ED-18**, 898–908 (1971).
11. G.S. Kino and T.M. Reeder. 'A normal mode theory for the Rayleigh wave amplifier', *IEEE Trans.*, **ED-18**, 909–920 (1971).
12. K.-Y. Hashimoto and M. Yamaguchi. 'Excitation and propagation of shear-horizontal-type surface and bulk acoustic waves', *IEEE Trans. Ultrason. Ferroelect. Freq. Contr.*, **48**, 1181–1188 (2001).
13. K. Hashimoto and M. Yamaguchi. 'Free software products for simulation and design of surface acoustic wave and surface transverse wave devices', *IEEE Annual Freq. Contr. Symp.*, 1996, pp. 300–307.

14. K.-Y. Hashimoto. *Surface Acoustic Wave Devices in Telecommunications – Modelling and Simulation*, Springer, 2000.

15. K. Bløtekjaer, K.A. Ingebrigtsen and H. Skeie. 'A method for analysing waves in structures consisting of metal strips on dispersive media', *IEEE Trans.*, **ED-20**, 1133–1138 (1973).

16. G.S. Kino. 'Acoustoelectric interactions in acoustic-surface-wave devices', *Proc. IEEE*, **64**, 724–748 (1976).

17. G.S. Kino and R.S. Wagers. 'Theory of interdigital couplers on non-piezoelectric substrates', *J. Appl. Phys.*, **44**, 1480–1488 (1973).

4

PROPAGATION EFFECTS AND MATERIALS

Previous chapters have considered acoustic waves and their excitation, assuming idealized propagation conditions. In practical devices several propagation effects, such as diffraction, propagation loss and temperature effects, are often significant. These are considered in this chapter. An extensive survey of work on these topics is given by Slobodnik [1, 2]. For practical devices the choice of materials is strongly influenced by these effects, in addition to the piezoelectric coupling, so the materials are also considered here.

4.1 DIFFRACTION AND BEAM STEERING

This section considers the diffraction of surface waves on an uniform surface, assuming that the only wave motion is a non-leaky surface wave. The wave field generated by an unapodized transducer has many similarities with the diffraction of light by a slit aperture. Near the transducer there is a 'near-field', or 'Fresnel', region in which the beam propagates with relatively little distortion. In the 'far-field', or 'Fraunhofer', region, the beam diverges with an angle determined by the transducer aperture and the wavelength. In many surface-wave devices the receiving transducer is in the near field of the launching transducer, so that the diffraction does not cause much distortion. Nevertheless, for good performance it is often necessary to assess diffraction and to compensate for it in the device design.

The main distinctions between surface-wave diffraction and conventional optical diffraction arise from the anisotropy of the materials. For an anisotropic material, a straight-crested wave, that is, one with a straight wavefront, has a velocity dependent on its propagation direction. Consequently, the diffracted field depends strongly on the substrate material and its orientation. The analysis here is based on the 'angular spectrum of plane waves' method, using Fourier synthesis [3, 4]. Alternatively, a Green's function approach has been developed [5, 6], generalizing Huygen's principle to the anisotropic case.

For the analysis, the surface-wave amplitude at a point (x, y) is represented by a scalar $\psi(x, y)$. This may be taken as the associated surface potential $\phi_s(x, y)$, or a component of the surface displacement $\mathbf{u}(x, y)$. This approach is a considerable simplification. For example, for a given power density the magnitudes of these quantities will generally vary with direction. Diffraction will also affect the charge density on a transducer, and the charges can be regarded as a surface-wave source. However, such factors lead to considerable complexity, so they will be ignored here. In this section, the scalar model is used and we consider diffraction on a uniform surface, ignoring perturbations due to metallization patterns such as transducer electrodes. This is generally adequate for practical devices.

4.1.1 Formulation using angular spectrum of plane waves

Taking the surface to be in the x–y plane, the amplitude of a straight-crested wave is proportional to $\exp[-j(xk_x + yk_y)]$, omitting a factor $\exp(j\omega t)$. Propagation loss is neglected. As shown in Fig. 4.1, k_x and k_y are the x- and y-components of the wave vector $\mathbf{k}(\phi)$, which makes an angle ϕ with the x-axis. Thus, $k_x = k(\phi) \cos \phi$ and $k_y = k(\phi) \sin \phi$, where $k(\phi) = |\mathbf{k}(\phi)|$. In addition, $k(\phi) = \omega/v(\phi)$, where $v(\phi)$ is the phase velocity for straight-crested waves with propagation direction ϕ. This may be calculated by the method in Chapter 2. The quantity $1/v(\phi)$ is the slowness, and a polar plot of $k(\phi)$ as a function of ϕ is the slowness curve, since the frequency ω is constant and does not affect the argument. Figure 4.1 shows schematically a slowness curve and its relation to the wave vector. The form of the slowness curve will depend on the substrate material and its orientation. For an isotropic material $k(\phi)$ and $v(\phi)$ are independent of ϕ, and the curve is a circle.

The x-axis is the main propagation direction, normal to the electrodes of a launching transducer. The angle θ in Fig. 4.1 relates this to some reference direction defined by the crystal lattice. The disturbance contains components propagating with angle ϕ relative to the main propagation direction. Thus, for constant θ the surface-wave velocity is a function of ϕ, and this is the same function of θ when $\phi = 0$. Initially, we take θ to be constant.

Since the system is linear, a general solution may be obtained by summing straight-crested waves of the form $\exp[-j(xk_x + yk_y)]$, allowing the propagation direction to vary [3]. It is assumed that k_x is determined by k_y, and that the total

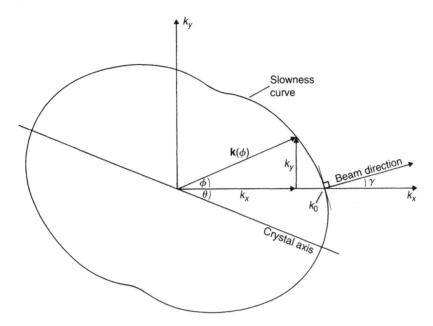

Figure 4.1 Diffraction analysis using the slowness curve.

field can be represented by a scalar $\psi(x, y)$ given by

$$\psi(x, y) = \int_{-\infty}^{\infty} \Psi(k_y) \exp[-j\{xk_x(k_y) + yk_y\}]dk_y \qquad (4.1)$$

where $\Psi(k_y)$ is the amplitude distribution of the component waves. Setting $x = 0$ shows that $\psi(0, y)$ is the Fourier transform of $\Psi(k_y)$, and hence $\Psi(k_y)$ is the inverse transform of $\psi(0, y)$, so that

$$\Psi(k_y) = \frac{1}{2\pi} \int_{-\infty}^{\infty} \psi(0, y') \exp(jy'k_y)dy' \qquad (4.2)$$

Hence, the disturbance at any point (x, y) can be obtained from the disturbance at $x = 0$ by transforming to give $\Psi(k_y)$ and then using eq. (4.1). The function $k_x(k_y)$ is given by

$$[k_x(k_y)]^2 = [k(\phi)]^2 - k_y^2. \qquad (4.3)$$

It can be assumed that, for $x \geq 0$, only positive solutions for k_x are required, and this usually determines k_x uniquely from k_y. This condition is necessary if the field is to be obtained from eq. (4.1). This is not always the case; if the slowness curve has a minimum near $\theta = \pm\pi/2$, some values of k_y can give more than one solution for k_x. However, $\Psi(k_y)$ is nearly always small in such regions, so the method remains valid. For large k_y, when $k_y > k$, k_x is imaginary and it is taken to be negative, so that the contribution to eq. (4.1) decays with x.

4.1.2   Beam steering in the near field

In the near field, the waves due to an aperture with uniform illumination exhibit little diffraction spreading. This is true for both isotropic and anisotropic materials. However, anisotropic materials can exhibit *beam steering*, which causes the beam to propagate in a direction that is not normal to the wavefronts. In surface-wave devices this phenomenon can be significant even when the receiving transducer is in the near field of the launching transducer.

Suppose that $\psi(0, y)$ represents a line source at $x = 0$ extending many wavelengths in the y-direction, and with phase independent of y. In this case, the transform $\Psi(k_y)$ will be significant only for k_y near to zero, so only a small region of the slowness curve is relevant. The phase xk_x in eq. (4.1) can be approximated using a Taylor expansion for $k_x(k_y)$, provided x is not too large. Defining $k_0 = k_x(0)$, equal to the value of $k(\phi)$ for $\phi = 0$, we have

$$k_x(k_y) \approx k_0 - k_y \tan \gamma \tag{4.4}$$

for $k_y \ll k_x$. Here quadratic and higher-order terms are ignored, and γ is defined by

$$\tan \gamma = -[dk_x/dk_y]_{\phi=0}. \tag{4.5}$$

Generally, γ depends on θ. Substituting eq. (4.4) into eq. (4.1), we find

$$\psi(x, y) \approx \exp(-jxk_0)\psi(0, y - x \tan \gamma). \tag{4.6}$$

Thus, in the near field the disturbance propagates with no distortion, but the propagation direction of the beam makes an angle γ with the x-axis, which is normal to the wavefronts. This phenomenon, known as beam steering, can occur only if the material is anisotropic. The beam direction is normal to the slowness curve, as shown in Fig. 4.1. The beam steering angle γ generally depends on the orientation angle θ. From eq. (4.5) we have

$$\tan \gamma = \left[\frac{1}{v} \frac{dv}{d\phi} \right]_{\phi=0} = \frac{1}{v} \frac{dv}{d\theta}. \tag{4.7}$$

Usually γ is small, so that $\tan \gamma \approx \gamma$.

Figure 4.2 illustrates beam steering for a wave launched by an unapodized interdigital transducer, where $\psi(0, y)$ is taken as the wave amplitude at the edge of the transducer. The wavefronts are parallel to the transducer electrodes. Owing to beam steering, only part of the beam overlaps a receiving transducer aligned with the launching transducer. The power available at the device output is proportional to the overlap. Beam steering has been confirmed experimentally by probing [1].

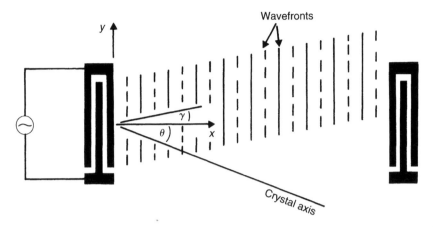

Figure 4.2 Schematic diagram of beam steering.

The substrate orientation θ is usually chosen such that the velocity $v(\phi)$ is symmetrical about $\phi = 0$ for small angles, giving $\gamma = 0$. However, beam steering can still arise in practice because of an error in θ due to misalignment. For small errors, γ can be found from its differential $d\gamma/d\theta$, which is found to be

$$\frac{d\gamma}{d\theta} = \frac{1}{v}\frac{d^2v}{d\theta^2}\cos^2\gamma - \sin^2\gamma \tag{4.8}$$

$$= \frac{1}{v}\frac{d^2v}{d\theta^2} \quad \text{if } \gamma = 0.$$

In the limit when the transducer aperture becomes infinite the disturbance becomes a straight-crested wave, with wave vector $\mathbf{k}$ directed along the x-axis. It is reasonable to conclude that the energy of this wave propagates in the beam direction, normal to the slowness curve, making an angle γ with $\mathbf{k}$. Thus, γ is often called the *power flow angle*. The energy velocity is $v/\cos\gamma$. A more rigorous derivation is obtained by defining a Poynting vector for a piezoelectric medium. This gives the energy flow direction at one point, but its application to surface waves is complicated because generally the magnitude and direction of the flow vary with depth. The overall energy flow can be obtained by integrating over the depth [7]. This generally agrees with the result found from the slowness curve, though there does not seem to be any proof that the methods should agree.

4.1.3 Minimal-diffraction orientations

It was assumed above that only small values of k_y give significant contributions to the integral of eq. (4.1), so that k_x could be approximated by eq. (4.4). Consider now a hypothetical case in which k_x is a constant, independent of k_y, giving $\tan \gamma = 0$ in eq. (4.4). The field $\psi(x, y)$ is then given by eq. (4.6) with $\tan \gamma = 0$, and this applies for all x, not just in the near field. In this case the beam propagates with no diffraction spreading. Physically, this occurs because the energy flow direction is parallel to the x axis, irrespective of the direction of the wave vector $\mathbf{k}$. For k_x to be constant, the phase velocity v must be proportional to $\cos \phi$. Noting that $dv/d\theta$ is equal to $dv/d\phi$ at $\phi = 0$, and using eq. (4.8), we have in this case

$$d\gamma/d\theta = -1. \tag{4.9}$$

In practice this is unrealistic because $k_y = k_x \tan \phi$ is infinite at $\phi = \pm\pi/2$ if k_x is constant. However, several materials have orientations with k_x almost constant for small ϕ values, so that $d\gamma/d\theta \approx -1$. These *minimal-diffraction orientations* give much less diffraction spreading than the isotropic case. An example is Y–Z lithium niobate, which gives $d\gamma/d\theta = -1.08$. Crabb *et al.* [6] have confirmed this by measuring the surface-wave distribution generated by a uniform transducer, using an electrostatic probe which responds to the electrical potential accompanying the wave.

4.1.4 Diffracted field in the parabolic approximation: scaling

We now consider the wave distribution for all x, including the far field. This can be obtained from the angular spectrum of plane waves method, eqs (4.1) and (4.2), in which $k_x(k_y)$ depends on the material and orientation. A simpler method, which applies for some substrates, is obtained using the *parabolic approximation*, in which k_x is taken to be a quadratic function of k_y. This method is more convenient to use. It also leads to the conclusion that the diffraction pattern for an anisotropic material is related to that of an isotropic material simply by scaling in the x-direction. The validity of the parabolic approximation depends on the nature of the slowness curve. For some materials, notably Y–Z lithium niobate, the approximation cannot be used.

Assuming the parabolic approximation to be valid, the function $k_x(k_y)$ is taken to have the form

$$k_x(k_y) \approx k_0 - ak_y - \frac{1}{2}bk_y^2/k_0, \tag{4.10}$$

where a, b and k_0 are constants for a given orientation, though they will generally vary with θ. Using eqs (4.5) and (4.8), a and b are given by

$$a = \tan \gamma, \tag{4.11a}$$

$$b = 1 + \frac{1}{v}\frac{d^2v}{d\theta^2} = \left(1 + \frac{d\gamma}{d\theta}\right)\sec^2 \gamma. \tag{4.11b}$$

Equation (4.11b) is found by evaluating $d^2 k_x / dk_y^2$ as a function of ϕ, at $\phi = 0$. For an isotropic material we have

$$k_x = \sqrt{k_0^2 - k_y^2} \approx k_0 - \frac{1}{2} k_y^2 / k_0,$$

where the approximate form applies for $k_y \ll k_0$. Thus, the isotropic case is given by eq. (4.10) with $a = 0$ and $b = 1$.

The field $\psi(x, y)$ is found by substituting eq. (4.10) into eq. (4.1), with $\Psi(k_y)$ determined by the source distribution. This shows that the effect of the parameter a can be expressed straightforwardly. If $\psi_0(x, y)$ is the field calculated with $a = 0$, it is found that

$$\psi(x, y) = \psi_0(x, y - ax). \tag{4.12}$$

This shows that the diffraction pattern can be calculated for $a = 0$, and then $y - ax$ is substituted for y. This is valid irrespective of the form of $\Psi(k_y)$. The linear term in eq. (4.10) skews the entire diffraction pattern. This beam steering was found for the near-field case in eq. (4.6). Equation (4.12) shows that it is true for all x, including the far field, if the parabolic approximation is valid.

Equation (4.10) also leads to a scaling theorem. Substitute into eq. (4.1) and take $a = 0$. This shows that, apart from a factor $\exp(-jxk_0)$, the x-variation of $\psi(x, y)$ can be expressed in terms of the product xb. Specifically, if $\psi_i(x, y)$ is the diffraction pattern for the isotropic case ($b = 1$), the pattern for the general case is given by

$$\psi(x, y) = \psi_i(xb, y) \exp[-jxk_0(1 - b)].$$

Aperture with uniform illumination

We now consider a uniformly illuminated aperture with width W, located at $x = 0$. This can refer to an unapodized launching transducer with few electrodes, so that diffraction within it can be ignored. For simplicity we assume $a = 0$, since the field for $a \neq 0$ is easily obtained from eq. (4.12). To find the field $\psi(x, y)$, we substitute eq. (4.10) into eq. (4.1), using eq. (4.2) for $\Psi(k_y)$. The field at the aperture is taken as $\psi(0, y) = 1$ for $|y| \leq W/2$ and $\psi(0, y) = 0$ elsewhere. We thus have, for $a = 0$,

$$\psi(x, y) = \frac{\exp(-jxk_0)}{\pi} \int_{-W/2}^{W/2} dy' \int_{-\infty}^{\infty} dk_y \exp\left[j \left\{ (y' - y)k_y + \frac{1}{2} xbk_y^2 / k_0 \right\} \right]. \tag{4.13}$$

This can be re-arranged using the standard integral [8]:

$$\int_{-\infty}^{\infty} \exp(jKt^2) dt = \sqrt{\frac{\pi}{|K|}} e^{\pm j\pi/4}, \tag{4.14}$$

where $K \neq 0$ is a real constant and the sign on the right is the same as the sign of K. Using eq. (4.14), eq.(4.13) can be written as

$$\psi(x, y) = \frac{1}{\sqrt{2}} \exp(-jxk_0 \pm j\pi/4) \int_{A_-}^{A_+} \exp\left(\mp \frac{1}{2} j\pi u^2\right) du, \qquad (4.15)$$

where the upper signs are used for $b > 0$ and the lower signs for $b < 0$, and the limits are

$$A_\pm = (y \pm W/2) \left[\frac{k_0}{\pi x |b|}\right]^{1/2}. \qquad (4.16)$$

Equation (4.15) is readily computed using Fresnel integrals. The anisotropy is expressed by the constant b, which occurs in the product $x|b|$. Thus, apart from the phase xk_0 in eq. (4.15), the diffraction pattern is the same as that for an isotropic material ($b = 1$), except for scaling in the x-direction by a factor $|b|$. In the limit $b \to 0$, k_x becomes a constant and there is no diffraction spreading, as in eq. (4.6).

In most practical cases, the orientation is chosen such that the phase velocity $v(\phi)$ is symmetrical about $\phi = 0$, so that $a = \gamma = 0$ and, from eq. (4.11b), $b = 1 + d\gamma/d\theta$. Thus, $d\gamma/d\theta$ determines the diffraction scaling. In addition, $d\gamma/d\theta$ gives the alignment accuracy required to keep the near-field beam steering within specified limits, as noted in Section 4.1.2, and minimal-diffraction orientations occur when $d\gamma/d\theta \approx -1$. Thus, $d\gamma/d\theta$ is a very significant parameter for any particular orientation. Its numerical values for several cases are given later.

In the far field, where $4\pi x |b| \gg k_0 W^2$, the limits $A_\pm$ in eq. (4.16) are close together and eq. (4.15) gives

$$|\psi(x, y)|^2 \approx \frac{\chi W}{\pi x} \left[\frac{\sin(\chi y/x)}{\chi y/x}\right]^2, \qquad (4.17)$$

where $\chi = W k_0 / |2b|$. Thus the diffraction pattern spreads out radially in this region. On the other hand, in the near field the beam propagates with little distortion or spreading, as shown in Section 4.1.2. It is useful to define a *Fresnel distance* x_F which gives approximately the demarcation between the two regions. Crabb *et al.* [6] have shown that this distance is given by

$$x_F \approx \frac{W^2 k_0}{10\pi |b|}. \qquad (4.18)$$

Table 4.1 Diffraction data for several materials.

Material	$d\gamma/d\theta$	Parabolic approximation valid?	Minimal diffraction?
Lithium niobate (Y–Z)	−1.08	No	Yes
Lithium niobate ($128°Y$–X)	−0.35	Yes	No
Quartz (ST–X)	0.38	Yes	No
Lithium tantalate (X–$112°Y$)	−0.29	No	No
Langasite*	−1.00	No	Yes

*Euler angles $0°$, $138.5°$, $26.6°$.

Measurements on Y-cut lithium niobate, with a variety of transducer orientations, gave good agreement with this formula [6], confirming the scaling of the diffraction pattern according to the value of $|b|$. The measurements were done with an electrostatic probe.

Validity of the parabolic approximation

The above analysis is valid if the slowness curve can be taken as a parabola, as in eq. (4.10). This approximation generally breaks down if x is large, or when the aperture is small. For many cases, the parabolic approximation is found to give the diffraction pattern with good accuracy for x well into the far-field region, provided the aperture is realistic. The approximation is then acceptable for practical purposes. This was established by comparison with experiments and with accurate calculations [9]. The validity was established for a variety of materials and orientations, and some of these are listed in Table 4.1.

For Y–Z lithium niobate there is very little diffraction spreading. The parabolic approximation is not valid so diffraction calculations must be done using a more accurate method, such as the angular spectrum of plane waves. Moreover, there is another difficulty because the pattern is very sensitive to errors in the velocity anisotropy. Velocities calculated from the measured bulk constants, using the method of Chapter 2, can give significant errors. However, this sensitivity can be exploited to improve the velocity accuracy [10]. A narrow-aperture transducer is used to generate waves spreading over a relatively wide angle, and the wave amplitude and phase are measured as functions of y at two x values. The analysis of eqs (4.1) and (4.2) then gives k_x as a function of k_y, and hence the slowness curve.

4.1.5 Two-transducer devices

The above analysis gives the field $\psi(x, y)$ at all points, due to an unapodized launching transducer with aperture W. Further development is needed to find the effect of diffraction on a surface-wave device. In this section, the parabolic approximation is used.

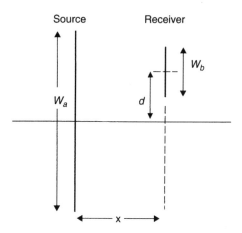

Figure 4.3 Line source and receiver for diffraction calculation.

We consider initially a device consisting of two unapodized transducers, both with few electrodes so that diffraction within them can be ignored. The diffraction can be analyzed using the simple configuration of Fig. 4.3. The transducers are represented by a line source with aperture W_a and a line receiver with aperture W_b, and the receiver is displaced in the y-direction by an amount d. More complicated cases, involving apodized transducers and diffraction within transducers, are considered later.

In Fig. 4.3, the field $\psi(x, y)$ due to the source can be found by the methods above. The response of the receiver is found by noting that, for an unapodized transducer, the short-circuit output current is proportional to the average surface-wave potential at the transducer input. This statement is justified later in Chapter 5, Section 5.6.2. Thus, the effect of diffraction is found by integrating $\psi(x, y)$ over the receiver aperture, and comparing with the diffractionless analysis [11]. The integral required is

$$R = \int_{d-W_b/2}^{d+W_b/2} \psi(x, y) dy, \tag{4.19}$$

where x gives the position of the receiver relative to the source. Assuming that the parabolic approximation is valid, $\psi(x, y)$ is given by eq. (4.15) with W replaced by W_a. For convenience we define the functions

$$F_{\pm}(t) = \int_0^t \exp(\pm j\pi u^2/2) du = C(t) \pm jS(t), \tag{4.20}$$

where $C(t)$ and $S(t)$ are the Fresnel integrals. In addition, functions $X_\pm(s)$ are defined as integrals of $F_\pm(t)$, so that

$$X_\pm(s) \equiv \int_0^s F_\pm(t)\mathrm{d}t = sF_\pm(s) \pm \frac{\mathrm{j}}{\pi}[\exp(\pm \mathrm{j}\pi s^2/2) - 1], \qquad (4.21)$$

where the final result follows using integration by parts. Using these formulae with eq. (4.15) for $\psi(x, y)$, the function R of eq. (4.19) is found to be

$$R = \frac{C}{\eta k_0}\left[X_\mp(B_1) - X_\mp(B_2) - X_\mp(B_3) + X_\mp(B_4)\right] \qquad (4.22)$$

where $\eta = 1/\sqrt{(\pi x|b|k_0)}$ and $C = (1/\sqrt{2})\exp(-\mathrm{j}xk_0 \pm \mathrm{j}\pi/4)$. Here, and in eq. (4.22), the upper sign applies for $b > 0$ (the usual case), and the lower sign for $b < 0$. The functions B_n are defined by

$$\begin{aligned}
B_1 &= [(W_a + W_b)/2 + d]\eta k_0; \quad B_2 = [(W_a - W_b)/2 + d]\eta k_0; \\
B_3 &= [(W_b - W_a)/2 + d]\eta k_0; \quad B_4 = [-(W_a + W_b)/2 + d]\eta k_0.
\end{aligned} \qquad (4.23)$$

For comparison, the response for no diffraction is obtained by setting $b \to 0$, and the value of R is taken as R_0 for this case. If the source transducer completely overlaps the receiver, this is found to give $R_0 = W_b \exp(-\mathrm{j}xk_0)$. If the transducers overlap by an amount w, the response for no diffraction is $R_0 = w \exp(-\mathrm{j}xk_0)$.

The effect of diffraction is given by a function $D = R/R_0$. If the transducers are coaxial, so that $d = 0$, the function D depends only on two normalized variables – the aperture ratio W_b/W_a and a normalized transducer separation

$$\hat{x} = \frac{|b|\,(x/\lambda)}{(W_a/\lambda)^2}, \qquad (4.24)$$

where $\lambda = 2\pi/k_0$ is the wavelength. Figure 4.4 shows the amplitude and phase of D for several values of W_b/W_a. For $W_b = 0$, D is simply the field $\psi(x, y)$ due to the source, evaluated on the center line $y = 0$, with the term $\exp(-\mathrm{j}xk_0)$ omitted.

Diffraction for long transducers and apodized transducers

In the above description, the transducers were assumed to be short so that they could be represented by a line source and a line receiver, as in Fig. 4.3. However, in general it is necessary to consider the sources and receivers individually. Each of the transducers thus becomes an array of line elements. This is illustrated in Fig. 4.5. To a good approximation, the lines can be identified as the active inter-electrode gaps. Element n in the source has polarity $C_n = \pm1$ determined by which of the adjacent electrodes is connected to the live bus bar, and similarly element m in the receiver has polarity C_m. For these two elements the output is

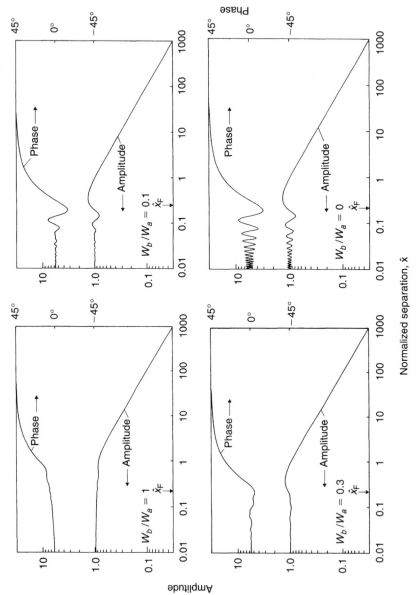

Figure 4.4 Diffraction factor D for a line source and receiver, taken to be coaxial ($d = 0$). Parabolic anisotropy, with $a = 0$ and $b > 0$.

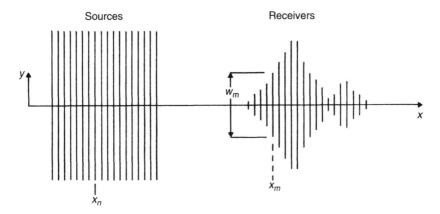

Figure 4.5 Diffraction analysis for extended and apodized transducers.

R_{nm}, given by eq. (4.22) with x as the separation of the elements. For the whole device, the response has the form

$$R_d \propto \sum_n \sum_m C_n C_m R_{nm}. \qquad (4.25)$$

If all the lines in the apodized receiver are overlapped by the sources, and if $b \to 0$ so that diffraction is negligible, this equation corresponds to a delta-function analysis. For a full analysis, it is necessary to include other factors that are present in the diffraction-free case, including the element factor and the transducer admittances. The device Y-matrix can then be calculated.

In practice, the computation of eq. (4.25) can be very time-consuming because the number of terms is the product of the numbers of sources in the two transducers. This product may well exceed 10^4. A fast calculation can be done by first tabulating the functions $X_{\pm}(s)$, and then using linear interpolation for the diffraction calculation. The accuracy of linear interpolation depends on the second differential which, as seen from the definition of $X_{\pm}(s)$, cannot exceed unity.

In the above method, the diffraction is analyzed as if the wave were propagating everywhere on a uniform surface. In practice this is not true, of course – the wave is subject to metal film structures such as transducers. However, it is usually a reasonable approximation to ignore the presence of these structures. Milsom [12] has given a formulation making allowance for the presence of transducer electrodes.

4.2 PROPAGATION LOSS AND NON-LINEAR EFFECTS

In addition to diffraction, the surface-wave amplitude is attenuated because of propagation loss, and this is often significant at high frequencies. Assuming a wide beam, so that diffraction can be ignored, a wave on a free surface is subject to several causes of loss. Surface imperfections such as cracks or scratches can cause attenuation. However, the work of Slobodnik [1] showed that this is of little significance, even at GHz frequencies, if the substrate is a crystal with an optical-quality finish. More fundamentally, the wave is attenuated because of interactions with thermal lattice vibrations, that is, phonons. From analysis of this phenomenon [13] the attenuation at frequency f, in dB per unit length, is expected to be proportional to f^2. This relation applies for a surface *in vacuo*. In air, there is additional loss due to generation of longitudinal acoustic waves in the air, and this is predicted to give an attenuation coefficient proportional to f [13]. Room-temperature measurements on various materials confirm these predictions [1]. For example, for $Y–Z$ lithium niobate the loss is found to be $\alpha = 0.19f + 0.88f^2$ dB/μs, while for ST–X quartz $\alpha = 0.47f + 2.62f^2$dB/μs, with f in GHz. Thus, at 1 GHz, $Y–Z$ lithium niobate gives $\alpha \approx 1$ dB/μs. The attenuation decreases rapidly if the temperature is reduced [1].

Practical devices have somewhat larger attenuation due to the presence of the metal electrodes in transducers. Room-temperature measurements on $Y–Z$ lithium niobate with *continuous* aluminum film gave [14] $\alpha \approx Kf^{2.2}$dB/μs, with $K = 3.0$ for a 50 nm (500 Å) film and $K = 5.2$ for a 200 nm (2000 Å) film.

Non-linear effects can be significant at relatively high power levels. This causes the excitation of waves at harmonic frequencies, and corresponding attenuation of the fundamental. Both effects can be of practical significance. The harmonics can be mixed together by the non-linearity so as to re-constitute the fundamental, so that the fundamental initially decreases with distance and then increases again. Thus the amplitudes of the various components are complicated functions of position. These phenomena have been studied in detail by optical probing. The wave interactions are limited, and made more complex, by a very small amount of dispersion which is found to be present in practical crystals. This is probably due to surface imperfections. Experiments on free-surface crystals confirm the theoretically expected behavior [1, 15].

For practical devices the presence of metal strips in transducers and gratings has a significant effect. Despite the complexity of the phenomenon, Williamson [16] found an empirical formula by examining a variety of devices, giving a guideline for practical cases. For devices using free-surface propagation on $Y–Z$ lithium niobate, a 1 dB depletion of the fundamental was found when the power

P_1 of the input wave was given by

$$P_1/(W\lambda) \approx 4 \text{ to } 8 \text{ W/mm}^2,$$

where λ is the wavelength and W is the beam width. However, the presence of a metal film reduces the non-linearity because it introduces extra dispersion. In many high-frequency devices, the limit on power handling is not the non-linearity of the substrate. Instead, it is due to acoustic migration of the electrodes (Chapter 11, Section 11.3), or electrical breakdown of the transducers.

4.3 TEMPERATURE EFFECTS AND VELOCITY ERRORS

Most surface-wave devices are required to meet precise specifications over a range of temperature which might be as large as $-30°C$ to $+100°C$. For this reason, temperature stability usually needs to be considered when a device is designed. This factor is also very important when assessing the suitability of a material for practical devices.

Temperature coefficient of delay

Temperature effects can be characterized by considering a line source and line receiver on a free surface, as in Fig. 4.3, which represents two short transducers fabricated on the surface. Diffraction, propagation loss and dispersion are ignored here. If the source and receiver are separated in the x-direction by a distance l, the delay is $T = l/v_f$. The velocity v_f can be calculated from the bulk constants of the material. If the temperature variations of these constants are known, the variation of v_f can be found using the method described in Chapter 2. In addition, l varies with temperature because of the thermal expansion of the material. The temperature coefficient of delay (TCD) is therefore given by

$$\alpha_T \equiv \frac{1}{T}\frac{dT}{d\Theta} = \frac{1}{l}\frac{dl}{d\Theta} - \frac{1}{v_f}\frac{dv_f}{d\Theta}, \tag{4.26}$$

where Θ is the temperature. Note that α_T is independent of the transducer separation l. Experimentally, α_T can be found from the frequency variation of a surface-wave delay-line oscillator [17], which is a sensitive measure of T.

In practice the fractional change of T is usually small, and T is often a linear function of Θ so that α_T is almost constant. Slobodnik [1] gives data for a variety of materials. For Y–Z lithium niobate, $\alpha_T = 94 \times 10^{-6}$ $(°C)^{-1}$, though most materials give values smaller than this. For quartz the two terms on the right of eq. (4.26) are generally both positive, and for some particular orientations

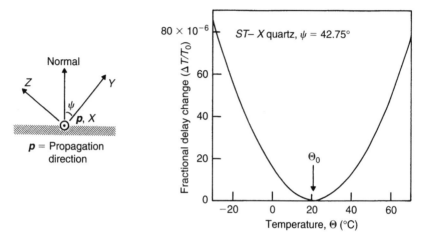

Figure 4.6 Temperature variation of delay for *ST–X* quartz.

they are equal at room temperature, so that $\alpha_T = 0$. One of these cases is often used for devices needing good temperature stability, and this is called *ST-cut quartz*. Propagation is in the crystal X-direction, so the orientation is '*ST–X* quartz'. As shown in Fig. 4.6, this is a rotated Y-cut, with rotation angle $\psi = 42.75°$. The delay T is approximately a quadratic function of temperature [17], given theoretically and experimentally by

$$T(\Theta) \approx T(\Theta_0)[1 + c(\Theta - \Theta_0)^2], \tag{4.27}$$

where $c = 32 \times 10^{-9}$ $(°C)^{-2}$ and $\Theta_0 = 21°C$. The delay is minimized at temperature Θ_0, which is known as the 'turn-over temperature'. Figure 4.6 shows the delay variation, where $\Delta T = T(\Theta) - T(\Theta_0)$ and $T_0 = T(\Theta_0)$. From eq. (4.27), the temperature coefficient for this case is $\alpha_T = 2c(\Theta - \Theta_0)$, and is therefore zero at the turn-over temperature.

The turn-over temperature Θ_0 depends on the rotation angle ψ, decreasing at a rate of about $10°C$ per degree of rotation, though this variation is not very linear [17]. In practical devices Θ_0 is also affected by transducers, which may depress it by $20–50°C$ [18]. It is common practice to reduce ψ to a value in the range $34–38°C$. This compensates for the reduction of Θ_0 due to the transducers. Another reason for this is that the center of the operational temperature range is often greater than $21°C$. The change of ψ has little effect on the other surface-wave properties, such as piezoelectric coupling and diffraction effects. With reduced ψ values, the orientation is still generally referred to as *ST–X* quartz.

Temperature effects in device responses

In a surface-wave device, the response will vary with temperature. Consider a two-transducer device at temperatures Θ_1 and Θ'_1, assuming that the response is ideal at temperature Θ_1. If T and T' are the delay between two points at temperatures Θ_1 and Θ'_1, we can define a small quantity ε such that

$$T' = (1 + \varepsilon)T. \tag{4.28}$$

Thus, if T varies linearly with Θ we have $\varepsilon = \alpha_T(\Theta'_1 - \Theta_1)$. For ST–X quartz, ε can be found from eq. (4.27), noting that Θ_1 is not generally equal to Θ_0. Equation (4.28) can be assumed to be valid for any two points on the surface. For a surface-wave device, the temperature changes modify the time scale of the impulse response, which changes from $h(t)$ at temperature Θ_1 to $h'(t)$ at Θ'_1. Noting that dispersion is negligible, we have

$$h'(t) = h\left(\frac{t}{1 + \varepsilon}\right). \tag{4.29}$$

These responses can be taken to refer to the short-circuit case, thus excluding temperature effects in terminating circuits. The corresponding frequency responses are $H_{sc}(\omega)$ and $H'_{sc}(\omega)$, the Fourier transforms of $h(t)$ and $h'(t)$. From the scaling theorem [eq. (A.5)] we have

$$H'_{sc}(\omega) \approx H_{sc}[\omega(1 + \varepsilon)], \tag{4.30}$$

neglecting a small amplitude change. If we write $H_{sc}(\omega) = A(\omega)\exp[j\phi(\omega)]$ and $H'_{sc}(\omega) = A'(\omega)\exp[j\phi'(\omega)]$, we have

$$A'(\omega) = A[\omega(1 + \varepsilon)], \quad \phi'(\omega) = \phi[\omega(1 + \varepsilon)]. \tag{4.31}$$

Thus, the amplitude and phase of the frequency response are simply scaled by a factor $(1 + \varepsilon)$. For example, the center frequency f_0 of a bandpass filter will change according to $\Delta f_0/f_0 = -\Delta T/T$, and its bandwidth will also change in the same ratio. A linear chirp filter (Chapter 7) has phase $\phi(\omega) = -(\omega - \omega_0)^2/(4\pi\mu)$, apart from constants, where μ is the chirp rate. Using eq. (4.31), we find that a temperature change modifies the chirp rate from μ to $\mu' = \mu/(1 + \varepsilon)^2$.

Velocity errors in device responses

If a device is designed assuming a wave velocity v and the actual velocity v' is different, this has an effect similar to a temperature change. If T is the intended delay between two points, and T' is the actual delay, we have $T' = (1 + \varepsilon)T$ as in eq. (4.28), but now ε is given by

$$\varepsilon \approx -(v' - v)/v. \tag{4.32}$$

With this value of ε, the effect of a velocity change is given by eqs (4.29)–(4.31) above. These equations determine the velocity accuracy needed for a particular application.

Velocity errors arise for several reasons. Errors due to misorientation of the crystal can be assessed by calculating the angular variation of velocity. It is often necessary to orient the substrate to an accuracy of $1°$ or better. Ferroelectric crystals often show significant sample-to-sample variations, such as the 1 in 10^3 velocity variation seen in lithium niobate [19]. Velocity changes are also caused by the presence of electrodes or other structures on the surface (Chapter 8).

4.4   MATERIALS FOR SURFACE-WAVE DEVICES

4.4.1   Orientation: Euler angles

Because surface-wave materials are anisotropic, it is essential to quote the orientation when considering a particular case. For the simpler cases it is sufficient to quote the crystal cut and the propagation direction. For example, $Y–Z$ lithium niobate has the crystal Y-axis normal to the surface and directed outward, with wave propagation in the crystal Z-direction. In some cases the substrate is a rotated Y-cut, an example being $ST–X$ quartz, Fig. 4.6, which has $42.75°$ rotation. This can be written as $42.75°Y–X$ quartz, since propagation is in the X-direction.

For general cases a more versatile description is needed, and a common method is to use Euler angles [1]. Define X, Y and Z as the crystal lattice axes and x, y and z as the device orientation axes. Here z is the outward-directed normal to the surface and x is the wave propagation direction. Initially, x, y and z are the same as X, Y and Z, and the Euler angles are zero. We define up to three rotations:

(1) Rotate the x, y and z axes anticlockwise about z through an angle λ (that is, x and y rotate anticlockwise as seen by an observer looking down the z-axis towards the origin). The new axes are called x_1, y_1 and z_1, and z_1 is the same as z and Z.
(2) Rotate the x_1, y_1 and z_1 axes anticlockwise about x_1 through an angle μ. The new axes are called x_2, y_2 and z_2.
(3) Rotate the x_2, y_2 and z_2 axes anticlockwise about z_2 through an angle θ. The final axes are called x_3, y_3 and z_3.

In the final orientation, z_3 is the surface normal and x_3 is the propagation direction. If there are less than three rotations we use x_j, y_j and z_j, with $j < 3$. The

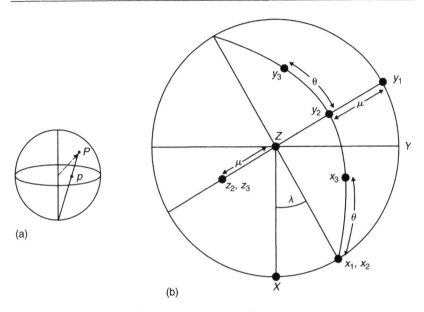

Figure 4.7 (a) Stereographic projection and (b) Euler angle rotations.

angles λ and μ define the substrate orientation, while θ defines the propagation direction in the surface.

Rotated orientations may be shown diagramatically using the stereographic projection familiar to crystallographers. This shows three-dimensional directions unambiguously on a two-dimensional diagram. The method is shown in Fig. 4.7a. Imagine a sphere with a straight line drawn from the origin in the direction to be represented. The line meets the surface at a point P. From P, draw a line to the south pole of the sphere, crossing the equatorial plane at a point p. The stereogram is a drawing of the equatorial plane with all the points p marked. If the point P is in the lower hemisphere, the line from P is drawn to the north pole instead of the south, and the point p is marked by an open circle instead of a closed one. A plane in three-dimensional space becomes a circular arc, or a straight line, on the stereogram. Figure 4.7b illustrates the stereogram for the Euler angles, after three rotations.

For a rotated Y-cut with rotation ψ, as in Fig. 4.6, the Euler angles are $\lambda = \theta = 0$ and $\mu = \psi - 90°$.

4.4.2 Single-crystal materials

The suitability of materials for device applications is determined by examining many factors including velocity, coupling constant ($K^2 = 2\Delta v/v$), TCD, diffraction effects, propagation loss, coupling to unwanted bulk waves, suitability for lithographic processing and availability. Many of these properties can be deduced from the wave velocity and its variation with direction and temperature, so a detailed examination of the velocity is a basic part of the assessment. The velocity can be calculated from the bulk elastic, dielectric and piezoelectric constants, as explained in Chapter 2. Temperature coefficients of these are also needed to obtain the surface-wave TCD. The bulk constants are derived from extensive measurements, and values are available for many materials such as quartz [36], lithium niobate and tantalate [10, 37] and many others [38, 39]. The need to vary three orientation angles, such as the Euler angles, makes the search very extensive.

Data for surface-wave materials are summarized in Table 4.2. The parameter ε_∞ is the effective permittivity at infinite wavenumber, $\varepsilon_s(\infty)$, defined in Chapter 3, Section 3.3; it governs the capacitance of an interdigital transducer as shown in Chapter 5, Section 5.5.2. Among the common materials, lithium niobate has relatively strong piezoelectric coupling but poor temperature stability; it is best suited to wide-band devices. The Y–Z orientation has minimal diffraction but it generates unwanted bulk waves rather badly. A multistrip coupler can be used to minimize bulk wave signals, and the absence of diffraction effects then leads to high-quality performance. The 128° rotated Y-cut gives much less bulk wave excitation so that a coupler is not necessary and the substrate area needed is less [21]. This is often preferred to the Y–Z case, though diffraction effects generally need to be compensated in the design.

At the other extreme lies ST–X quartz, with weak piezoelectric coupling but good temperature stability. The weak coupling generally limits it to narrow-band applications such as resonators and some bandpass filters, for which good temperature stability is usually needed. Lithium tantalate, with orientation X–112°Y, gives intermediate coupling and temperature stability. Other materials, with zero TCD and with coupling stronger than that of ST–X quartz, include berlinite, gallium phosphate and lithium tetraborate. Several new materials show properties like quartz but with stronger coupling; these are langasite, langatate and langanite. The langasite orientation shown in the table exhibits the N-SPUDT effect, described in Chapter 8, Section 8.2.3. New orientations of quartz give coupling stronger than that of the ST–X case, in addition to either minimal diffraction [30] or increased electrode reflectivity [40].

Bismuth germanium oxide is notable for its unusually small surface-wave velocity, reducing the size needed to obtain a given acoustic delay. Another

Table 4.2 Single-crystal surface-wave materials.

Material	Euler angles λ, μ, θ (°)	v_f (m/s)	$\Delta v/v$ (%)	$\varepsilon_\infty/\varepsilon_0$	TCD[a] (ppm/°C)	Reference
(A) Materials in common use						
Lithium niobate, LiNbO$_3$, Y–Z	0, −90, −90	3488	2.4	46	94	[20]
LiNbO$_3$, 128°Y–X	0, 38, 0	3979	2.7	56	75	[21]
Quartz, SiO$_2$, ST–X (42.75°Y–X)	0, −47.25, 0	3159	0.06	5.6	0(32)	[17]
Lithium tantalate, LiTaO$_3$, X–112°Y	90, 90, 112	3300	0.35	48	18	[22]
Lithium tetraborate, Li$_2$B$_4$O$_7$, 45°X–Z	135, 90, −90	3350	0.45	11	0(270)	[23]
LiNbO$_3$, 64°Y–X (LSAW)	0, −26, 0	4742	5.5	52	80	[24]
LiNbO$_3$, 41°Y–X (LSAW)	0, −49, 0	4792	8.5	63	80	[24]
LiTaO$_3$, 36°Y–X (LSAW)	0, −54, 0	4212	2.4	50	32	[25, 26]
Quartz, 36°Y–X+90° (STW)	0, −54, 90	5100	–	5.6	0(60)	[25]
(B) Other established materials						
Bismuth germanium oxide, Bi$_{12}$GeO$_{20}$ (001, 110)	0, 0, 45	1681	0.7	46	120	[1]
Ditto	45, 40.04, 90	1827	0.3	46	–	
Berlinite, AlPO$_4$, 114.5°Y–X	0, 24.5, 0	2741	0.25		0	[27]
Gallium phosphate, GaPO$_4$	0, 54.5, 0	2342	0.15		0	[28]
Quartz, LST cut, −75°Y–X (LSAW)	0, −165, 0	3950	0.05	5.5	0(−6)	[29]
Quartz[b]	0, 43, 23.7	3162	0.07	5.5	0(50)	[30]
Langasite, La$_3$Ga$_5$SiO$_{14}^{c}$	0, 138.5, 26.6	2730	0.16		0	[31]
Langanate, La$_3$Ga$_{5.5}$Nb$_{0.5}$O$_{14}$	0, 146, 23.5	2668	0.26		−10	[28]
Langatate, La$_3$Ga$_{5.5}$Ta$_{0.5}$O$_{14}$	0, 151.2, 24	2618	0.35		0	[32]
Potassium niobate, KNbO$_3$, 60°Y–X	0, −30, 0	4000	17		0(1830)	[33]
Lithium tetraborate (LLSAW)	0, 47.3, 90	6790	0.6		−3	[34]
SiO$_2$/LiNbO$_3$ 128°Y–X	0, 38, 0	3990	3		0	[35]

All cases shown have the power flow angle γ equal to zero.
LSAW: leaky surface acoustic wave; LLSAW: longitudinal LSAW.
[a] Numbers in brackets are $c \times 10^9$ (°C)$^{-2}$, with c defined in eq. (4.32).
[b] Minimal-diffraction cut with the N-SPUDT effect.
[c] This orientation of langasite shows the N-SPUDT effect.

interesting case is potassium niobate, which gives exceptionally high coupling with zero TCD.

Leaky waves

Entries marked 'LSAW' in Table 4.2 refer to leaky surface acoustic waves. These are surface-wave solutions which have velocities higher than that of the slowest shear wave. The velocity is generally higher than that of conventional Rayleigh surface waves. They exist for specific propagation directions such that coupling to the shear wave is negligible, and consequently propagation loss due to such coupling is also negligible. These waves are discussed later in Chapter 11, Section 11.4. They are usually shear-horizontal waves, with the elastic displacement normal, or almost normal, to the sagittal plane. They usually have stronger piezoelectric coupling than Rayleigh-type surface waves. In quartz, the LST and STW orientations shown give leaky waves, as do many cases in the langasite family [41]. In some materials it is possible to find a longitudinal leaky wave, denoted by LLSAW, in which the displacement is along, or almost along, the propagation direction. These generally have very high velocities, attractive for high-frequency devices. The table includes an orientation of lithium tetraborate, and others are mentioned later in Chapter 11, Section 11.4.

4.4.3 Thin films

Thin films can function in surface-wave devices in several ways:

(a) Conducting films, usually of aluminum, are needed for components such as transducers and gratings.
(b) A non-piezoelectric dielectric film can be used to modify the surface-wave properties. The prime example of this is an isotropic silicon oxide (SiO_2) film, which for many substrates can be used to improve the temperature stability. Table 4.2 includes data for a SiO_2 film on $128°Y$–X lithium niobate, showing that the film minimizes the TCD but has little other effect.
(c) A piezoelectric film can be deposited on a non-piezoelectric substrate, or a weakly piezoelectric substrate, so that interdigital transducers can be used for wave generation and reception. In particular, a diamond or sapphire substrate, which is non-piezoelectric, gives a high surface-wave velocity and is therefore attractive for high-frequency operation. Piezoelectric films of zinc oxide (ZnO) or aluminum nitride (AlN) have been widely used.
(d) Piezoelectric films can also be used for semiconducting substrates such as silicon and gallium arsenide. This introduces the possibility of integrating surface-wave devices on the same substrate as semiconductor circuits. It has also been used to study semiconductor properties.

This section gives some discussion of piezoelectric films.

Piezoelectric zinc oxide films have been used for many years for bulk wave generation. They are convenient at microwave frequencies because typical film thicknesses, in the region of 1 μm, are comparable to the acoustic wavelength.

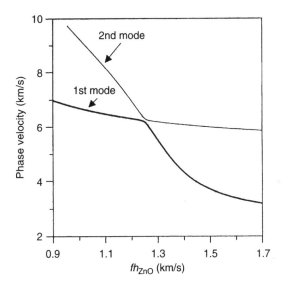

Figure 4.8 Dispersion for a zinc oxide film on diamond substrate. The abscissa shows frequency multiplied by zinc oxide film thickness. From Wu and Chen [49], copyright *IEEE*, 2002, reproduced with permission.

For surface-wave devices the application is rather different because the thickness is not critically related to the frequency of use. In both cases the film deposition needs to be done with considerable care in order to obtain optimal piezoelectric coupling and minimal losses. Moreover, the conditions needed for surface-wave devices were found to be more critical than those for bulk-wave devices. Consequently, early work in the 1970s and 1980s was mainly restricted to low-frequency devices, particularly television intermediate frequency (IF) filters using glass substrates with frequencies around 40 MHz. Technological aspects are described by Kadota *et al.* [47, 48], demonstrating that the coupling can be close to the theoretical ideal and that filters at 1 GHz are realizable.

Zinc oxide films have been used with various substrates, including diamond [42, 49]. This material is non-piezoelectric but its very high surface-wave velocity of 12 000 m/s makes it potentially very attractive for high frequencies. Generally, the diamond is grown as a film on a silicon substrate, so the system actually has two layers. However, the diamond is usually made thick enough for the silicon to have little effect on waves propagating at the upper surface. Figure 4.8 shows theoretical results for velocity. Here the surface normal is the (001) axis of the zinc oxide and the diamond, and propagation is in the [110] direction of the diamond. In contrast to a single-crystal material, there is substantial dispersion and a sequence of modes exists, numbered 1, 2, 3, . . . , in

Table 4.3 Systems using piezoelectric layers.

Material layer/substrate	v_{f0} (m/s)	Mode number (a)	fh (m/s)	v_f (m/s)	K^2 (%)	Reference
ZnO/diamond	12000	1	621	7800	0.4	[42]
		2	875	11000	1.5	
ZnO/sapphire	5000	1	455	4880	3	[43]
		2	870	5800	6	
AlN/diamond	12000	1	1464	9200	–	[44]
		2	4775	10000	2.5	
AlN/sapphire	5700	1	1833	5760	1.5	[45]
ZnO/GaAs	2870	1	288	2880	0.7	[46]

(a) 1: fundamental mode (Rayleigh) and 2: first higher mode (Sezawa).

order of increasing velocity. The second mode, with index 2, is often called the Sezawa wave. In many cases this shows stronger piezoelectric coupling than the fundamental, and is therefore the preferred mode for device applications. For film thicknesses up to about 0.2 wavelengths, the system offers velocities much higher than other materials, with a reasonable coupling factor, and it thus appears attractive for practical devices. It has also been shown that a silicon oxide (SiO_2) film can be added to reduce the TCD, ideally to zero [42]. Most of the features of zinc oxide on diamond, in particular the multiple modes, are also shown by other layered cases.

Several examples using piezoelectric films are summarized in Table 4.3. For zinc oxide on diamond the transducer position is taken to be between the film and the diamond substrate, but for other cases the transducer position is above the piezoelectric layer. The properties depend on the film thickness, of course, and the table refers to some representative thicknesses, expressed as the frequency-thickness product fh. The parameter v_{f0} is the velocity for a free substrate, in the absence of the piezoelectric film ($h = 0$). Also, v_f is the wave velocity for $h \neq 0$, with no electrical shorting plane at the transducer location.

Values for the coupling constant K^2 are from measured Q-values of interdigital transducers, using the relation $Q = \omega C_t/G_a = 4K^2 N_p/\pi$, where N_p is the number of periods. This relation was given by the early crossed-field network model. It applies for non-reflective single-electrode transducers with metallization ratio 0.5. The quasi-static theory of Chapter 5 shows that this definition gives $K^2 = 2.26\Delta v/v$, though it is also common to define $K^2 = 2\Delta v/v$. Here $\Delta v/v = (v_f - v_m)/v_f$, where v_m is the velocity when a shorting plane is present at the intended transducer location. The two velocities vary with frequency, of course. The theory in Chapter 3, leading to $\Delta v/v$ as a coupling parameter, applies when the frequency ω is held constant.

Another non-piezoelectric substrate giving high velocities is sapphire. Data for propagation on the R-plane of sapphire with a zinc oxide film are shown in the table [43]. Again, there is a sequence of modes, but these are less dispersive than the diamond case because the sapphire alone has a lower surface-wave velocity of 5000 m/s. Koike *et al.* [50] have demonstrated a 1.5 GHz interdigitated inter-digital transducer (IIDT) filter using this material, giving 1.3 dB insertion loss, 60 MHz bandwidth and 30 dB stop-band rejection. Aluminum nitride films on sapphire have been used for 5 GHz filters [45].

A zinc oxide film on quartz has been shown to give both better coupling and better temperature stability than ST–X quartz [51]. Table 4.3 also includes zinc oxide on gallium arsenide, where the substrate is {001} cut with <110> propagation. With no zinc oxide, gallium arsenide gives a free-surface velocity of 2870 m/s and it is weakly piezoelectric, giving $\Delta v/v = 0.035\%$.

As mentioned earlier, single-crystal materials sometimes show leaky wave solutions with attractive properties, such as high velocity. Leaky waves have also been found in layered systems. For example, in zinc oxide on diamond a longitudinal-type leaky wave has been observed, with a phase velocity of about 16 000 m/s [52].

REFERENCES

1. A.J. Slobodnik. 'Materials and their influence on performance', in A.A. Oliner (ed.), *Acoustic Surface Waves*, Springer, 1978, pp. 226–303.
2. A.J. Slobodnik. 'Surface acoustic waves and SAW materials', *Proc. IEEE*, **64**, 581–595 (1976).
3. M.S. Kharusi and G.W. Farnell. 'Diffraction and beam steering for surface-wave comb structures on anisotropic substrates', *IEEE Trans.*, **SU-18**, 34–42 (1971).
4. I.M. Mason and E.A. Ash. 'Acoustic surface wave beam diffraction on anisotropic substrates', *J. Appl. Phys.*, **42**, 5343–5351 (1971).
5. N.R. Ogg. 'A Huygens principle for anisotropic media', *J. Phys. A*, **4**, 382–388 (1971).
6. J.C. Crabb, J.D. Maines and N.R. Ogg. 'Surface-wave diffraction in $LiNbO_3$', *Electron. Lett.*, **7**, 253–255 (1971).
7. G.W. Farnell. 'Properties of elastic surface waves', in W.P. Mason and R.W. Thurston (eds.), *Physical Acoustics*, Vol. 6, Academic Press, 1970, pp. 109–166.
8. M. Abramowitz and I.A. Stegun. *Handbook of Mathematical Functions*, Dover, 1968.
9. T.L. Szabo and A.J. Slobodnik. 'The effect of diffraction on the design of acoustic surface wave devices', *IEEE Trans.*, **SU-20**, 240–251 (1973).
10. G. Kovacs, M. Anhorn, H.E. Engan, G. Visintini and C.C.W. Ruppel. 'Improved material constants for $LiNbO_3$ and $LiTaO_3$', *IEEE Ultrason. Symp.*, **1**, 435–438 (1990).
11. T.L. Szabo and A.J. Slobodnik. 'Diffraction compensation in periodic apodised acoustic surface wave filters', *IEEE Trans.*, **SU-21**, 114–119 (1974).
12. R.F. Milsom. 'A diffraction theory for SAW filters on non-parabolic high-coupling orientations', *IEEE Ultrason. Symp.*, **1**, 827–833 (1997).

13. K. Dransfeld and E. Salzmann. 'Excitation, detection and attenuation of high-frequency elastic surface waves', in W.P. Mason and R.N. Thurston (eds.), *Physical Acoustics*, Vol. 7, Academic Press, 1970, pp. 219–272.

14. K.L. Davies and J.F. Weller. 'SAW attenuation in metal film coated delay lines', *IEEE Ultrason. Symp.*, 1979, pp. 659–662.

15. Y. Nakagawa, K. Yamanouchi and K. Shibayama. 'Control of nonlinear effects in acoustic surface waves', *J. Appl. Phys.*, **45**, 2817–2822 (1974).

16. R.C. Williamson. 'Problems encountered in high-frequency surface wave devices', *IEEE Ultrason. Symp.*, 1974, pp. 321–328.

17. J.F. Dias, H.E. Karrer, J.A. Kusters, J.H. Matsinger and M.B. Schulz. 'The temperature coefficient of delay time for X-propagating acoustic surface waves on rotated Y-cuts of alpha-quartz', *IEEE Trans.*, **SU-22**, 46–50 (1975).

18. S.J. Kerbel. 'Design of harmonic surface acoustic wave oscillators without external filtering and new data on the temperature coefficient of quartz', *IEEE Ultrason. Symp.*, 1974, pp. 276–281.

19. J. Temmyo, I. Kotaka, T. Inamura and S. Yoshikawa. 'Precise measurement of SAW propagation velocity on lithium niobate', *IEEE Trans.*, **SU-27**, 218–219 (1980).

20. J.J. Campbell and W.R. Jones. 'A method for estimating optimal crystal cuts and propagation directions for excitation of piezoelectric surface waves', *IEEE Trans.*, **SU-15**, 209–217 (1968).

21. K. Shibayama, K. Yamanouchi, H. Sato and T. Meguro. 'Optimum cut for rotated Y-cut $LiNbO_3$ crystal used as the substrate of acoustic-surface-wave filters', *Proc. IEEE*, **64**, 595–597 (1976).

22. S. Takahashi, T. Kodama, F. Miyashiro, B. Suzuki, A. Onoe, T. Adachi and K. Fujiwara. 'SAW IF filter on $LiTaO_3$ for colour TV receivers', *IEEE Trans. Consumer Electron.*, **CE-24**, 337–348 (1978).

23. B. Lewis, N.M. Shorrocks and R.W. Whatmore. 'An assessment of lithium tetraborate for SAW applications', *IEEE Ultrason. Symp.*, 1982, pp. 389–393.

24. K. Yamanouchi and M. Takeuchi. 'Applications for piezoelectric leaky surface waves', *IEEE Ultrason. Symp.*, **1**, 11–18 (1990).

25. M. Lewis. 'Surface-skimming bulk waves, SSBW', *IEEE Ultrason. Symp.*, 1977, pp. 744–752.

26. K. Nakamura, M. Kazumi and H. Shimizu. 'SH-type and Rayleigh type surface waves on rotated Y-cut $LiTaO_3$', *IEEE Ultrason. Symp.*, 1977, pp. 819–822.

27. R. O'Donnell and P.H. Carr. 'High piezoelectric coupling temperature-compensated cuts of berlinite ($AlPO_4$) for SAW applications', *IEEE Trans.*, **SU-24**, 376–384 (1977).

28. M.P. DaCunha and S. de A. Fagundes. 'Investigation of recent quartz-like materials for SAW applications', *IEEE Trans. Ultrason. Ferroelec. Freq. Contr.*, **46**, 1583–1590 (1999).

29. Y. Shimizu, M. Tanaka and T. Watanabe. 'A new cut of quartz with extremely small temperature coefficient by leaky surface waves', *IEEE Ultrason. Symp.*, **1**, 233–236 (1985).

30. B.P. Abbot and L. Solie. 'A minimal diffraction cut of quartz for high performance SAW filters', *IEEE Ultrason. Symp.*, **1**, 235–240 (2000).

31. N. Naumenko and L. Solie. 'Optimal cuts of langasite, $La_3Ga_5SiO_{14}$, for SAW devices', *IEEE Trans. Ultrason. Ferroelec. Freq. Contr.*, **48**, 530–537 (2001).

32. M.P. DaCunha, D.C. Malocha, E.L. Adler and K.J. Casey. 'Surface and pseudo surface acoustic waves in langatate: predictions and measurements', *IEEE Trans. Ultrason. Ferroelec. Freq. Contr.*, **49**, 1291–1299 (2002).

33. K. Yamanouchi and H. Odagawa. 'Super high electromechanical coupling and zero temperature coefficient surface acoustic wave substrates in $KNbO_3$ single crystal', *IEEE Trans. Ultrason. Ferroelec. Freq. Contr.*, **46**, 700–705 (1999).

34. T. Sato and H. Abe. 'Propagation properties of longitudinal leaky surface waves on lithium tetraborate', *IEEE Trans. Ultrason. Ferroelec. Freq. Contr.*, **45**, 136–151 (1998).

35. K. Yamanouchi, H. Sato, T. Meguro and Y. Wagatsuma. 'High temperature stable GHz-range low-loss wide band transducers and filter using SiO_2/$LiNbO_3$, $LiTaO_3$', *IEEE Trans. Ultrason. Ferroelec. Freq. Contr.*, **42**, 392–396 (1995).

36. J. Kushibiki, I. Takanaga and S. Nishiyama. 'Accurate measurements of the acoustical physical constants of synthetic α-quartz for SAW devices', *IEEE Trans. Ultrason. Ferroelec. Freq. Contr.*, **49**, 125–135 (2002).

37. I. Takanaga and J. Kushibiki. 'A method of determining acoustical physical constants for piezoelectric materials by line-focus-beam acoustic microscopy', *IEEE Trans, Ultrason. Ferroelec. Freq. Contr.*, **49**, 893–904 (2002).

38. J.G. Gualtieri, J.A. Kosinski and A. Ballato. 'Piezoelectric materials for acoustic wave applications', *IEEE Trans. Ultrason. Ferroelec. Freq. Contr.*, **41**, 53–59 (1994).

39. J.A. Kosinski. 'New piezoelectric substrates for SAW devices', in C.C.W. Ruppel and T.A. Fjeldy (eds.), *Advances in Surface Acoustic Wave Technology, Systems and Applications*, Vol. 2, World Scientific, 2001, pp. 151–202.

40. S. Ballandras, W. Steichen, E. Briot, M. Solal, M. Doisy and J.-M. Hode. 'A new triply rotated quartz cut for the fabrication of low loss IF SAW filters', *IEEE Trans. Ultrason. Ferroelec. Freq. Contr.*, **51**, 121–126 (2004).

41. M.P. DaCunha, D.C. Malocha, D. Puccio, J. Thiele and T. Pollard. 'High coupling zero TCD SH wave on LGX', *IEEE Ultrason. Symp.*, **1**, 381–384 (2002).

42. H. Nakahata, K. Higaki, S. Fujii, A. Hachigo, H. Kitabayashi, K. Tanabe, Y. Seki and S. Shikata. 'SAW devices on diamond', *IEEE Ultrason. Symp.*, **1**, 361–370 (1995).

43. N.W. Emanetoglu, G. Patounakis, S. Liang, C.R. Gorla, R. Wittstruck and Y. Lu. 'Analysis of SAW properties of epitaxial ZnO films grown on R-Al_2O_3 substrates', *IEEE Trans. Ultrason. Ferroelec. Freq. Contr.*, **48**, 1389–1394 (2001).

44. O. Elmazria, V. Mortet, M. El Hakiki, M. Nesladek and P. Alnot. 'High velocity SAW using aluminium nitride film on unpolished nucleation side of free-standing CVD diamond', *IEEE Trans. Ultrason. Ferroelec. Freq. Contr.*, **50**, 710–715 (2003).

45. K. Uehara, C.-M. Yang, T. Shibata, S.-K. Kim, S. Kameda, H. Nakase and K. Tsubouchi. 'Fabrication of 5 GHz-band SAW filter with atomically-flat-surface AlN on sapphire', *IEEE Ultrason. Symp.*, **1**, 203–206 (2004).

46. Y. Kim, W.D. Hunt, F.S. Hickernell, R.J. Higgins and C.-K. Jen. 'ZnO films on {001}-cut <110>-propagating GaAs substrates for surface acoustic wave device applications', *IEEE Trans. Ultrason. Ferroelec. Freq. Contr.*, **42**, 351–361 (1995).

47. M. Kadota, T. Kasanami and M. Minakata. 'Characterisation of zinc oxide films on glass substrates deposited by RF-mode electron cyclotron resonance sputtering system', *Jpn. J. Appl. Phys.*, **32**, 2341–2345 (1993).

48. M. Kadota and M. Minakata. 'Piezoelectric properties of zinc oxide films on glass substrates deposited by RF-magnetron-mode electron cyclotron resonance sputtering system', *IEEE Trans. Ultrason. Ferroelec. Freq. Contr.*, **42**, 345–350 (1995).

49. T.-T. Wu and Y.-Y. Chen. 'Exact analysis of dispersive SAW devices on ZnO/diamond/Si layered systems', *IEEE Trans. Ultrason. Ferroelec. Freq. Contr.*, **49**, 142–149 (2002).

50. J. Koike, K. Shimoe and H. Ieki. '1.5 GHz low-loss surface acoustic wave filter using ZnO/sapphire substrate', *Jpn. J. Appl. Phys.*, **32**, 2337–2340 (1993).

51. M. Kadota and H. Kando. 'Small and low-loss IF SAW filters using zinc oxide film on quartz substrates', *IEEE Trans. Ultrason. Ferroelec. Freq. Contr.*, **51**, 464–469 (2004).

52. D.L. Dreifus, R.J. Higgins, R.B. Henard, R. Almar and L.P. Solie. 'Experimental observation of high velocity pseudo-SAWs in ZnO/diamond/Si multilayers', *IEEE Ultrason. Symp.*, **1**, 191–194 (1997).

5

NON-REFLECTIVE TRANSDUCERS

Interdigital transducers are of fundamental importance to all surface-wave devices, and often the transducer performance is the main factor determining device performance. This chapter describes methods for transducer analysis, and for analysis of devices comprising two transducers. This forms the basis for the device design described in Chapter 6. It is also used in developing the analysis for reflective transducers, described in Chapter 8. Most of the analysis here is based on the Green's function developed in Chapter 3. A simplified version called the delta-function model was presented in Chapter 1. It is assumed that there are no losses, thus excluding propagation loss and diffraction. Thus, the waves propagate with no attenuation.

The term 'non-reflective' refers to transducers that do not reflect surface waves when they are shorted. In terms of the well-known P-matrix, this means that $P_{11} = P_{22} = 0$. Generally, a shorted transducer can reflect the waves because individual electrodes reflect, and these reflections can be significant if the electrode pitch p is a multiple of the half-wavelength $\lambda/2$. However, this phenomenon is easily avoided by using double-electrode transducers with electrode spacing $\lambda/4$, as noted in Chapter 1, Section 1.3. Hence double-electrode transducers are non-reflective. Single-electrode transducers have an electrode pitch of $\lambda/2$, so generally they are reflective, with $P_{11} \neq 0$. For this case, the methods in this chapter are inadequate, though methods in Chapter 8 can be used. However, short single-electrodes transducers can have small reflectivity if the electrode reflectivity is small, and for this case the non-reflective analysis is approximately valid. A quantitative discussion is given in Chapter 8. Transducers with p less than $\lambda/2$ are non-reflective even if the electrodes reflect quite strongly, because the electrode reflections do not add coherently. The analysis here applies to such cases, though we assume that electrical and mechanical perturbations are not too strong.

114

The main relevance of this topic is to surface-wave transversal filters. These common high-performance devices are described in Chapter 6. They use regular arrays of electrodes, that is, the electrode width and pitch are constant. A simple extension applies to interdigital chirp filters, as shown in Chapter 7.

In this chapter we first develop the theory for electrode arrays with general geometry. This is general enough to cover some important applications such as unidirectional transducers, considered in Chapter 9. However, most surface-wave devices use regular electrodes. Section 5.3 has a summary of results from the earlier sections, and it is followed by sections mainly concerned with regular electrodes. The reader interested in devices with regular electrodes may find it helpful to omit Sections 5.1 and 5.2 on a first reading.

5.1   ANALYSIS FOR A GENERAL ARRAY OF ELECTRODES

Here we consider a general array of electrodes on a piezoelectric half-space, illustrated in Fig. 5.1a. The analysis is based on the Green's function of Chapter 3, and therefore excludes mechanical loading. The fields are assumed to be invariant in the y-direction, so that distortions near the electrode ends can be neglected and diffraction can be ignored. It is assumed that the only wave present is a piezoelectric Rayleigh wave, and that the electrodes have negligible resistivity. The electrode voltages all have the same frequency ω, but are otherwise arbitrary.

5.1.1   The quasi-static approximation

In Chapter 3, Section 3.4 it was shown that the potential $\phi(x, \omega)$ and charge density $\sigma(x, \omega)$ at the surface of the piezoelectric are related by the Green's function $G(x, \omega)$, at frequency ω. We assume that the only wave present is a piezoelectric Rayleigh wave, ignoring other waves such as bulk waves. The Green's function can be approximated as a sum of an electrostatic term $G_e(x)$ and a surface-wave term $G_s(x, \omega)$ as in eq. (3.48), so that

$$\phi(x, \omega) = [G_e(x) + G_s(x, \omega)] * \sigma(x, \omega) \tag{5.1}$$

where the asterisk indicates convolution with respect to x. Here the piezoelectric occupies the region $z < 0$, and the waves propagate in the $\pm x$-directions. The two components of the Green's function are

$$G_e(x) = -\frac{\ln|x|}{\pi \varepsilon_\infty} \tag{5.2}$$

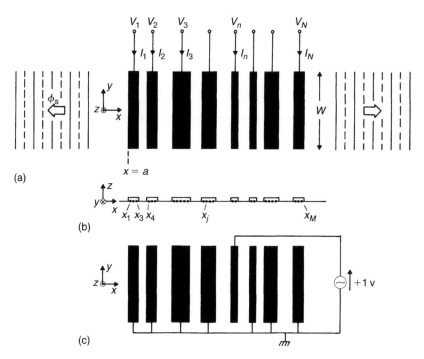

Figure 5.1 Generalized array of electrodes. (a) Electrode array. (b) Elevation, showing points x_j. (c) Array with one live electrode.

and

$$G_s(x, \omega) = j\Gamma_s \exp(-jk_f|x|) \tag{5.3}$$

where k_f is the wavenumber for surface-wave propagation on a free surface at frequency ω. The constant ε_∞ is the effective permittivity at infinite wavenumber, $\varepsilon_s(\infty)$. The constant Γ_s is, from eq. (3.31),

$$\Gamma_s \approx \frac{1}{\varepsilon_\infty} \frac{v_f - v_m}{v_f} \equiv \frac{\Delta v/v}{\varepsilon_\infty} \tag{5.4}$$

where v_f and v_m are the wave velocities for a free surface and metallized surface, respectively. The notation $\Delta v/v \equiv (v_f - v_m)/v_f$ denotes a commonly used measure of the piezoelectric coupling. It is also common to define a surface-wave coupling constant K^2 as

$$K^2 = 2\Delta v/v. \tag{5.5}$$

However, in this book the coupling will usually be expressed in terms of $\Delta v/v$.

Since the electrodes have negligible resistivity, the potential $\phi(x, \omega)$ will be constant over the surface of each electrode, and equal to the electrode voltage. The charge density $\sigma(x, \omega)$ must be zero in all unmetallized regions. Using these boundary conditions with the equations above, the potential and charge density are determined everywhere if the electrode voltages are specified.

At points remote from the electrodes, the potential is primarily the potential $\phi_s(x, \omega)$ associated with the surface waves generated by the structure. For example, if the electrodes exist only for $x > a$, as in Fig. 5.1a, we have $\sigma(x, \omega) = 0$ for $x < a$, so the potential due to the electrostatic term $G_e(x)$ decays with distance in this region. Thus, the surface-wave potential can be identified as $G_s(x, \omega) * \sigma(x, \omega)$. Defining $\overline{\sigma}(\beta, \omega)$ as the Fourier transform of $\sigma(x, \omega)$, with ω held constant in the Fourier integral, the surface-wave potential is found to be

$$\phi_s(x, \omega) = j\Gamma_s \overline{\sigma}(k_f, \omega) \exp(jk_f x) \quad \text{for } x < a. \tag{5.6}$$

Here the modulus sign in eq. (5.3) has disappeared because $\sigma(x, \omega) = 0$ for $x < a$. Thus, the surface-wave potential is basically the Fourier transform of the charge density. Although bulk-wave excitation has been ignored here, eq. (5.6) is still valid when bulk waves are excited, as shown in Appendix B, eq. (B.31).

The surface potential $\phi_s(x, \omega)$ is regarded here as the amplitude of the wave. Of course, another parameter, such as one of the displacement components, could alternatively be used to specify the amplitude. A definition related to the wave power is often used in connection with the P-matrix, as described in Chapter 5, Section 5.3.

It should be noted that $\overline{\sigma}(k_f, \omega)$ is not the Fourier transform of $\sigma(x, \omega)$ in the usual sense, because k_f is not an independent parameter. If ω is changed $\sigma(x, \omega)$ will change, so the Fourier integral must be calculated separately for each ω. For clarity, β is used here as the independent variable in the transform, while k, with various subscripts, refers to wavenumbers of propagating waves.

Although the above equations can be solved for the potential and charge density, the calculation is generally very complex. It will be considered later in Chapter 8. Here the *quasi-static approximation* [1] is introduced to simplify the problem. The charge density $\sigma(x, \omega)$ is assumed to be dominated by an electrostatic term $\sigma_e(x, \omega)$, defined as the charge density obtained when acoustic wave excitation is ignored. Thus, from eq. (5.1), $\sigma_e(x, \omega)$ is the solution of

$$\phi_e(x, \omega) = G_e(x) * \sigma_e(x, \omega). \tag{5.7}$$

Here $\phi_e(x, \omega)$ is a new surface potential, equal to the specified voltage when x is on an electrode, and $\sigma_e(x, \omega) = 0$ on all unmetallized regions. Note that $\sigma_e(x, \omega)$

is independent of ω if the voltages are independent of ω, and it is zero if the voltages are all the same. The total charge density is taken to be the sum

$$\sigma(x, \omega) \approx \sigma_e(x, \omega) + \sigma_a(x, \omega) \tag{5.8}$$

where $\sigma_a(x, \omega)$ is a small contribution due to the piezoelectric effect, arising from the presence of the acoustic waves. The surface potential is given by eq. (5.1), using eq. (5.8) for $\sigma(x, \omega)$. On the right there are now four terms, but a term $G_s(x, \omega) * \sigma_a(x, \omega)$ is small and can be omitted; both of the functions involved are zero if there is no piezoelectricity. We thus have, in the quasi-static approximation,

$$\phi(x, \omega) = [G_e(x) + G_s(x, \omega)] * \sigma_e(x, \omega) + G_e(x) * \sigma_a(x, \omega). \tag{5.9}$$

Given the function $\sigma_e(x, \omega)$, this equation, together with the boundary conditions, determines the acoustic charge density $\sigma_a(x, \omega)$, as discussed further in Section 5.1.3. Also, $\sigma_e(x, \omega)$ is determined by eq. (5.7). Hence the total charge density, eq. (5.8), and the surface potential, eq. (5.9), are determined. The solution is approximate because one term was omitted in deriving eq. (5.9). However, we define $\sigma_a(x, \omega)$ such that eq. (5.9) is exactly true; it then follows that the charge density is not exactly the sum of $\sigma_e(x, \omega)$ and $\sigma_a(x, \omega)$, so that eq. (5.8) is approximate.

As explained earlier, the potential $\phi_s(x, \omega)$ of a surface wave generated by the electrodes can be obtained by omitting terms arising from the electrostatic Green's function $G_e(x)$. Using eq. (5.9), the surface-wave potential for $x < a$, to the left of the structure, is

$$\phi_s(x, \omega) = j\Gamma_s \overline{\sigma}_e(k_f, \omega) \exp(jk_f x) \quad \text{for } x < a \tag{5.10}$$

where we define $\overline{\sigma}_e(\beta, \omega)$ as the Fourier transform of $\sigma_e(x, \omega)$. This is the same as the more accurate result of eq. (5.6) except that $\overline{\sigma}(k_f, \omega)$ is replaced by $\overline{\sigma}_e(k_f, \omega)$. This change simplifies the analysis considerably. For example, consider surface-wave excitation by a two-terminal transducer, taking the electrode voltages to be independent of ω. The electrostatic charge density $\sigma_e(x, \omega)$ will be independent of ω, so its transform $\overline{\sigma}_e(\beta, \omega)$ is also independent of ω. Hence, it is sufficient to do the transformation at $\omega = 0$, and then use eq. (5.10) to obtain $\phi_s(x, \omega)$ for all ω. For this reason, the approximation is described as 'quasi-static'.

5.1.2 Electrostatic equations and charge superposition

The electrostatic charge density $\sigma_e(x, \omega)$ is determined by eq. (5.7), with the boundary conditions that $\phi_e(x, \omega)$ must be equal to the electrode voltage whenever x is on an electrode, and $\sigma_e(x, \omega)$ is zero on all unmetallized regions. Techniques for calculating $\sigma_e(x, \omega)$ are considered below and in Section 5.4.

Here we consider a simple method which is not very practical, but which serves for developing the analysis. We define a set of M points x_j, as shown in Fig. 5.1b. The points exist only on the electrodes, where they have a small spacing Δx. The electrostatic equation (5.7) is written as

$$\phi_e(x_i, \omega) = \sum_{j=1}^{M} A_{ij}\sigma_e(x_j, \omega). \tag{5.11}$$

This becomes identical to eq. (5.7) in the limit when $\Delta x \to 0$ and $M \to \infty$. The obvious choice for the A_{ij} is $G_e(x_i - x_j)\Delta x$, but this is infinite when $i = j$, as seen from eq. (5.2). A more suitable choice is

$$A_{ij} = (\Delta x/2)[G_e(x_i - x_j + \Delta x/2) + G_e(x_i - x_j - \Delta x/2)]. \tag{5.12}$$

This matrix takes account of the electrode geometry, since the points x_j exist only on the electrodes. Since $G_e(x)$ is symmetrical, A_{ij} is also symmetrical, so that $A_{ji} = A_{ij}$. We now invert eq. (5.11) to express $\sigma_e(x, \omega)$ in terms of $\phi_e(x, \omega)$, giving

$$\sigma_e(x_i, \omega) = \sum_{j=1}^{M} B_{ij}\phi_e(x_j, \omega) \tag{5.13}$$

where B_{ij} is the reciprocal of the matrix A_{ij}. Since A_{ij} is symmetrical, B_{ij} will be as well, so that $B_{ji} = B_{ij}$.

In eq. (5.13), all the $\phi_e(x_j, \omega)$ are equal to known electrode voltages. The potential in the interelectrode gaps, which is not known initially, is not required. The condition that $\sigma_e(x, \omega)$ should be zero in the gaps is implied by eq. (5.11), because the summation excludes points not on the electrodes.

Charge superposition

The determination of $\sigma_e(x, \omega)$ is often simplified by superposition, illustrated in Fig. 5.1c. Here the electrodes are as in Fig. 5.1a, but now electrode n has unit voltage while all other electrodes are grounded (that is, they have zero voltage). The electrostatic charge density for this case is denoted by $\rho_{en}(x)$, which is independent of frequency. Generally, there will be charges on all electrodes, not just on electrode n. To find $\rho_{en}(x)$, we define an electrode polarity function $\hat{p}_n(x)$ such that $\hat{p}_n(x) = 1$ if x is on electrode n, and $\hat{p}_n(x) = 0$ for other x. For a point x_j on any electrode, the potential in Fig. 5.1c is $\hat{p}_n(x_j)$, and from eq. (5.13) the electrostatic charge density is

$$\rho_{en}(x_i) = \sum_{j=1}^{M} B_{ij}\hat{p}_n(x_j) \tag{5.14}$$

with $\rho_{en}(x) = 0$ on the unmetallized regions.

The charge superposition principle states that, when some arbitrary set of voltages V_n is applied to the N electrodes, as in Fig. 5.1a, the charge density is given by the linear sum

$$\sigma_e(x, \omega) = \sum_{n=1}^{N} V_n \rho_{en}(x). \tag{5.15}$$

This follows directly from eq. (5.13). It has considerable practical value because $\sigma_e(x, \omega)$ can be found for any set of electrode voltages, once the functions $\rho_{en}(x)$ are known. The charge density is zero if the surface potential is independent of x, since for this case there is no electric field. It follows that $\sigma_e(x, \omega)$ is unaffected if a constant is added to all the V_n in eq. (5.15). It is therefore determined by the voltage differences.

Although the above derivation considered the electrostatic case, eq. (5.15) is actually much more general because any Green's function could have been used. In particular, the theorem remains valid if various acoustic waves are excited, including Rayleigh waves, bulk waves and leaky waves.

Numerical evaluation of the electrostatic charge density

The form of the function $\sigma_e(x, \omega)$ is fundamental to the operation of surface-wave transducers, and it has been given very extensive attention. Section 3.1 of Chapter 3 gives some basic relations. It is common to assume that the film thickness h is negligible because finite h values introduce extra complexities without affecting the results substantially. Exact algebraic formulae are known only for a few special cases. Fortunately, these include the common case of regular electrodes, described in Section 5.4. There are also formulae for transducers using either two or three electrodes [2]. In general, numerical calculations are needed, and they are not easy to do. One reason for this is that the charge density has a $1/\sqrt{x}$ dependence near each electrode edge, so that it is infinite at the edge [2]. Milsom *et al.* [3] used unequal point spacing to deal with this problem. Alternatively [4], a function that behaves as $1/\sqrt{x}$ for small x can be integrated numerically using the substitution $x = t^2$. Another complication is that the logarithmic Green's function of eq. (5.2) is not sufficient – it is necessary to add the condition that the total charge on the transducer is zero.

In early work, Hartmann and Secrest [5] used numerical calculations to evaluate end effects, showing the distortion of the charge density for electrodes near the transducer ends. A useful form for the charge density on electrode n is the expression

$$\sigma_{en}(x) = \sum_m \alpha_{mn} \frac{T_m(x)}{\sqrt{1 - x^2}} \tag{5.16}$$

where $T_m(x)$ is a Chebychev polynomial and the α_{mn} are coefficients to be solved for. This form automatically incorporates the infinities at the electrode edges, and the inclusion of the Chebychev polynomial has useful mathematical properties. The form in eq. (5.16) needs to be scaled according to the width and position of electrode n, parameters that can be specified arbitrarily. Often, only three or four values of m are sufficient to give the required accuracy, so the number of unknown coefficients is three or four times the number of electrodes. Many authors have used this type of formulation [6–9].

A useful theorem relates $\sigma_e(x)$ to the solution obtained for electrodes in a vacuum, with the substrate absent. For a vacuum, the fields obey Laplace's equation everywhere (except for discontinuities at the electrode plane $z = 0$), and it can be shown [2, 10] that $F(x, z) = E_x - jE_z$ is analytic function of the variable $Z = x + jz$. This implies that the fields $F(x, 0)$ on the plane $z = 0$ determine the fields everywhere. At $z = 0$, $F(x, 0)$ is imaginary on the electrodes, where $E_x = 0$. Also, at $z = 0$, $F(x, 0)$ is real in the unmetallized regions, where $E_z = 0$ and there are no charges. Suppose the charge density is $\sigma_v(x)$, on each side of the electrodes. When a dielectric substrate is present, the total charge density is $\sigma_e(x) = (1 + \varepsilon_p/\varepsilon_0)\sigma_v(x)$. Hence it is sufficient to find the function $F(x, 0)$.

A very effective method is given by Biryukov and Polevoi [10]. For $z = 0$, the solution for the vacuum case is

$$F'(x, 0) = (-1)^{N+m} \frac{\displaystyle\sum_{n=1}^{N-1} \gamma_n x^{n-1}}{\sqrt{\displaystyle\prod_{n=1}^{N} |(x - a_n)(x - b_n)|}} \tag{5.17}$$

where $n = 1, 2, \ldots, N$ is the electrode number. The left and right edges of electrode n are at $x = a_n$ and b_n, respectively, and m is defined such that x is between a_m and a_{m+1}, taking $a_0 = -\infty$ and $a_{N+1} = \infty$. For a given electrode geometry, the fields are determined by the $N - 1$ voltage differences. The real coefficients γ_n are defined such that integrals of E_x between the electrodes give the specified voltage differences. This requirement gives $N - 1$ equations in the $N - 1$ unknown coefficients γ_n. The function $F'(x, 0)$ is a modified form of $F(x, 0)$, equal to E_x in the gaps. On the electrodes, the charge density on both sides is $\sigma_v(x) = F'(x, 0)$.

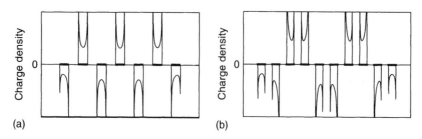

Figure 5.2 Electrostatic charge density. (a) Single-electrode transducer. (b) Double-electrode transducer.

Figure 5.2 shows charge densities calculated by this method, for single-electrode and double-electrode transducers. It can be seen that end effects cause the charges on the end electrodes to be reduced.

For long transducers, such methods are impractical because they involve inversion of large matrices. However, other methods can be used for infinite-length periodic transducers, as explained in Chapter 9. These methods can be applied to long finite transducers if end effects can be ignored.

5.1.3 Current entering one electrode

The current entering one electrode can be found from the charge density on it. We first consider the latter in more detail, showing that $\sigma_a(x, \omega)$ can be expressed in terms of $\sigma_e(x, \omega)$. The electrostatic term $\sigma_e(x, \omega)$ is defined as the solution of eq. (5.7), which involves the potential $\phi_e(x, \omega)$. The actual potential $\phi(x, \omega)$ is given by eq. (5.9), which includes a term equal to $\phi_e(x, \omega)$. Since $\phi_e(x, \omega) = \phi(x, \omega)$ when x is on an electrode, the remaining terms in eq. (5.9) must be zero at such points, so that

$$[G_s(x, \omega) * \sigma_e(x, \omega) + G_e(x) * \sigma_a(x, \omega)]_{x_j} = 0. \qquad (5.18)$$

This equation relates $\sigma_a(x, \omega)$ to $\sigma_e(x, \omega)$. It is convenient to define an acoustic potential by the expression

$$\phi_a(x, \omega) = G_s(x, \omega) * \sigma_e(x, \omega). \qquad (5.19)$$

This potential is zero if the substrate material is not piezoelectric. Equation (5.18) can be written as

$$[G_e(x) * \sigma_a(x, \omega)]_{x_j} = -\phi_a(x_j, \omega). \qquad (5.20)$$

This has the same form as eq. (5.7), so it can be expressed in a form similar to eq. (5.11). Comparing with eq. (5.13), the acoustic charge density is

$$\sigma_a(x_i, \omega) = -\sum_{j=1}^{M} B_{ij} \phi_a(x_j, \omega), \qquad (5.21)$$

where B_{ij} is the reciprocal of the matrix A_{ij}. For x values on the electrodes, eq. (5.21) gives $\sigma_a(x, \omega)$ exactly, when $\Delta x \to 0$. For other x, $\sigma_a(x, \omega)$ is zero. Thus, $\sigma_a(x, \omega)$ can be deduced from $\sigma_e(x, \omega)$, using eq. (5.19) for $\phi_a(x, \omega)$.

The current I_n flowing into electrode n is the time differential of the total charge on it. Thus,

$$I_n = j\omega W \int_n [\sigma_e(x, \omega) + \sigma_a(x, \omega)]dx \qquad (5.22)$$

where the integral is taken over electrode n. W is the aperture, as shown in Fig. 5.1. The current is the sum of an electrostatic term I_{en} and an acoustic term I_{an}, so that

$$I_n = I_{en} + I_{an}. \qquad (5.23)$$

The term I_{en} is the part of eq. (5.22) dependent on $\sigma_e(x, \omega)$. The acoustic contribution can be written as

$$I_{an} = j\omega W \int_{-\infty}^{\infty} \hat{p}_n(x)\sigma_a(x, \omega)dx, \qquad (5.24)$$

where, as before, $\hat{p}_n(x)$ is unity when x is on electrode n, and zero for other x. Since the integrand here is zero on unmetallized regions, the integral can be expressed as a sum over the points x_j, so that

$$I_{an} = j\omega W \sum_{j=1}^{M} \hat{p}_n(x_j)\sigma_a(x_j, \omega)\Delta x$$

$$= -j\omega W \Delta x \sum_{i=1}^{M} \phi_a(x_i, \omega) \sum_{j=1}^{M} \hat{p}_n(x_j)B_{ji}, \qquad (5.25)$$

where eq. (5.21) has been used and the summations have been re-ordered. Now, since B_{ij} is symmetrical, comparison with eq. (5.14) shows that the sum over j can be identified as $\rho_{en}(x_i)$. Taking the limit as $\Delta x \to 0$, and noting that $\rho_{en}(x)$ is zero in unmetallized regions, eq. (5.25) becomes

$$I_{an} = -j\omega W \int_{-\infty}^{\infty} \rho_{en}(x)\phi_a(x, \omega)dx. \qquad (5.26)$$

5.1.4 Evaluation of the acoustic potential

The acoustic potential $\phi_a(x, \omega)$ in eq. (5.26) is defined in terms of $\sigma_e(x, \omega)$ by eq. (5.19). Here we derive an alternative expression involving the Fourier transform of $\sigma_e(x, \omega)$, denoted by $\overline{\sigma}_e(\beta, \omega)$. The surface-wave Green's function $G_s(x, \omega)$ is given by eq. (5.3), and substitution into eq. (5.19) gives

$$\phi_a(x, \omega) = j\Gamma_s \int_{-\infty}^{x} \sigma_e(x', \omega) \exp[-jk_f(x - x')]dx'$$

$$+ j\Gamma_s \int_{x}^{\infty} \sigma_e(x', \omega) \exp[jk_f(x - x')]dx'. \tag{5.27}$$

If the electrodes are present only for $x > a$, as in Fig. 5.1, this equation gives $\phi_a(x, \omega) = j\Gamma_s \overline{\sigma}_e(k_f, \omega) \exp(jk_f x)$ for the region $x < a$ to the left of the transducer. Comparison with eq. (5.10) shows that $\phi_a(x, \omega)$ is equal to the surface-wave potential $\phi_s(x, \omega)$ for $x < a$, and this is also found to be true for the unmetallized region to the right of the structure.

Equation (5.27) may be re-arranged using the step function $U(x)$, which is equal to 1 for $x > 0$ and to zero for $x < 0$. Thus,

$$\phi_a(x, \omega) = j\Gamma_s e^{-jk_f x} \int_{-\infty}^{\infty} \sigma_e(x', \omega) U(x - x') e^{jk_f x'} dx'$$

$$+ j\Gamma_s e^{jk_f x} \int_{-\infty}^{\infty} \sigma_e(x', \omega) U(x' - x) e^{-jk_f x'} dx', \tag{5.28}$$

where the limits are now $\pm\infty$. These integrals can be evaluated by Fourier methods, taking x as a constant. The first integral is the transform of $[\sigma_e(x', \omega) U(x - x')]$, from the x' domain to the β domain, with the result evaluated at $\beta = -k_f$. A similar method is used for the second integral. The relationships needed are given in Appendix A. The transform of $U(x')$ is, from eq. (A.29),

$$U(x') \leftrightarrow \pi\delta(\beta) - j/\beta. \tag{5.29}$$

The shifting and scaling theorems, eqs (A.6) and (A.5), are used to obtain the transforms of $U(x - x')$ and $U(x' - x)$, and the products in eq. (5.28) are transformed using the convolution theorem, eq. (A.14). With $\overline{\sigma}_e(\beta, \omega)$ defined as the Fourier transform of $\sigma_e(x, \omega)$, the result is

$$\phi_a(x, \omega) = \frac{1}{2}j\Gamma_s e^{-jk_f x}[\overline{\sigma}_e(-k_f, \omega) + jF(-k_f)/\pi]$$

$$+ \frac{1}{2}j\Gamma_s e^{jk_f x}[\overline{\sigma}_e(k_f, \omega) - jF(k_f)/\pi], \tag{5.30}$$

where the function $F(\beta)$ is defined by

$$F(\beta) = \overline{\sigma}_e(\beta, \omega) * \frac{\exp(-j\beta x)}{\beta}.$$

Here the terms $\overline{\sigma}_e(\pm k_f, \omega)$ arise because of the delta function in eq. (5.29). It will be seen later that these give rise to the acoustic conductance $G_a(\omega)$ of a transducer. The terms $F(\pm k_f)$, involving a convolution, are due to the j/β term in eq. (5.29) and will be found to give the acoustic susceptance $B_a(\omega)$.

5.2 QUASI-STATIC ANALYSIS OF TRANSDUCERS

We now apply the results of Section 5.1 to a two-terminal unapodized transducer such as that in Fig. 5.3, where each electrode is connected to one of the two bus bars. It is assumed that the bus bars have no effect other than providing electrical connections, with no resistance. The method uses the quasi-static approximation of Section 5.1.1, and this implies that electrode reflections are neglected, so that a shorted transducer does not reflect incident surface waves. Another consequence is that waves are predicted to travel through the transducer with the free-surface velocity v_f. In practice the electrodes cause small changes to the wave velocity, as discussed in Chapter 8.

5.2.1 Launching transducer

We first consider the transducer of Fig. 5.3a, which is taken to be isolated so that there are no incident acoustic waves. The charge density on the electrodes is

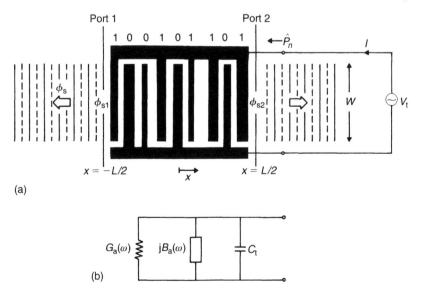

(a)

(b)

Figure 5.3 Two-terminal transducer. (a) Wave generation and (b) equivalent circuit.

assumed to be unaffected by the presence of any other conductors in the vicinity. As before, the wave amplitudes are expressed in terms of the surface potentials $\phi_s(x, \omega)$. A voltage V_t is applied, and the electrostatic part of the resulting charge density is $\sigma_e(x, \omega)$, with Fourier transform $\overline{\sigma}_e(\beta, \omega)$. In the quasi-static approximation, the potential $\phi_s(x, \omega)$ of the wave radiated in the $-x$-direction is, from eq. (5.10),

$$\phi_s(x, \omega) = j\Gamma_s\overline{\sigma}_e(k_f, \omega)\exp(jk_fx).$$

We define $\rho_e(x) = \sigma_e(x, \omega)/V_t$ as the electrostatic charge density for unit applied voltage. This function is real and independent of frequency. If $\overline{\rho}_e(\beta)$ is the Fourier transform of $\rho_e(x)$, we then have

$$\phi_s(x, \omega) = j\Gamma_s V_t\overline{\rho}_e(k_f)\exp(jk_fx). \tag{5.31}$$

The transducer has two acoustic ports which are taken to be at $x = \pm L/2$. The ports are simply reference locations for specifying the wave amplitudes. They are taken to be near the ends of the transducer, though their precise locations are not significant. With this definition, the origin for x is midway between the ports. Actually, the position of the origin is arbitrary. The position affects the x-dependent functions and the x-values at the ports, but it does not affect the transducer properties referred to the ports.

Defining $\phi_{s1}(\omega)$ as the potential of the wave launched at port 1, where $x = -L/2$, we have

$$\phi_{s1}(\omega) = j\Gamma_s V_t\overline{\rho}_e(k_f)\exp(-jk_fL/2). \tag{5.32}$$

The power P_s carried by this wave is, from eq. (3.27),

$$P_s = \frac{1}{4}\omega W|\phi_{s1}(\omega)|^2/\Gamma_s. \tag{5.33}$$

Similarly, the potential of the wave launched in the $+x$-direction, measured at port 2 where $x = L/2$, is found to be

$$\phi_{s2}(\omega) = j\Gamma_s V_t\overline{\rho}_e(-k_f)\exp(-jk_fL/2). \tag{5.34}$$

Since $\rho_e(x)$ is real we have $\overline{\rho}_e(-k_f) = \overline{\rho}_e^*(k_f)$, so that $\phi_{s2}(\omega)$ is essentially the conjugate of $\phi_{s1}(\omega)$. Clearly, the two waves have the same power.

The electrostatic charge density can be evaluated by methods discussed in Section 5.1.2. For a two-terminal transducer we define an electrode polarity vector by

$$\hat{P}_n = 1 \quad \text{if electrode } n \text{ is connected to the upper bus}$$
$$= 0 \quad \text{if electrode } n \text{ is connected to the lower bus} \tag{5.35}$$

as shown in Fig. 5.3a. For unit voltage across the transducer, the charge density is found by taking the voltage of electrode n to be $\hat{P}_n$, and hence, using eq. (5.15),

$$\rho_e(x) = \sum_{n=1}^{N} \hat{P}_n \rho_{en}(x) \tag{5.36}$$

where N is the number of electrodes and $\rho_{en}(x)$ is the charge density for unit voltage on electrode n, defined in Section 5.1.2.

5.2.2 Transducer admittance

When a voltage V_t is applied to an isolated transducer, as in Fig. 5.3a, the transducer current is I and the ratio I/V_t is the transducer admittance, Y_t. Usually, a major part of the admittance arises from the electrostatic charge density $\sigma_e(x, \omega)$, which gives a capacitive contribution to Y_t. This is usually written explicitly and denoted by C_t. The admittance is written as

$$Y_t(\omega) = G_a(\omega) + jB_a(\omega) + j\omega C_t. \tag{5.37}$$

Here $G_a(\omega)$ and $B_a(\omega)$ are the real and imaginary contributions due to the acoustic charge density $\sigma_a(x, \omega)$. They are respectively the acoustic conductance and the acoustic susceptance. The admittance may be represented as an equivalent circuit with these three contributions in parallel, as in Fig. 5.3b. The admittance $Y_t(\omega)$ can be regarded as the Fourier transform of a time-domain function $y_t(t)$ which must be real. It follows that $G_a(\omega)$ must be symmetric, so that $G_a(-\omega) = G_a(\omega)$, and $B_a(\omega)$ is antisymmetric so that $B_a(-\omega) = -B_a(\omega)$.

If I_n is the current entering electrode n, the transducer current I is given by

$$I = \sum_n \hat{P}_n I_n = j\omega W \sum_n \hat{P}_n \int_n [\sigma_e(x, \omega) + \sigma_a(x, \omega)] dx, \tag{5.38}$$

where eq. (5.22) has been used for I_n, and the integral is over the x-region of electrode n. The electrostatic contribution is due to $\sigma_e(x, \omega)$ and is equal to $j\omega C_t V_t$. Since $\sigma_e(x, \omega) = V_t \rho_e(x)$, we have

$$C_t = W \sum_n \hat{P}_n \int_n \rho_e(x) dx, \tag{5.39}$$

which is simply the sum of the electrostatic charges connected to one bus, for unit applied voltage.

The acoustic part of the current, due to the $\sigma_a(x, \omega)$ term in eq. (5.38), is denoted by I_a. The contribution due to electrode n is I_{an}, given by eq. (5.26), and hence

$$I_a = \sum_n \hat{P}_n I_{an} = -j\omega W \int_{-\infty}^{\infty} \sum_n \hat{P}_n \rho_{en}(x) \phi_a(x, \omega) dx. \tag{5.40}$$

Using eq. (5.36), this can be expressed as

$$I_a = -j\omega W \int_{-\infty}^{\infty} \rho_e(x)\phi_a(x,\omega)dx. \tag{5.41}$$

This is equal to $V_t[G_a(\omega) + jB_a(\omega)]$, by definition. The acoustic potential $\phi_a(x,\omega)$ is given by eq. (5.30). The two terms involving $\overline{\sigma}_e(\pm k_f, \omega)$ are found to give the real part of I_a, equal to $V_t G_a(\omega)$. The terms involving $F(\pm k_f)$ give the imaginary part, $jV_t B_a(\omega)$. Noting that $\overline{\sigma}_e(k_f, \omega) = V_t \overline{\rho}_e(k_f)$ and $\overline{\rho}_e(-k_f) = \overline{\rho}_e^*(k_f)$, the conductance is found to be

$$G_a(\omega) = \omega W \Gamma_s |\overline{\rho}_e(k_f)|^2. \tag{5.42}$$

More directly, the conductance can be obtained from the surface-wave power generated, using eqs (5.32)–(5.34). Since we assume no power losses, this is equal to the power extracted from the voltage source, which is $V_t^2 G_a(\omega)/2$, and this gives eq. (5.42).

The acoustic susceptance $B_a(\omega)$ is found by substituting the $F(\pm k_f)$ terms of eq. (5.30) into eq. (5.41). After some manipulation, the result obtained is

$$B_a(\omega) = -\frac{\omega W \Gamma_s}{\pi}\left[|\overline{\rho}_e(k_f)|^2 * \frac{1}{k_f}\right]. \tag{5.43}$$

where the asterisk indicates convolution with respect to k_f. Now, the Hilbert transform of a function is obtained by convolution with $-1/(\pi\omega)$. Noting that $k_f = \omega/v_f$, we find from eq. (5.43) that $B_a(\omega)/\omega$ is the Hilbert transform of $G_a(\omega)/\omega$. Explicitly, this transform can be written as the sum of two integrals of the form $\int d\omega' G_a(\omega')/(\omega - \omega')$ and $\int d\omega' G_a(\omega')/\omega'$. The latter integral is zero because $G_a(\omega)$ is a symmetric function, so that the argument is antisymmetric. This leads to the conclusion that $B_a(\omega)$ is the Hilbert transform of $G_a(\omega)$, so that

$$B_a(\omega) = G_a(\omega) * \frac{-1}{\pi\omega} \equiv -\int_{-\infty}^{\infty} \frac{G_a(\omega')}{\pi(\omega - \omega')}d\omega'. \tag{5.44}$$

This result has been derived within the quasi-static approximation, but it is in fact much more general. The general proof follows from the fact that the relation between I and V_t must be causal, that is, if the voltage is zero for $t < 0$ then the current must also be zero for $t < 0$. It follows that the real and imaginary parts of $Y_t(\omega)$ are related by the Hilbert transform [11, 12]. The proof assumes the admittance to be zero at infinite ω, so the capacitance must be excluded.

5.2.3 Receiving transducer

We now consider a transducer with a surface wave incident on it. As shown in Fig. 5.4a, the transducer geometry is as before, but now the two bus bars are

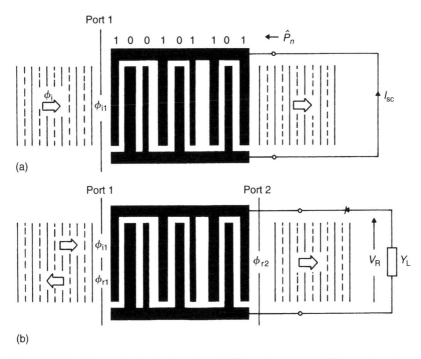

Figure 5.4 Reception by a two-terminal transducer. (a) Shorted transducer and (b) transducer with finite load.

shorted and a current I_{sc} flows between them. The incident surface wave has a potential $\phi_i(x, \omega)$, which is taken to be $\phi_{i1}(\omega)$ at port 1, where $x = -L/2$. In the absence of the transducer, the potential of the wave would be

$$\phi_i(x, \omega) = \phi_{i1}(\omega) \exp(-jk_f x) \exp(-jk_f L/2). \tag{5.45}$$

The electrodes can be taken to be at zero potential, so there must be charges present such that $\phi_i(x, \omega)$ is canceled at the electrode locations.

To find the charge distribution we return to eq. (5.9), which is the quasi-static relation between potential and charge density. The charge density is $\sigma_e(x, \omega) + \sigma_a(x, \omega)$, but here the electrostatic term $\sigma_e(x, \omega)$ is zero because the electrode voltages are zero. Hence the charge density is $\sigma_a(x, \omega)$, related to the surface potential by

$$\phi(x, \omega) = G_e(x) * \sigma_a(x, \omega). \tag{5.46}$$

The potential of the incident wave, $\phi_i(x, \omega)$, is an additional term which can be considered to be due to some remote source whose charge density is not included in eq. (5.46). To obtain zero potential on the electrodes, we therefore have

$$\phi(x_j, \omega) = [G_e(x) * \sigma_a(x, \omega)]_{x_j} = -\phi_i(x_j, \omega), \tag{5.47}$$

where the points x_j exist only on the electrodes, as before.

A similar relation was found before when evaluating the electrode current. Equation (5.47) has the same form as eq. (5.20). It follows that the current I_{an} entering electrode n is given by eq. (5.26), as before, but with $\phi_a(x, \omega)$ replaced $\phi_i(x, \omega)$ so that

$$I_{an} = -j\omega W \int_{-\infty}^{\infty} \rho_{en}(x)\phi_i(x, \omega)dx. \tag{5.48}$$

The transducer short-circuit current, I_{sc}, is therefore

$$I_{sc} = \sum_n \hat{P}_n I_{an} = -j\omega W \int_{-\infty}^{\infty} \rho_e(x)\phi_i(x, \omega)dx, \tag{5.49}$$

where $\rho_e(x)$ has been introduced using eq. (5.36). Finally, substituting eq. (5.45) for $\phi_i(x, \omega)$, we find

$$I_{sc} = -j\omega W \phi_{i1}(\omega)\overline{\rho}_e(k_f) \exp(-jk_f L/2). \tag{5.50}$$

Comparing this with eq. (5.32) gives a reciprocity relation for the processes of launching and receiving waves at port 1. We find

$$\left[\frac{I_{sc}}{\phi_{i1}(\omega)}\right]_{receive} = -\frac{\omega W}{\Gamma_s}\left[\frac{\phi_{s1}(\omega)}{V_t}\right]_{launch}. \tag{5.51}$$

A more general derivation of this is given in Appendix B, Section B.5, showing that it is valid when electrode reflections are significant. It is also valid when the transducer couples to bulk waves as well as surface waves, in which case eq. (5.51) refers to the surface-wave components.

5.3 SUMMARY AND *P*-MATRIX FORMULATION

This section summarizes the above results. They are expressed in terms of the *P*-matrix, a common form of scattering matrix introduced by Tobolka [13]. This matrix is convenient for analyzing a wide variety of surface-wave devices. Its properties are summarized in Appendix D, which also gives methods for cascading it for device analysis. It is assumed here that only one acoustic wave, a piezoelectric Rayleigh wave, is present, thus excluding leaky waves and bulk waves. Losses have been ignored. The transducer is assumed to be non-reflective, so that it does not reflect the waves when shorted. This applies,

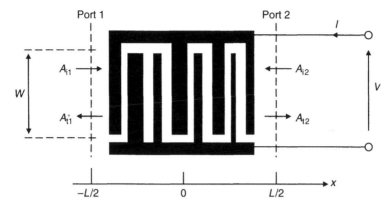

Figure 5.5 Parameters used for transducer P-matrix.

for example, to a double-electrode transducer. In the theory, this condition is obtained by the use of the quasi-static approximation.

For the P-matrix, the amplitudes of waves incident at ports 1 and 2 are denoted by A_{i1} and A_{i2}, respectively. The amplitudes of waves leaving the transducer, at these ports, are denoted by A_{t1} and A_{t2}. The ports are simply reference points for the wave amplitudes, near the transducer edges. The transducer current is I and the voltage is V. Any three of these six variables can be expressed in terms of the other three. The P-matrix is defined such that

$$\begin{bmatrix} A_{t1} \\ A_{t2} \\ I \end{bmatrix} = \begin{bmatrix} P_{11} & P_{12} & P_{13} \\ P_{21} & P_{22} & P_{23} \\ P_{31} & P_{32} & P_{33} \end{bmatrix} \cdot \begin{bmatrix} A_{i1} \\ A_{i2} \\ V_t \end{bmatrix}. \tag{5.52}$$

Here P_{33} equals the transducer admittance Y_t. Other terms in the last column and last row depend on the definition of the wave amplitudes A, which are subscripted as above. Figure 5.5 illustrates the parameters used.

In previous sections, the surface potential ϕ_s has been used to specify the wave amplitude. It is convenient to use an electrical variable because this relates strongly to the behavior of a transducer, particularly when there is negligible mechanical loading. It is also convenient if the variable gives the expected symmetry, and the surface potential satisfies this requirement. For example, when a symmetrical transducer generates surface waves, the potentials of the two waves are the same at equal distances from the transducer center.

For the P-matrix, the amplitude A is usually defined such that the surface-wave power is $|A|^2/2$. It is convenient to take the phase of A to be the same as that of ϕ_s.

The surface-wave power is related to ϕ_s by eq. (5.33), and this gives the amplitude A as

$$A = \phi_s\sqrt{\omega W/(2\Gamma_s)}, \tag{5.53}$$

where W is the beam width, shown in Fig. 5.5. With this definition, the reciprocity relation of eq. (5.51) gives $P_{31} = -2P_{13}$ and $P_{32} = -2P_{23}$, these being general reciprocity relations as shown in Appendix D. Reciprocity also requires that $P_{21} = P_{12}$.

Using the results of Section 5.2, the P-matrix components for a lossless non-reflective transducer are

$$P_{11} = P_{22} = 0 \tag{5.54a}$$

$$P_{12} = P_{21} = \exp(-jkL) \tag{5.54b}$$

$$P_{13} = -P_{31}/2 = j\bar{\rho}_e(k)\sqrt{\omega W\Gamma_s/2}\exp(-jkL/2) \tag{5.54c}$$

$$P_{23} = -P_{32}/2 = j\bar{\rho}_e(-k)\sqrt{\omega W\Gamma_s/2}\exp(-jkL/2) \tag{5.54d}$$

$$P_{33} = Y_t(\omega) = G_a(\omega) + jB_a(\omega) + j\omega C_t. \tag{5.54e}$$

For these equations, we define a function $\rho_e(x)$ as the electrostatic charge density on the electrodes when unit voltage is applied. The Fourier transform of $\rho_e(x)$ is $\bar{\rho}_e(k)$, and this governs the frequency response of the transducers, as seen in the expressions for P_{13} and P_{23}. Equations (5.54c) and (5.54d) are derived from eqs (5.32) and (5.34). Since $P_{11} = P_{22} = 0$, incident waves are not reflected if the transducer is shorted. Equation (5.54b) expresses the fact that an incident wave travels through a shorted transducer with no reflection, and with velocity $v = \omega/k$. The velocity is taken to be independent of frequency, a good approximation in practice. The origin for x has been taken to be midway between the ports, which are at $x = \pm L/2$. Changing the position of the origin will change the phase of $\bar{\rho}_e(k)$ and the x values at the ports, but the P-matrix remains the same. The constant Γ_s is defined as $\Gamma_s = (\Delta v/v)/\varepsilon_\infty$, where ε_∞ equals the effective permittivity $\varepsilon_s(\beta)$ at $\beta \to \infty$. It will be seen later that the capacitance of a single-electrode transducer with aperture W and with N_p periods is simply $N_p W\varepsilon_\infty$, if the metallization ratio a/p is 0.5.

We can also define a transducer response $H_t(\omega)$ such that, when a voltage V_t is applied, the potential of the wave emerging at port 1 is $\phi_s = jV_tH_t(\omega)[\Gamma_s/(\omega W)]^{1/2}$. This expression, used in Chapter 4 of the earlier book [14], is useful for a two-transducer device where each transducer has port 1 facing the other transducer. Using eq. (5.54c) shows that $H_t(\omega) = -j\sqrt{2}P_{13}$.

Appendix D derives some general transducer properties for a non-reflective transducer. In particular, the transducer is *bidirectional*, so that $|P_{23}| = |P_{13}|$. Hence, when a voltage is applied the transducer generates surface waves of equal amplitude in the two directions. This can also be seen in eqs (5.54c) and (5.54d), because the fact that $\rho_e(x)$ is real implies that $\overline{\rho}_e(k)$ and $\overline{\rho}_e(-k)$ have equal magnitude. From power conservation the conductance is given by $G_a(\omega) = 2|P_{13}|^2$. The susceptance $B_a(\omega)$ is the Hilbert transform of this, as shown by eq. (5.44). Appendix D shows that there are various other constraints associated with reciprocity and power conservation, and these are satisfied by eq. (5.54).

If the transducer is connected to an electrical load with admittance Y_L, we have $I/V_t = -Y_L$. For a wave incident on port 1, part of the incident wave power is delivered to the load. The ratio of the load power P_L to the incident wave power P_s is the conversion coefficient, C_p. The quantity $-20 \log C_p$, in decibels, is often called the conversion loss. From the P-matrix, we find

$$C_p \equiv P_L/P_s = 2G_a G_L/|P_{33} + Y_L|^2 \tag{5.55}$$

where G_L is the real part of the load admittance Y_L. The same ratio applies for a wave incident on port 2. It also applies when a voltage is applied to the transducer, giving the ratio of wave power (in one direction) to available input power. The conversion coefficient is maximized if the transducer is electrically matched, that is, if $Y_L = P_{33}^*$. This condition usually requires an inductor to tune out the transducer capacitance. When matched, eq. (5.55) leads to $C_p = 1/2$, so the conversion loss is 3 dB. Thus half of the input power is inaccessible, a consequence of bidirectionality. For the same condition, the power acoustic reflection coefficient is 1/4, as shown in Appendix D. In a two-transducer device, this gives rise to a familiar problem – large unwanted reflections causing triple- and multiple-transit signals. These cause ripples in the frequency response. In Chapter 6 we will consider transversal filters, in which this problem is dealt with by mismatching the transducers, so that the ripples are minimized at the expense of high insertion loss. Other techniques, giving acceptable ripple and low loss, are considered later in Chapters 9 and 11.

The quasi-static theory of Section 5.2 leads to the conclusion that the waves travel through the transducer with the free-surface velocity $v_f = \omega/k_f$. In practice, the electrodes modify the velocity a little, as discussed later in Chapter 8. It is therefore necessary to take account of this change, and to bear in mind that different regions of a device may have different velocities. In the above equations, and in the following sections, the wavenumber is denoted by k for simplicity.

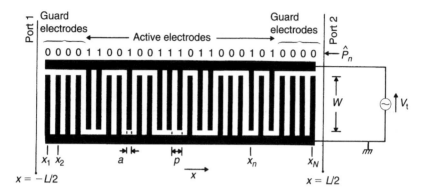

Figure 5.6 Transducer with regular electrodes.

5.4 TRANSDUCERS WITH REGULAR ELECTRODES: ELEMENT FACTOR

Many surface-wave devices make use of transducers with *regular* electrodes, that is, electrodes with constant width and spacing. This constraint simplifies the analysis substantially, and the rest of this chapter concentrates on this case. The description here includes uniform transducers, which have a repetitive sequence of electrode polarities. These are considered further in Section 5.5. However, we first consider a more general case in which the polarities are irregular. This includes withdrawal-weighted transducers.

A typical transducer under consideration is shown in Fig. 5.6. The electrodes are regular, with the same width a and constant pitch p. The ratio a/p is the *metallization ratio*, usually with a nominal value of 1/2. Electrode n has polarity $\hat{P}_n$, defined as 1 or 0 for an electrode connected to the upper or lower bus, respectively.

The analysis is substantially simplified by using superposition. Suppose we apply unit voltage to electrode n and zero voltage to all the other electrodes, and define $\rho_{en}(x)$ as the resulting electrostatic charge density. Then charge superposition states that when voltages V_n are applied to the electrodes, the electrostatic charge density is $\sum_n V_n \rho_{en}(x)$. For unit voltage applied across the bus bars, we have $V_n = \hat{P}_n$, and hence $\rho_e(x) = \sum_n \hat{P}_n \rho_{en}(x)$. A proof of this is given in Section 5.1.2.

If the electrodes are regular the functions $\rho_{en}(x)$ are all very similar, except for electrodes near the ends where end effects can be significant. To exploit this feature we consider another array, shown in Fig. 5.7a. Here there is only

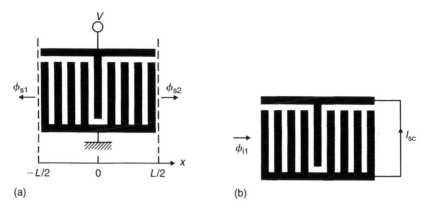

Figure 5.7 Transducer with only one live electrode. (a) Launching and (b) reception.

one live electrode, centered at $x = 0$, all other electrodes being grounded. We define $\rho_f(x)$ as the electrostatic charge density assuming that the number of electrodes, N, is infinite. In practice N is finite, of course, but we assume that N is large and the live electrode is not near either end, so that end effects are negligible. For the transducer of Fig. 5.6 we have $\rho_{en}(x) \approx \rho_f(x - x_n)$, where x_n is the center location of electrode n and end effects are ignored. Thus, for this general transducer we have

$$\rho_e(x) = \sum_{n=1}^{N} \hat{P}_n \rho_f(x - x_n). \qquad (5.56)$$

The transducer of Fig. 5.7 has a P-matrix described by eq. (5.54), with $\overline{\rho}_e(k)$ replaced by $\overline{\rho}_f(k)$, where the latter is the Fourier transform of $\rho_f(x)$. Because $\rho_f(x)$ is real and symmetrical, it follows that $\overline{\rho}_f(k)$ is also real and symmetrical, so that $\overline{\rho}_f(-k) = \overline{\rho}_f(k)$. An algebraic formula for this function is given below. When a voltage V_t is applied, the potentials of the waves generated, measured at ports 1 and 2, are given by

$$\phi_{s1} = \phi_{s2} = j\Gamma_s V_t \overline{\rho}_f(k) \exp(-jkL/2) \qquad (5.57)$$

where the ports are taken to be at $x = \pm L/2$. If these potentials are referred to the center of the live electrode, the exponential in eq. (5.57) disappears. If the transducer is shorted and a surface wave is incident, as in Fig. 5.7b, the current I_{sc} in the short circuit can be obtained from P_{31} or P_{32}. If the incident wave has potential ϕ_i, the current is given by

$$I_{sc} = -j\omega W \phi_i \overline{\rho}_f(k). \qquad (5.58)$$

In this formula, the phase of the incident wave potential, ϕ_i, is referred to the center of the live electrode. The same formula applies for incident waves in either direction.

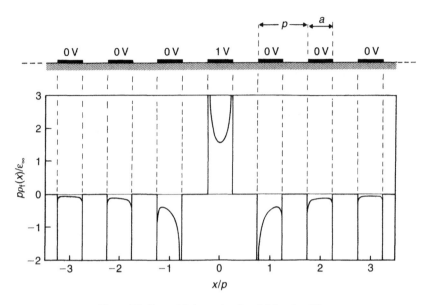

Figure 5.8 Elemental charge density $\rho_f(x)$, for $a/p = 1/2$.

The function $\rho_f(x)$ is considered in Appendix C, and it is plotted in Fig. 5.8. For many purposes, the Fourier transform $\overline{\rho}_f(\beta)$ is required, and for this there is an algebraic formula derived by Peach [2] and Datta and Hunsinger [15]. Appendix C shows that this is given by

$$\overline{\rho}_f(\beta) = \varepsilon_\infty \frac{2 \sin \pi s}{P_{-s}(-\cos \Delta)} P_m(\cos \Delta), \quad \text{for } m \leq \frac{\beta p}{2\pi} \leq m + 1, \qquad (5.59)$$

where $s = \beta p/(2\pi) - m$, so that $0 \leq s \leq 1$, and $\Delta = \pi a/p$. Also, $P_{-s}(-\cos \Delta)$ is a Legendre function and $P_m(\cos \Delta)$ is a Legendre polynomial. The Legendre function and polynomial are easily evaluated using a series expansion in Appendix C. Figure 5.9 shows the function $\overline{\rho}_f(k)$. Noting that k is proportional to ω, the horizontal axis has been expressed as frequency, and normalized to the sampling frequency f_s. The latter is the frequency at which the spacing p equals the wavelength, so that $f_s = v/p$ and $k = 2\pi/p$. The function consists of a series of lobes, all with the same shape, with relative amplitudes determined by the polynomials $P_m(\cos \Delta)$. These amplitudes are strongly affected by the metallization ratio a/p. To clarify this, Fig. 5.10 shows the values of the peaks, plotted against a/p. The peaks occur at $\beta = M\pi/p$ with M odd, and the values are given by eq. (5.59) with $m = (M - 1)/2$ and $s = 1/2$.

It is noted that $\overline{\rho}_f(k)$ is zero at frequency f_s. This can be understood by considering a wave incident on a periodic array of disconnected electrodes.

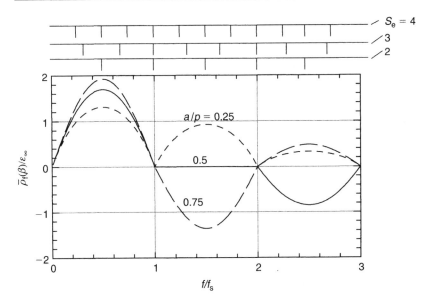

Figure 5.9 The function $\bar{\rho}_f\,(\beta)$, as a function of f/f_s, where f_s is the sampling frequency. Upper scales indicate the fundamental and harmonic frequencies of transducers with $S_e = 2$, 3 and 4 electrodes per period. The transducers do not respond at all of the harmonics (see text).

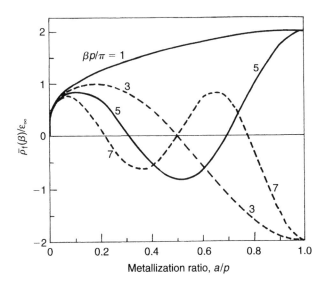

Figure 5.10 Elemental charge density: values at the peaks of the lobes, as functions of metallization ratio.

At frequency f_s, the potential of the wave has the same periodicity as that of the electrodes, so the electrodes will all have the same voltage. Hence, connecting the electrodes to bus bars will have no effect on the fields and no current can be produced. A transducer cannot therefore respond to the wave, so $\overline{\rho}_f(k)$ must be zero. A similar argument applies at multiples of f_s.

We can now return to the analysis of a general transducer, illustrated in Fig. 5.6. For an applied voltage, the electrostatic charge density is given by eq. (5.56). The Fourier transform of this is needed to obtain the frequency response. Using the shifting theorem, eq. (A.6), the transform is simply

$$\overline{\rho}_e(\beta) = \sum_{n=1}^{N} \hat{P}_n \overline{\rho}_f(\beta) \exp(-j\beta x_n). \tag{5.60}$$

This determines the responses for wave generation, given by P_{13} and P_{23} in eqs (5.54), and the responses for reception. It corresponds to the delta-function analysis described in Chapter 1. Here $\overline{\rho}_f(\beta)$ is basically the *element factor*, representing the response of one electrode. The function $\sum_n \hat{P}_n \exp(-j\beta x_n)$ is the *array factor*. Using eq. (5.54c) we have

$$P_{13} = j\sqrt{\omega W \Gamma_s/2}\,\overline{\rho}_f(k) \sum_{n=1}^{N} \hat{P}_n \exp(-jkx_n) \tag{5.61}$$

taking $x = 0$ at port 1. The acoustic conductance is $G_a(\omega) = \omega W \Gamma_s |\overline{\rho}_e(k)|^2$, from eq. (5.42), and the susceptance $B_a(\omega)$ is given by eq. (5.44).

The capacitance is found by summing the electrostatic charges on electrodes connected to one bus. For regular electrodes, it is convenient to consider the net charges on the electrodes for a transducer with a single live electrode, as in Fig. 5.7a. For unit voltage and unit aperture, these charges are denoted by Q_m, where m is the electrode number and $m = 0$ for the live electrode. Thus we have

$$Q_m = \int_{mp-a/2}^{mp+a/2} \rho_f(x)dx. \tag{5.62}$$

The sum total of the Q_m is zero, since the total charge is zero. For the general transducer of Fig. 5.6, the charge density is given by eq. (5.56). The net charge on electrode n of this transducer, for unit voltage, is thus

$$q_n = W \sum_{m=1}^{N} \hat{P}_m Q_{m-n}.$$

This sum is dominated by electrodes near electrode n, that is for m near to n. The total charge on electrodes connected to the upper bus gives the capacitance C_t, so that

$$C_{\mathrm{t}} = \sum_{n=1}^{N} \hat{P}_n q_n = W \sum_{n=1}^{N} \sum_{m=1}^{N} \hat{P}_n \hat{P}_m Q_{m-n}. \tag{5.63}$$

Thus, the capacitance is readily found from the net charges Q_m. These are functions of the metallization ratio a/p, and they are given by eq. (C.23) of Appendix C. In the particular case of $a/p = 1/2$ they are given by the simple formula

$$Q_m = \frac{4\varepsilon_\infty}{\pi(1 - 4m^2)}, \quad \text{for } a/p = 1/2. \tag{5.64}$$

End effects

In the above description it was assumed that ends effects could be ignored, so that the electrostatic charge for each electrode is given by $\rho_{\mathrm{f}}(x)$, as in eq. (5.56). In practice, the charge density is distorted for electrodes near the ends of the transducer, as can be seen in Fig. 5.2. In many cases this affects only a small proportion of the electrodes, and the distortion may be acceptable. The analysis can be extended to include this effect by calculating the charge density numerically, as discussed in Section 5.1.2. Alternatively, a simple practical measure is to include 'guard' electrodes, with zero voltage, at the transducer ends as in Fig. 5.6. For the guard electrodes the polarity $\hat{P}_n$ is zero, so they do not contribute to the charge density in eq. (5.56) and the distortions do not affect the transducer response.

5.5 ADMITTANCE OF UNIFORM TRANSDUCERS

Uniform transducers are defined here as transducers using regular electrodes and with a repetitive polarity sequence $\hat{P}_n$. For example, a single-electrode transducer has a polarity sequence $0, 1, 0, 1, \ldots$ and a double-electrode transducer has a polarity sequence $0, 0, 1, 1, 0, 0, 1, 1, \ldots$ These are shown in Fig. 5.11, together with a type with three electrodes per period. Surface-wave devices often use such transducers, and the repetitive sequence gives rise to some simple expressions for the main properties. Since we assume no internal reflectivity ($P_{11} = 0$), the results apply to a single-electrode transducer only if the electrode reflections are minimized by, for example, recessing the electrodes. Engan [16] has given an analysis based on the theory of multistrip couplers. The method here is based on the above quasi-static analysis and applies for any repetitive sequence of polarities. As before, end effects are ignored. This is

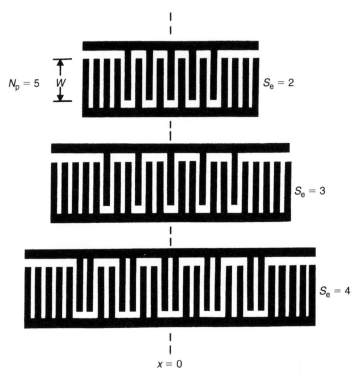

$N_p = 5$ W $S_e = 2$

$S_e = 3$

$S_e = 4$

$x = 0$

Figure 5.11 Common types of uniform transducers.

indicated in Fig. 5.11 by including guard electrodes, though these are not usually necessary.

5.5.1 Acoustic conductance and susceptance

The response is determined by the electrostatic charge density $\rho_e(x)$, as given by eq. (5.56). For a uniform transducer it is convenient to express this in terms of an array factor

$$A(x) = \sum_{n=1}^{N} \hat{P}_n \delta(x - x_n) \tag{5.65}$$

where the delta functions exist for all electrodes and the polarities $\hat{P}_n$ are used to select the live electrodes. In terms of this function, the electrostatic charge density is

$$\rho_e(x) = A(x) * \rho_f(x). \tag{5.66}$$

We can express $A(x)$ as $A_1(x) * A_N(x)$, where $A_1(x)$ is an array factor for one period, and $A_N(x)$ is used to give a repetition such that there are N_p periods. The array factor for one period, with S_e electrodes, is

$$A_1(x) = \delta(x) \qquad\qquad \text{for } S_e = 2 \text{ or } 3$$
$$ = \delta(x + p/2) + \delta(x - p/2) \quad \text{for } S_e = 4. \tag{5.67}$$

The array factor $A_N(x)$ is written as the symmetrical function

$$A_N(x) = \sum_{n=1}^{N_p} \delta[x - (2n - N_p - 1)pS_e/2] \tag{5.68}$$

where N_p is the number of periods. This function is the same for all the uniform transducers, and its Fourier transform $\overline{A}_N(\beta)$ is easily found to be

$$\overline{A}_N(\beta) = \frac{\sin(N_p \beta p S_e/2)}{\sin(\beta p S_e/2)}. \tag{5.69}$$

The Fourier transform of $A_1(x)$ is

$$\overline{A}_1(\beta) = 1 \qquad\qquad \text{for } S_e = 2 \text{ or } 3$$
$$\phantom{\overline{A}_1(\beta)} = 2\cos(\beta p/2) \quad \text{for } S_e = 4. \tag{5.70}$$

For the transducer as a whole, the electrostatic charge density is $\rho_e(x) = A_1(x) * A_N(x) * \rho_f(x)$, so $\overline{\rho}_e(\beta)$ is the product of the three transforms. We thus have

$$\overline{\rho}_e(k) = \overline{A}_1(k)\overline{A}_N(k)\overline{\rho}_f(k) \tag{5.71}$$

and the conductance is

$$G_a(\omega) = \omega W \Gamma_s |\overline{\rho}_e(k)|^2, \tag{5.72}$$

from eq. (5.42).

For $S_e = 2$ or 3 we have $\overline{A}(k) = \overline{A}_N(k)$, given by eq. (5.69). This function has maxima when $k = 2\pi M/(pS_e)$, with $M = 0, 1, 2, \ldots$ The frequencies of these maxima are at $\omega = M\omega_c$, where $\omega_c = 2\pi v/(pS_e)$ is the center frequency of the fundamental response. However, the transducers do not respond at all the harmonic frequencies, because of the presence of the elemental charge density. Figure 5.9 has marks indicating the fundamental and harmonic frequencies for these transducers. For a single-electrode transducer, with $S_e = 2$, harmonics with $M = 2, 4, 6, \ldots$ are absent because $\overline{\rho}_f(k)$ is zero at these points. In addition, the third harmonic is absent if $a/p = 1/2$. Similar remarks apply for $S_e = 3$, where there is no response if M is a multiple of 3. For the double-electrode transducer, with $S_e = 4$, the situation is complicated by the factor $\overline{A}_1(\beta)$, eq. (5.70), and this suppresses harmonics with $M = 2, 6, 10, \ldots$ In this case, harmonic responses occur only for odd values of M.

Some general points can be deduced from Fig. 5.9. The element factor $\bar{\rho}_f(\beta)$ is zero at the sampling frequency f_s and at its multiples, as noted in Section 5.4. Consequently, all periodic transducers have zero response for harmonic numbers that are multiples of S_e. Thus, for $S_e = 2$ there is no response at the 2nd, 4th, ... harmonics, and for $S_e = 4$ there is no response at the 4th, 8th, ... harmonics. For $S_e = 3$ there is no response at the 3rd, 6th, ... harmonics. In addition, the element factor is zero in the region $f_s < f < 2f_s$ if $a/p = 1/2$. For $S_e = 2$, this suppresses the 3rd harmonic. For $S_e = 4$, the 5th, 6th and 7th harmonics are suppressed.

In the fundamental passband, where ω is close to ω_c, the array factor of eq. (5.69) can be approximated as

$$\bar{A}_N(k) \approx -(-1)^{N_p} \frac{N_p \sin X}{X}, \tag{5.73}$$

where $X = \pi N_p(\omega - \omega_c)/\omega_c$. This function varies rapidly with ω. The conductance, eq. (5.72), contains other frequency-dependent terms, but if N_p is not too small the variation is mainly due to $\bar{A}_N(k)$. Hence, for ω close to ω_c the conductance can be approximated as

$$G_a(\omega) \approx G_a(\omega_c) \left[\frac{\sin X}{X} \right]^2. \tag{5.74}$$

The constant $G_a(\omega_c)$ is evaluated below. The acoustic susceptance $B_a(\omega)$ is the Hilbert transform of $G_a(\omega)$. This requirement follows from causality, as discussed in Section 5.2.2. Section A.6 of Appendix A shows that, using eq. (5.74), $B_a(\omega)$ is given by

$$B_a(\omega) \approx G_a(\omega_c) \frac{\sin(2X) - 2X}{2X^2}. \tag{5.75}$$

These approximate expressions are plotted in Fig. 5.12. In many practical cases the susceptance $B_a(\omega)$ is of little consequence because the total susceptance, $B_a(\omega) + \omega C_t$, is dominated by the capacitive term. Experimental observations [16, 17] agree well with eqs (5.74) and (5.75). There are often additional contributions due to bulk-wave generation. However, if N_p is not too small these are generally confined to a frequency band above the surface-wave response. Distortions due to electrode reflections will be considered later in Chapter 8.

For $M \le S_e$ the conductance at the fundamental center frequency and harmonics is given by

$$G_a(M\omega_c) = \alpha M \omega_c N_p^2 W \Gamma_s \left[\frac{2\varepsilon_\infty \sin \pi s}{P_{-s}(-\cos\Delta)} \right]^2, \tag{5.76}$$

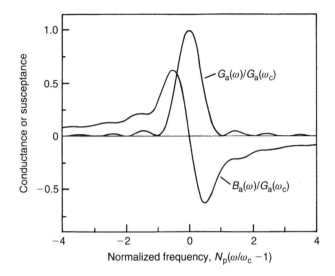

Figure 5.12 Acoustic conductance and susceptance for uniform transducers.

where $s = M/S_e$ and $\Delta = \pi a/p$. For $S_e = 2$ or 3, $\alpha = 1$. For $S_e = 4$, $\alpha = 2$ if $M = 1$ or 3 and $\alpha = 0$ for $M = 2$. Equation (5.76) follows from eq. (5.72) with the array factors of eqs (5.69) and (5.70), and with eq. (5.59) for $\bar{\rho}_f(\beta)$.

5.5.2 Capacitance

The capacitance is given by eq. (5.63) in terms of the net charges q_n on the electrodes. For a uniform transducer the q_n can be taken to be the same in all periods, assuming that end effects are small. This is generally reasonable if N_p is not too small. In this case, eq. (5.63) can be written as

$$C_t \approx N_p \sum_{n=1}^{S_e} \hat{P}_n q_n \tag{5.77}$$

where the q_n are evaluated for one period of a transducer of infinite length. These charges are given by

$$q_n = W \sum_{r=-\infty}^{\infty} \sum_{i=1}^{S_e} \hat{P}_i Q_{i+rS_e-n} \tag{5.78}$$

where Q_m is the charge on electrode n associated with the voltage on electrode $(n - m)$, as defined in eq. (5.62). The sum over r in eq. (5.78) can be done using eq. (C.30) of Appendix C, and the remaining summations are straightforward.

For the transducers considered here, the result is

$$C_t = \gamma W N_p \varepsilon_\infty \frac{\sin(\pi/S_e)}{P_\nu(-\cos\Delta)} P_\nu(\cos\Delta), \qquad (5.79)$$

where $\gamma = 1$, 4/3 or 2 for $S_e = 2$, 3 or 4, respectively, and $\nu = -1/S_e$.

5.5.3 Comparative performance

When comparing the performance of transducers, an important quantity is the electrical Q-factor, denoted by Q_t. This is defined as the ratio of susceptance to conductance, at the center of the band. At this frequency $B_a(\omega)$ is zero, so $Q_t = \omega_c C_t / G_a(\omega_c)$ for the fundamental. Allowing for harmonic frequencies, where $\omega = M\omega_c$, the Q-factor is

$$Q_t = M\omega_c C_t / G_a(M\omega_c). \qquad (5.80)$$

This generally depends on the harmonic number, M. The reciprocal of Q_t is a measure of the surface-wave coupling strength, taking account of the transducer geometry and the substrate parameters. This parameter has a strong bearing on the bandwidth obtainable when the transducer is matched for minimum conversion loss.

The basic properties of the three transducers are summarized in Fig. 5.13, and Table 5.1 shows data for $a/p = 1/2$. A normalized capacitance $\tilde{C}_t$, conductance $\tilde{G}_{aM}$ and Q-factor $\tilde{Q}_t$ have been defined, such that

$$C_t = W N_p \varepsilon_\infty \tilde{C}_t \qquad (5.81)$$

$$G_a(M\omega_c) = M\omega_c \varepsilon_\infty^2 N_p^2 W \Gamma_s \tilde{G}_{aM} \qquad (5.82)$$

$$Q_t = \tilde{Q}_t / (N_p \varepsilon_\infty \Gamma_s). \qquad (5.83)$$

The normalized parameters are independent of the substrate properties ε_∞ and Γ_s, and of the number of periods N_p and the aperture W. In Fig. 5.13 the capacitance and Q-factor are plotted as functions of the metallization ratio a/p. It can be seen that the metallization ratio is not very critical, and that the coupling is maximized at $a/p = 1/2$ for all three transducer types. Table 5.1 shows that for $a/p = 1/2$ the coupling is strongest for the single-electrode transducer ($S_e = 2$). For $S_e = 3$ or 4 there is a harmonic with coupling strength equal to that of the fundamental. The capacitance is smallest for $S_e = 2$. Engan [16] has confirmed these results experimentally.

The last line of Table 5.1 refers to another type of transducer with $S_e = 4$ electrodes per period. This is strongly related to a unidirectional transducer called the DART (distributed acoustic reflection transducer), considered later in Chapter 9.

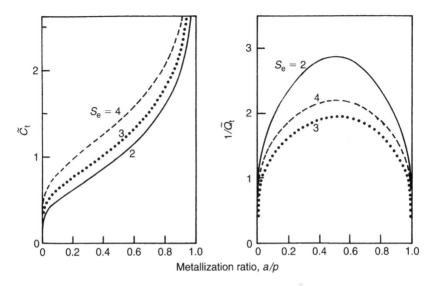

Metallization ratio, a/p

Figure 5.13 Uniform transducers: normalized capacitance and Q-factor, as functions of metallization ratio.

Table 5.1 Normalized data for uniform transducers, with $a/p = 1/2$.

S_e	Polarity sequence	Capacitance $\tilde{C}_t$	Harmonic M	Conductance $\tilde{G}_{aM}$	Coupling $1/\tilde{Q}_t$
2	0 1 0 1 0 ...	1	1	2.871	2.871
3	0 0 1 0 0 1 0 ...	1.155	1	2.231	1.932
3	0 0 1 0 0 1 0 ...	1.155	2	2.231	1.932
4	0 1 1 0 0 1 1 0 0 ...	1.414	1	3.111	2.200
4	0 1 1 0 0 1 1 0 0 ...	1.414	2	0	0
4	0 1 1 0 0 1 1 0 0 ...	1.414	3	3.111	2.200
4	0 1 0 0 0 1 0 0 0 1 0 ...	1.207	1	1.556	1.288

The transducer geometry is shown in Fig. 9.1b, and the electrostatic solution is almost the same as that of a DART. This transducer has responses at the second and third harmonics, as well as the fundamental.

5.6 TWO-TRANSDUCER DEVICES

This section considers a device consisting of two transducers, a common configuration for surface-wave filters. As before the transducers are assumed to be non-reflective, so that they do not reflect the waves when they are shorted. This assumption is valid for many high-performance transversal filters, considered in Chapter 6. Reflections arise in the device when the transducers are connected to finite source and load impedances, giving multiple-transit signals. We consider unapodized transducers initially, and the extension to apodized transducers is considered in Section 5.6.2.

5.6.1 Device using unapodized transducers

For unapodized transducers, the analysis is conveniently based on the P-matrix of Section 5.3, which gives simple relationships for the Y-matrix of the device. A general device, with transducers denoted A and B, is shown in Fig. 5.14. The transducer ports are denoted by 1 and 2, and each transducer has port 1 facing the other transducer. This is done for reasons of symmetry. Correspondingly, each transducer needs an x-axis directed away from the other transducer in order to define the charge density $\rho_e(x)$, so separate axes are used. The device ports will be described as port A at the left and port B at the right. The currents at these ports are I_a and I_b, and the voltages are V_a and V_b. The Y-matrix is defined by the equations

$$I_a = Y_{11}V_a + Y_{12}V_b; \quad I_b = Y_{21}V_a + Y_{22}V_b. \tag{5.84}$$

It can be seen that Y_{11} is the admittance seen looking into port A when $V_b = 0$, that is, when port B is shorted. Since the transducers are non-reflective, transducer B does not reflect in this situation. Hence the admittance seen at port A is simply Y_t^a, the admittance shown by transducer A when no waves are incident. Similarly, Y_{22} equals the admittance Y_t^b of transducer B. Thus,

$$Y_{11} = Y_t^a; \quad Y_{22} = Y_t^b. \tag{5.85}$$

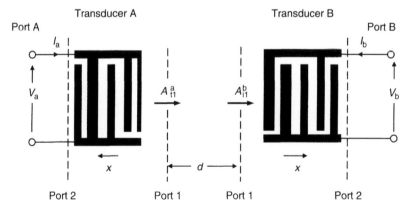

Figure 5.14 Two-transducer device with unapodized transducers.

To find the transadmittance Y_{21}, suppose that a voltage V_a is applied to port A, and port B is shorted so that $V_b = 0$. The wave generated to the right by transducer A has amplitude $A_{t1}^a = P_{13}^a V_a$, taking P_{ij}^a as the P-matrix of transducer A. The wave amplitude A_{t1}^a is related to surface potential by eq. (5.53). For transducer B, an incident wave A_{i1}^b at port 1 gives rise to a short-circuit current $I_b = A_{i1}^b P_{31}^b$, and we also have $P_{31}^b = -2P_{13}^b$ by reciprocity. In addition, the wave amplitudes are related by $A_{i1}^b = A_{t1}^a \exp(-jkd)$, where d is the distance between the transducer ports. Since we assumed $V_b = 0$ the ratio I_b/V_a equals Y_{21}, and thus we have

$$Y_{21} \equiv I_b/V_a = -2P_{13}^a P_{13}^b \exp(-jkd) = H_t^a(\omega)H_t^b(\omega)\exp(-jkd) \qquad (5.86)$$

Here H_t^a and H_t^b are the transducer responses, as defined in Section 5.3. For this derivation we have ignored reflections, and this is valid because both transducers are connected to electrical impedances of zero. A similar argument, applying a voltage to port B, gives the same expression for Y_{12}, so we have $Y_{21} = Y_{12}$ as expected. The magnitude of Y_{21} is related to the conductances. Since the transducers are bidirectional, they both give $G_a = 2|P_{13}|^2$, by power conservation. If $G_a^a(\omega)$ and $G_a^b(\omega)$ are the conductances of the two transducers, eq. (5.86) gives

$$|Y_{21}| = \sqrt{G_a^a(\omega)G_a^b(\omega)}. \qquad (5.87)$$

Response for finite source and load impedances

For finite terminating impedances, the device response is given by the transmission coefficient S_{21}. As shown in Fig. 5.15, each transducer is connected via a circuit whose details need not be specified here. The analysis here is valid if one transducer is apodized, as shown. The source has open-circuit voltage V_G and impedance R_G, and the load has impedance R_L and voltage V_L. To define S_{21} the device can be considered to be connected between transmission lines, with impedances R_G, R_L. The wave arriving on the input transmission line has amplitude $a_i = V_G/\sqrt{R_G}$, and the wave on the output line has amplitude

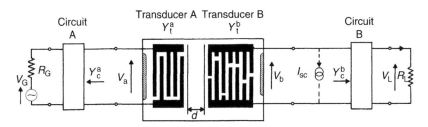

Figure 5.15 Device analysis allowing for terminating circuits.

$a_t = V_L/\sqrt{R_L}$. By definition $S_{21} = a_t/a_i$, and we also have $S_{12} = S_{21}$. The wave powers are $2|a_i|^2$ and $2|a_t|^2$. Formulae for S_{21} and other components of the scattering matrix S_{ij} are given in Appendix D.

In the present case, the fact that the transducers are non-reflective can be used to derive some simpler results. For transducer A, a circuit factor is defined by the equation

$$F_c^a = [V_a/V_G]_{V_b=0}. \tag{5.88}$$

This ratio is defined for the case where no waves are incident, so we set $V_b = 0$ to obtain this condition. Similarly, a circuit factor for the output transducer is defined by

$$F_c^b = [I_L/I_{sc}]_{V_a=0} \tag{5.89}$$

where I_L is the load current. I_{sc} is the transducer current I_b when transducer B is shorted, as shown by the broken line. For identical transducers with the same loading, the two circuit factors are the same.

The main response (first-transit response) is obtained by ignoring reflections. From the Y-matrix we have $I_{sc} = Y_{21}V_a$, and using eqs (5.88) and (5.89) gives

$$I_L/V_G = Y_{21}F_c^a F_c^b. \tag{5.90}$$

This simple expression ignores multiple-transit signals, but it is otherwise correct. If the power available from the generator is P_G and the power delivered to the load is P_L, the power ratio is $P_L/P_G = |Y_{21}|^2 R_G R_L$. The insertion loss, in decibels, is $-10\log(P_L/P_G)$. If the transducers are electrically matched, this loss is found to have its minimum value of 6 dB, corresponding to $P_L/P_G = 1/4$. This arises from a conversion loss of 3 dB at each transducer. In this case, there is generally large distortion due to the multiple-transit signal, of course. However, eq. (5.90) is still of some value because it incorporates the distortion due to the circuits, that is, the circuit effect. This formulation is convenient when designing bandpass filters, in which the effect needs to be compensated.

To evaluate the multiple-transit signals we consider the Y-matrix of eq. (5.84), where $Y_{11} = Y_t^a$, $Y_{22} = Y_t^b$ and $Y_{12} = Y_{21}$ are given by eq. (5.86). Suppose that the admittance of the load, as seen by transducer B, is Y_c^b, and the admittance seen by transducer A is Y_c^a. The source can be considered as a voltage generator, with voltage V_0, say, in series with an admittance Y_c^a. We then have the equations

$$I_a = (V_0 - V_a)Y_c^a; \quad I_b = -Y_c^b V_b. \tag{5.91}$$

Using these with eq. (5.84), the response can be expressed in the form

$$\frac{V_b}{V_0} = \frac{r_1}{1 + r_1 Y_{12}/Y_c^a} = r_1[1 + X + X^2 + \cdots] \qquad (5.92)$$

where

$$r_1 = -Y_{12} Y_c^a / [(Y_t^a + Y_c^a)(Y_t^b + Y_c^b)]$$

and $X = -r_1 Y_{12}/Y_c^a$. The series in eq. (5.92) is simply a binomial expansion. In these expressions, Y_{12}, r_1 and X are the only terms involving the delay due to the transducer separation. The parameter X is proportional to Y_{12}^2, corresponding to two device transits. Hence, the terms in the series correspond to the 1st, 3rd and 5th transits, and so on. For the main response, the ratio of V_b/V_0 is r_1. The ratio of the triple-transit to the first-transit signal is X. The ratios V_G/V_0 and V_L/V_b are given by standard circuit analysis, and with eq. (5.92) these give the ratio V_L/V_G which is the response allowing for the circuits.

5.6.2 Device using an apodized transducer

Here we consider a device using one apodized transducer and one unapodized, as indicated on Fig. 5.16. As before, each transducer has an x-axis directed away from the other transducer. It is assumed that both transducers have regular electrodes, though the principles can also be applied if the electrodes are irregular.

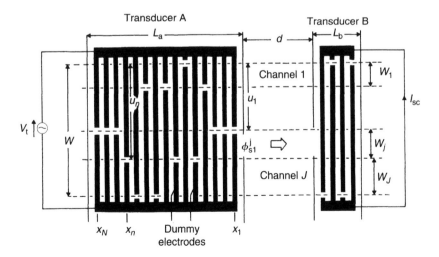

Figure 5.16 Two-transducer device with one transducer apodized.

A basic method of analysis is to divide the device into parallel channels, as originally suggested by Tancrell and Holland [18]. In Fig. 5.16 the channels are indicated by the horizontal broken lines. The channel edges are defined by the electrode breaks. Each channel behaves as an unapodized device, and it may be analyzed using the method of Section 5.6.1 above. The overall response is then obtained by connecting the channels in parallel electrically. If $Y_{ij}^{(m)}$ is the admittance matrix for channel m, the matrix for the whole device is easily found to be $Y_{ij} = \sum_m Y_{ij}^{(m)}$.

The channeling method has general validity. However, for the case of non-reflective transducers considered here there is a simpler approach. Consider first transducer B, which is unapodized and produces a short-circuit current I_{sc}. If a uniform surface-wave beam is incident, with potential ϕ_{i1}^u at port 1, the current is denoted by I_{sc}^u. From eq. (5.50) we have

$$I_{sc}^u = -j\phi_{i1}^u \omega W \overline{\rho}_e^b(k) \exp(-jkL_b/2) \tag{5.93}$$

with $x = 0$ at the center of the transducer. The subscript and superscript b refer to transducer B. Now consider that a narrow sub-beam of width ΔW and potential ϕ_{i1}^j is incident, producing an output current I_{sc}^j. Assuming that there is no resistance in the electrodes, I_{sc}^j is independent of the position of this beam, provided it is within the active aperture of the transducer. We have $I_{sc}^j = \alpha \phi_{i1}^j$, where α is some constant. If we now have a non-uniform beam of width W, composed of adjacent sub-beams each of width ΔW, the total current will be $I_{sc} = \sum_j I_{sc}^j = \sum_j \alpha \phi_{i1}^j$ by superposition. The constant α is found by taking the ϕ_{i1}^j to be the same and equal to ϕ_{i1}^u, and using eq. (5.93). We then find

$$I_{sc} = -j\langle \phi_{i1} \rangle \omega W \overline{\rho}_e^b(k) \exp(-jkL_b/2) \tag{5.94}$$

where $\langle \phi_{i1} \rangle = \sum_j \phi_{i1}^j \Delta W/W$ is the average potential at port 1. The current is therefore the same as for a uniform beam, except that the average potential is used. The argument remains valid if $\Delta W \to 0$, so that there is an infinite number of sub-beams and the incident potential varies smoothly over the transducer aperture.

The device response can be written in terms of transducer responses. This is more convenient than the P-matrix, because the latter is best suited to unapodized transducers. The response of an *unapodized* transducer is $H_t(\omega)$, defined such that when a voltage V_t is applied the potential of the wave emerging at port 1 is $\phi_{s1} = jV_t H_t(\omega)[\Gamma_s/(\omega W)]^{1/2}$. If the transducer is shorted and a wave is incident on port 1, with potential ϕ_{i1} at this port, the current produced is $I_{sc} = -j\phi_{i1} H_t(\omega)[\omega W/\Gamma_s]^{1/2}$. This follows from reciprocity, eq. (5.51). For a two-transducer device, this gives the response as in eq. (5.86) above. For an apodized transducer the response is defined in the same way, but with ϕ_{s1} replaced by its average $\langle\phi_{s1}\rangle$. Thus, when a voltage V_t is applied the average potential at port 1 is

$$\langle\phi_{s1}\rangle = jV_tH_t(\omega)[\Gamma_s/(\omega W)]^{1/2}. \tag{5.95}$$

This formula applies for both apodized and unapodized transducers. In Fig. 5.16, transducer A is apodized with response $H_t^a(\omega)$ and transducer B is unapodized with response $H_t^b(\omega)$, and each transducer has port 1 facing the other transducer. These ports are separated by a distance d, so the average potential at port 1 of transducer B is $\langle\phi_{i1}\rangle = \langle\phi_{s1}\rangle \exp(-jkd)$. From these equations, the device response can be written as

$$Y_{21} = I_{sc}/V_t = H_t^a(\omega)H_t^b(\omega) \exp(-jkd). \tag{5.96}$$

This has the same form as eq. (5.86) above. If both transducers are unapodized we have $\langle\phi_{s1}\rangle = \phi_{s1}$ and $\langle\phi_{i1}\rangle = \phi_{i1}$. Note that W is the aperture of the *unapodized* transducer, which is assumed to fully overlap the beam produced by the apodized transducer. The response for finite terminating impedances can be obtained as in Section 5.6.1 above, after evaluating the transducer admittances.

For regular electrodes, the response of transducer A is conveniently obtained by considering the electrodes individually. From eq. (5.57) the wave potential at port 1, due to the voltage on electrode n only, is $\phi_{s1} = j\Gamma_s V_t \overline{\rho}_f(k) \exp(-jkx_n)$, where x_n is the center location of electrode n relative to port 1. This potential is present only over the extent u_n of the electrode. The average value of ϕ_{s1}, over the aperture W of the unapodized transducer, is $<\phi_{s1}> = \phi_{s1} u_n/W$. We now sum over all the electrodes in transducer A, using superposition. The total beam has average potential

$$\langle\phi_{s1}\rangle = j\Gamma_s V_t \overline{\rho}_f(k) \sum_{n=1}^{N} \frac{u_n}{W} \exp(-jkx_n). \tag{5.97}$$

This corresponds to the delta-function analysis in Chapter 1. The response corresponds to that of a transversal filter, and consequently it has very flexible design capability as described in Chapter 6. The term $\overline{\rho}_f(k)$, given by eq. (5.59), is essentially the element factor, and the summation is the array factor. Comparing with eq. (5.95), the transducer response is

$$H_t^a(\omega) = \sqrt{\omega W \Gamma_s} \overline{\rho}_f(k) \sum_{n=1}^{N} \frac{u_n}{W} \exp(-jkx_n) \tag{5.98}$$

with $x=0$ at port 1. This agrees with the result for the unapodized case, eq. (5.61), with $\hat{P}_n$ replaced by u_n/W. As before, W is the aperture of the unapodized transducer, even though eq. (5.98) is the response of the apodized transducer.

The above frequency response is defined specifically for the case where the apodized transducer is used in conjunction with an unapodized transducer. In addition, it is necessary that the active gaps in the apodized transducer are completely overlapped by the unapodized transducer. Restrictions such as these are needed if the frequency response of the apodized transducer is to have any meaning.

Gap element formulation

For an alternative formulation, note that the array factor in eq. (5.98) can be written as

$$\overline{A}_f(k) = \sum_{n=1}^{N} \frac{u_n}{W} \exp(-jkx_n) = \sum_{n=1}^{N-1} \frac{u_{n+1}}{W} \exp[-jk(x_n + p)].$$

Here the second form follows because $x_n + p = x_{n+1}$ and we assume end effects to be negligible. Multiplying the second form by $\exp(jkp)$ and subtracting from the first form gives an array factor involving $u_{n+1} - u_n$. This enables $H_t^a(\omega)$ to become

$$H_t^a(\omega) = \sqrt{\omega W \Gamma_s} \overline{\rho}_g(k) \sum_{n=1}^{N-1} \frac{u_{n+1} - u_n}{W} \exp[-jk(x_n + p/2)]. \qquad (5.99)$$

This expression uses gaps as elements. Each term in the array factor has amplitude proportional to the overlap of adjacent electrodes, and it refers to a position in the center of the gap. The element factor is $\overline{\rho}_g(k) = -0.5j\overline{\rho}_f(k) \operatorname{cosec}(kp/2)$.

The gap element method is valid because an array of electrodes with the same voltage does not produce any surface-wave amplitude. This is associated with the fact that adding a constant voltage to all electrodes does not change the charge density.

5.6.3 Admittance of apodized transducers

As stated in Section 5.6.2, the Y-matrix of the device can be obtained by dividing into parallel channels. In the present case we assume the transducers to be non-reflective, giving some convenient simplifications. The previous section derived the transadmittance Y_{21} of the device. It remains to find the transducer admittances, which are equal to Y_{11} and Y_{22}. Section 5.4 gives the method for an unapodized transducer, so here an apodized transducer is considered. As before, the electrodes are taken to be regular.

Capacitance

First consider the charge on electrode m due to a voltage on electrode n. The parts of these electrodes connected to the upper bus overlap by an amount $\min(u_n, u_m)$. Suppose that the upper part of electrode n has unit voltage, and all other electrodes are grounded. The electrostatic charge on the upper part of electrode m is $Q_{m-n} \min(u_n, u_m)$, where Q_n is defined in Section 5.4. When a voltage is applied to the upper parts of all electrodes, the total charge on the upper part of electrode m is found by summing over n. The total charge connected to the upper bus is obtained by summing over m. The capacitance is therefore

$$C_t = \sum_{m=1}^{N} \sum_{n=1}^{N} Q_{m-n} \min(u_n, u_m). \tag{5.100}$$

Conductance and susceptance

To calculate G_a, we consider the device split into parallel channels, as discussed in Section 5.6.2. Each channel behaves as an unapodized device. For an unapodized transducer the conductance is $G_a(\omega) = \omega W \Gamma_s |\overline{\rho}_e(k)|^2$, with $\overline{\rho}_e(k) = \sum_n \hat{P}_n \overline{\rho}_f(k) \exp(-jkx_n)$, from eq. (5.60). For the apodized transducer we apply this to channel j, which has width W_j. For channel j, we find

$$G_a^j = \omega \Gamma_s W_j [\overline{\rho}_f(k)]^2 \sum_{n=1}^{N} \sum_{m=1}^{N} \hat{P}_n \hat{P}_m \exp[-jk(x_m - x_n)]$$

where $\hat{P}_n$ are electrode polarities for channel j. This expression needs to be summed over the channel number j to find the total conductance G_a. Some simplification is obtained by collecting terms with the same value of $x_m - x_n$. Analysis in terms of the gap elements used in eq. (5.99) gives [19]

$$G_a(\omega) = E(\omega) \sum_{s=0}^{N-2} A_s \cos(s\omega\tau) \tag{5.101}$$

where $\tau = kp/\omega$ is the electrode spacing in time units and $E(\omega) = \omega \Gamma_s |\overline{\rho}_g(k)|^2$. The coefficients A_s are

$$A_s = a_s \sum_{m=s+1}^{N-2} \alpha(m-s, m)$$

where $a_s = 2$ for $s > 0$ and $a_s = 1$ for $s = 0$, and

$$\alpha(n, m) = \min(u_{n+1}, u_{m+1}) - \min(u_n, u_{m+1}) - \min(u_{n+1}, u_m) + \min(u_n, u_m).$$

The susceptance $B_a(\omega)$ is the Hilbert transform of $G_a(\omega)$. Usually, the element factor $E(\omega)$ in eq. (5.101) varies slowly compared with other terms, so it can be regarded approximately as a constant. The Hilbert transform of $\cos(s\omega\tau)$ is $-\sin(s\omega\tau)$, giving the convenient formula

$$B_a(\omega) \approx -E(\omega) \sum_{s=0}^{N-2} A_s \sin(s\omega\tau). \qquad (5.102)$$

5.6.4 Two-transducer device using a multistrip coupler

The apodized transducer analysis of Section 5.6.2 is valid only if one of the two transducers is unapodized. If both are apodized, the device response is not a simple product of transducer responses. However, a simple product can be obtained by using a multistrip coupler (m.s.c). This consists of a parallel set of disconnected metal electrodes spanning two tracks [20], as in Fig. 5.17. With both transducers apodized there is an additional degree of design flexibility. Another advantage is that unwanted bulk-wave signals are reduced, because the bulk waves are transferred from one track to the other less efficiently than surface waves. This is important for devices using Y–Z lithium niobate substrates, with either or both of the transducers apodized. The number of strips needed for a full transfer of a surface wave between tracks is in the region of $3/(\Delta v/v)$. Consequently, the coupler is only suitable for strongly piezoelectric substrates such as lithium niobate.

Physically, it can be expected that a wave incident on the upper track will induce voltages on the electrodes, and these voltages will cause wave generation in the lower track. The voltages will have appropriate phase increments for wave

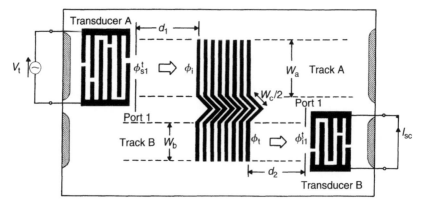

Figure 5.17 Interdigital device using a multistrip coupler.

generation so the process has a wide bandwidth, in contrast to an interdigital transducer whose bandwidth decreases with the length. Since the structure is periodic, it will give strong reflectivity at frequencies where the pitch is $\lambda/2$ or a multiple of this. However, practical devices use couplers with pitch smaller than $\lambda/2$, so that the reflections are not significant. The inclined strips in Fig. 5.17 allow the tracks to be separated, and there is little wave generation here because the strip spacing does not correspond to the wave velocity.

If a uniform beam is incident on the upper track, with potential ϕ_i at the m.s.c. input port, a uniform beam will appear at the output in the lower track with potential ϕ_t, say. We can define an m.s.c. response H_m by the equation $\phi_t = H_m\phi_i$. We now argue that if the input beam is non-uniform the output beam, which is uniform, is given by the mean value $\langle\phi_i\rangle$, so that

$$\phi_t = H_m\langle\phi_i\rangle. \tag{5.103}$$

The justification for this is almost identical to the argument for an unapodixed receiving transducer, given in Section 5.6.2. At the output of the device, we have a uniform beam incident on an apodized transducer. The short-circuit current produced by this transducer is easily shown to be given by the response defined by eq. (5.95). Without giving details here, the short-circuit device response is found to be

$$Y_{21} = I_{sc}/V_t = H_t^a H_t^b H_m \exp[-jk(d_1 + d_2)] \tag{5.104}$$

where d_1 and d_2 are the two spaces between the transducer ports and the m.s.c. ports. The device response is thus proportional to the product of the two apodized transducer responses. The averages are defined as in eq. (5.97), where W is taken as the m.s.c. aperture (with different W values for the two tracks in the case shown in Fig. 5.17). Because the m.s.c. has wide bandwidth, it is often adequate to regard H_m as a constant in the above equation.

Some theory for the m.s.c. is described in Appendix E.

REFERENCES

1. D.P. Morgan. 'Quasi-static analysis of generalised SAW transducers using the Green's function method', *IEEE Trans.*, **SU-27**, 111–123 (1980).
2. R.C. Peach. 'A general approach to the electrostatic problem of the SAW interdigital transducer', *IEEE Trans.*, **SU-28**, 96–105 (1981).
3. R.F. Milsom, N.H.C. Reilly and M. Redwood. 'Analysis of generation and detection of surface and bulk acoustic waves by interdigital transducers', *IEEE Trans.*, **SU-24**, 147–166 (1977).
4. W.H. Press, B.P. Flannery, S.A. Teukolsky and W.T. Vetterling. *Numerical Recipes – The Art of Scientific Computing* (Fortran version), Cambridge University Press, 1989.

5. C.S. Hartmann and B.G. Secrest. 'End effects in interdigital surface wave transducers', *IEEE Ultrason. Symp.*, 1972, pp. 413–416.
6. W.R. Smith and W.F. Peddler. 'Fundamental and harmonic frequency circuit-model analysis of interdigital transducers with arbitrary metallisation ratios and polarity sequences', *IEEE Trans.*, **MTT-23**, 853–864 (1975).
7. P. Ventura, J.M. Hodé and B. Lopes. 'Rigorous analysis of finite SAW devices with arbitrary electrode geometries', *IEEE Ultrason. Symp.*, **1**, 257–262 (1995).
8. K.-Y. Hashimoto. *Surface Acoustic Wave Devices in Telecommunications – Modelling and Simulation*, Springer, 2000, p. 55.
9. E. Bausk, E. Kolosvsky, A. Kozlov and L. Solie. 'Optimization of broadband uniform beam profile interdigital transducers weighted by assignment of electrode polarities', *IEEE Trans. Ultrason. Ferroelect. Freq. Contr.*, **49**, 1–10 (2002).
10. S.V. Biryukov and V.G. Polevoi. 'The electrostatic problem for the SAW interdigital transducer in an external electric field – Part I: A general solution for a limited number of electrodes', *IEEE Trans. Ultrason. Ferroelect. Freq. Contr.*, **43**, 1150–1159 (1996).
11. A. Papoulis. *The Fourier Integral and Its Applications*, McGraw-Hill, 1962.
12. A.L. Nalamwar and M. Epstein. 'Immittance characterisation of acoustic surface wave transducers', *Proc. IEEE*, **60**, 336–337 (1972).
13. G. Tobolka. 'Mixed matrix representation of SAW transducers', *IEEE Trans.*, **SU-26**, 426–428 (1979).
14. D.P. Morgan. *Surface-Wave Devices for Signal Processing,* Elsevier, 1991, Chapter 4.
15. S. Datta and B.J. Hunsinger. 'Element factor for periodic transducers', *IEEE Trans.*, **SU-27**, 42–44 (1980).
16. H. Engan. 'Surface acoustic wave multi-electrode transducers', *IEEE Trans.*, **SU-22**, 395–401 (1975).
17. W.R. Smith, H.M. Gerard, J.H. Collins, T.M. Reeder and H.J. Shaw. 'Analysis of interdigital surface wave transducers by use of an equivalent circuit model', *IEEE Trans.*, **MTT-17**, 856–864 (1969).
18. R.H. Tancrell and M.G. Holland. 'Acoustic surface wave filters', *Proc. IEEE*, **59**, 393–409 (1971).
19. D.P. Morgan. 'Admittance calculations for non-reflective SAW transducers', *IEEE Ultrason. Symp.*, **1**, 131–134 (1996).
20. F.G. Marshall, C.O. Newton and E.G.S. Paige. 'Theory and design of the surface acoustic wave multistrip coupler', *IEEE Trans.*, **MTT-21**, 206–215 (1973).

6

BANDPASS FILTERING USING NON-REFLECTIVE TRANSDUCERS

In this chapter we consider the surface-wave bandpass filter, whose basic function is to pass signals with frequencies within a specified band, the passband, and to reject signals outside this band. This is one of the commonest applications of surface waves, as shown by the very large quantity production of filters for domestic television receivers and for mobile telephony. There are several classes of such filters. In this chapter we consider filters using non-reflective transducers, which normally make use of apodization or withdrawal weighting in order to obtain the required frequency response. These filters, also called 'transversal filters', became established for television receivers and other applications in the 1970s, when surface-wave technology was new. However, for many applications, such as mobile telephony, there is a need for low insertion losses which cannot be met using the non-reflective types of device. For this purpose, devices using reflective components were developed. As described later, these include single-phase unidirectional transducers (Chapter 9), transversely coupled resonator filters (Chapter 10) and impedance element filters (Chapter 11). All of these devices also fall in a class known as 'linear filters', with general behavior described in Appendix A.

Bandpass filtering is commonly done using combinations of resonators, which might be inductor–capacitor combinations or transmission line sections or dielectric resonators. For this type of filter the response is expressed in terms of poles, and the design problem is concerned with finding suitable locations for the poles in the complex frequency plane. In contrast, the response of a transversal filter has no poles – it is determined by zeros, that is, points in the complex frequency plane where the response is zero. The design problem is therefore quite different. The methods used can give a very high degree of flexibility, so that an almost arbitrary frequency response can be realized. Moreover, this can be obtained in an optimized manner, taking account of required tolerances which can be specified to vary with frequency.

Suitable methods for analysis of interdigital bandpass filters are given in Chapter 5, and relevant propagation effects are described in Chapter 4. This chapter is therefore concerned with the design and performance of the devices. Internal reflections in the transducers are assumed to be negligible. In the case of single-electrode transducers, this assumes that the electrodes are recessed to minimize their reflection coefficients, or a short transducer is used so that its reflection coefficient is small. More commonly, the transducers may be the double-electrode type.

Section 6.1 describes some basic properties of uniform transducers, applicable to many devices. Section 6.2 considers the relation between an apodized transducer and a transversal filter, and discusses sampling. This reduces the design problem to that of designing transversal filters, and methods for this are described in Section 6.3. Section 6.4 then gives filter design techniques and performance.

6.1 BASIC PROPERTIES OF UNIFORM TRANSDUCERS

Uniform transducers are present in many surface-wave devices, and here we consider some basic properties, particularly bandwidth. Although this relates to uniform transducers, many of the conclusions also apply qualitatively if the transducers are weighted, for example, by apodization.

Figure 6.1a shows the equivalent circuit for a uniform transducer, as derived in Section 5.5. It consists of a capacitance C_t in parallel with an acoustic conductance G_a and susceptance B_a, with G_a approximated as

$$G_a(\omega) \approx G_a(\omega_c)[(\sin X)/X]^2, \qquad (6.1)$$

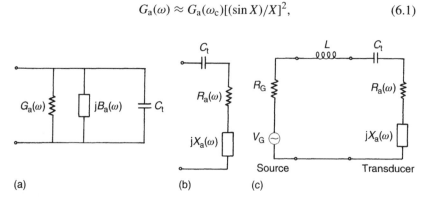

(a)   (b)   (c)

Figure 6.1 (a, b) Equivalent circuits for transducer. (c) Tuned transducer connected to source.

where ω_c is the center frequency, $X = \pi N_p (\omega - \omega_c)/\omega_c$ and N_p is the number of periods. When a voltage is applied, the surface-wave power is proportional to $G_a(\omega)$. The 1.5 dB points are where $G_a = G_a(\omega_c)/\sqrt{2}$, the bandwidth $\Delta\omega$ between these points is found to be

$$\Delta\omega/\omega_c = 0.638/N_p. \tag{6.2}$$

The Q-factor is Q_t, defined as

$$Q_t = \omega_c C_t/[G_a(\omega_c)] = \tilde{Q}_t/[N_p \Delta v/v], \tag{6.3}$$

where $\tilde{Q}_t$ is a normalized Q-factor given in Table 5.1. For single-electrode transducers, $\tilde{Q}_t = 1/2.871$.

Series equivalent circuit and matching

Some properties of the transducer can be appreciated from a series equivalent circuit shown in Fig. 6.1b. The capacitance C_t is included explicitly, and the remaining terms are expressed as an acoustic resistance $R_a(\omega)$ and reactance $X_a(\omega)$. Equating the impedances of the two circuits, we find that $R_a(\omega)$ is given by

$$R_a(\omega) = \frac{G_a(\omega)}{[G_a(\omega)]^2 + [B_a(\omega) + \omega C_t]^2}. \tag{6.4}$$

The reactance $X_a(\omega)$ has a form similar to $B_a(\omega)$ and is small at the center frequency. For many transducers the impedance is dominated by the capacitive term, so that the G_a and B_a terms in the denominator of eq. (6.4) are small. Thus, $R_a(\omega)$ has approximately the same form as $G_a(\omega)$. At the center frequency we have $R_a(\omega_c) \approx G_a(\omega_c)/(\omega_c C_t)^2$. It was shown in Chapter 5, Section 5.5, that $G_a(\omega_c)$ is proportional to $\omega_c N_p^2 W$, where W is the transducer aperture and C_t is proportional to $N_p W$. Hence $R_a(\omega_c)$ is proportional to $1/(\omega_c W)$ and is independent of the number of periods, N_p. Moreover, transducers can often be designed to give $R_a(\omega_c) \approx 50\,\Omega$. The transducer is then well matched to a $50\,\Omega$ source if a series inductor is used to tune out the capacitance, as in Fig. 6.1c. For single-electrode transducers at the fundamental frequency, the aperture needed to obtain $R_a(\omega_c) = 50\,\Omega$ is 111 wavelengths for Y–Z lithium niobate, and 42 wavelengths for ST–X quartz. Corresponding values for a double-electrode transducer are 60 and 23 wavelengths. These values assume a metallization ratio of $a/p = 1/2$.

In some cases the transducer cannot be designed such that $R_a(\omega_c)$ is close to $50\,\Omega$. However, it is still possible to match to a $50\,\Omega$ source, at a specified frequency, if a two-component matching circuit is used. Consider a transducer with impedance $Z_0 = R_0 + jX_0$ connected to a series reactance X_1 and then a parallel reactance X_2. The impedance seen looking into this circuit is to be a real quantity R_G. From circuit analysis, X_1 can be shown to be given by

$$(X_0 + X_1)^2 = R_0(R_G - R_0), \qquad (6.5)$$

and then $X_2 = -R_0 R_G/(X_0 + X_1)$. This scheme is valid provided $R_0 \leq R_G$. The reactances X_1 and X_2 can be realized using inductors or capacitors, depending on their signs. For another possible circuit, a parallel reactance X_1 is connected across the transducer, followed by a series reactance X_2. The transducer impedance is written as $Z_0 = 1/(G_0 + jB_0)$, and X_1 is given by

$$(B_0 - 1/X_1)^2 = -G_0(G_0 - 1/R_G). \qquad (6.6)$$

For X_2 we have $X_2 = (B_0 - 1/X_1)R_G/G_0$. This scheme is valid if $G_0 \leq 1/R_G$.

Bandwidth for minimum loss

If a lossless transducer is electrically matched to a source, at some frequency, then all of the available electrical power is converted into surface waves at this frequency. The conversion loss is ideally 3 dB owing to the bidirectionality. One penalty paid for this is that the bandwidth may be limited because of the electrical Q of the matching circuit [1, Chapter 7].

Consider the series-tuned circuit of Fig. 6.1c, matched so that $R_G = R_a(\omega_c)$. If N_p is small we can take $R_a(\omega) \approx R_a(\omega_c)$ and the circuit has resistance $R_a(\omega_c) + R_G$. The Q-factor is $Q = Q_t/2$ and the fractional 1.5 dB bandwidth is found to be $\Delta\omega/\omega_c = 0.644/Q$, which is proportional to N_p. On the other hand, if N_p is large the electrical Q is not relevant, and the bandwidth is determined by the transducer bandwidth. Equation (6.2) gave $\Delta\omega/\omega_c = 0.638/N_p$. However, there is an additional complication because transducer impedance causes the transducer voltage to vary with frequency. It is found that the bandwidth is $\Delta\omega/\omega_c \approx 1.14/N_p$. Hence the bandwidth varies as N_p when N_p is small, and as $1/N_p$ when N_p is large. An intermediate value of N_p, called N_{opt}, gives the maximum bandwidth. This can be estimated [1, Chapter 7] by equating the above expressions for $\Delta\omega/\omega_c$, giving

$$N_{opt} \approx 0.94\sqrt{\tilde{Q}_t/(\Delta v/v)}, \qquad (6.7)$$

and the corresponding bandwidth is $\Delta\omega/\omega_c \approx 1/N_{opt}$. For a single-electrode transducer with $a/p = 1/2$ we have $\tilde{Q}_t = 0.348$ and $N_{opt} = 0.55/\sqrt{\Delta v/v}$. This gives $N_{opt} = 4$ for Y–Z lithium niobate, and $N_{opt} = 21$ for ST–X quartz. In many devices this matching would give unacceptable triple-transit signals, but there are special methods for dealing with this, for example those in Chapter 9, Section 9.6.

Self-resonant transducers

Another consideration is the bandwidth obtainable if the transducer is directly connected to a resistive source or load, with no inductor. Low loss is possible here because the transducer can be 'self-resonant'. The susceptance $B_a(\omega)$, seen in Fig. 5.12, is negative at some frequencies, tending to cancel the capacitive susceptance ωC_t. Since B_a and C_t are proportional to N_p^2 and N_p, respectively, it is possible to choose N_p such that these susceptances cancel at one frequency. The transducer admittance is then real, and it can be matched directly to a resistive load. This is found to require $N_p \approx \pi \tilde{Q}_t/(2\Delta v/v)$, and the bandwidth is typically $\Delta\omega/\omega_c \approx 1/N_p$. Device bandwidths are limited to approximately this value if low loss is needed and inductors cannot be used. For example, a double-electrode transducer on $128° Y–X$ lithium niobate gives $N_p = 26$. Some practical applications are mentioned in Section 9.6 of Chapter 9.

Although the above has considered uniform transducers, the conclusions remain roughly true for weighted and reflective transducers.

Electrode resistance

The sheet resistance of aluminum film is typically $\rho = 0.04/h$ ohms per square, where h is the film thickness in μm. This can be a significant source of loss. The effect on a transducer is rather complicated, since it causes the voltages to vary along the lengths of the electrodes. However, if the losses are small the effect can be estimated by assuming that the transducer equivalent circuit is ideal except for an additional series resistance R_E, which accounts for the loss. Consider an unapodized transducer with N regular electrodes and aperture W, and define r_E as the resistance of one electrode per unit length. Then R_E is given by [1, Chapter 7]

$$R_E = \frac{r_E W}{3} \left[\sum_{n=1}^{N} q_n^2 \right] \bigg/ \left[\sum_{n=1}^{N} \hat{P}_n q_n \right]^2 \tag{6.8}$$

where $\hat{P}_n = 0$ or 1 are the electrode polarities and q_n are the electrostatic charges of Section 5.4 as shown in Chapter 5. In eq. (6.8) the polarity sequence $\hat{P}_n$ is arbitrary. In particular, for uniform single-electrode transducers we find $R_E = 2r_E W/(3N_p)$. This formula neglects the resistance in the bus bars, which can also cause significant loss.

6.2 APODIZED TRANSDUCER AS A TRANSVERSAL FILTER

We here consider a filter consisting of two transducers, one apodized and the other unapodized, as shown in Fig. 6.2. The transducers have regular

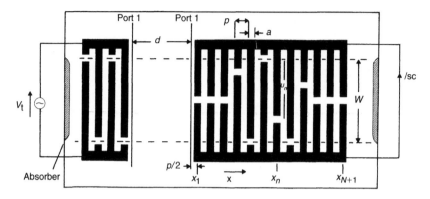

Figure 6.2 Bandpass filter using apodized transducer with regular electrodes.

electrodes and are non-reflective, a condition normally met by using double electrodes. Propagation loss and diffraction are neglected. As shown in Chapter 5, Section 5.6.2, the transadmittance Y_{21} is essentially the product of the two-transducer responses. For finite load impedances the response has additional terms due to multiple transits, as discussed in Section 5.6.2 of Chapter 5 and Section D.1 of Appendix D. These signals arise because the transducers have finite reflectivity. For practical filters it is usual to use a small load impedance, such that the multiple-transit signals are reduced to a low level where they can be considered as minor perturbations. Hence, the basic design is done with these signals ignored, concentrating on the transducer design. A consequence of this is that the device insertion loss is relatively high, typically 20 dB or more.

The electrodes in the apodized transducer extend to the bus bars on either side, even though the active interelectrode gaps do not generally extend over the entire aperture. The design thus includes 'dummy' electrodes, such as the upper part of the fourth electrode, which have electrodes on either side with the same potential. The dummy electrodes are included because they improve the uniformity of the wave velocity, thus improving the device performance. They also make the electrode geometry regular. This improves the accuracy when the convenient element factor of Section 5.4 in Chapter 5 is used.

The frequency response $H_t(\omega)$ of a non-reflective apodized transducer, with regular electrodes, was derived in Section 5.6.2 of Chapter 5. In terms of gap elements, eq. (5.99) shows that the response is

$$H_t(\omega) = E(\omega) \sum_{n=1}^{N} v_n \exp[-jk(x_n - x_1 + p)]. \tag{6.9}$$

The device response Y_{21} is the product of the transducer responses defined in this way, with an additional phase factor corresponding to the transducer separation (eq. 5.96). Electrode n extends a distance u_n from the upper bus and is centered at x_n. All electrodes have width a and the pitch p is constant. The electrode numbering starts from the electrode nearest to the other transducer. N is the number of gaps, so that the number of electrodes is $N + 1$. A few guard electrodes (Chapter 5, Section 5.4) are included at each end in order to minimize end effects. We define $v_n = (u_{n+1} - u_n)/W$ as a normalized form of the overlap between adjacent electrodes. The element factor $E(\omega)$ is defined as

$$E(\omega) = (\omega W \Gamma_s)^{1/2} \overline{\rho}_g(k). \tag{6.10}$$

In these equations, the aperture W refers to the *unapodized* transducer, which is assume to completely overlap the active parts of the apodized transducer. The function $\overline{\rho}_g(k)$ is the Fourier transform of the elemental charge density for gap elements, given by

$$\overline{\rho}_g(k) = -0.5\mathrm{j}\overline{\rho}_f(k)\operatorname{cosec}(kp/2), \tag{6.11}$$

with $\overline{\rho}_f(k)$ given by eq. (5.59). The term $x_n - x_1 + p$ in eq. (6.9) is the distance between element n, located at $x = x_n + p/2$, and port 1, which is taken to be just outside the transducer, at $x = x_1 - p/2$.

Equation (6.9) applies to both transducers. For the unapodized transducer the lengths u_n are equal to 0 or W. The two transducers may have different electrode pitches p, in which case the element factors will be different. The element factors do not affect the frequency response very much, but it is necessary to account for them if accurate results are to be obtained.

Transversal filter analogy

Define $\tau_s = p/v_e$ as the delay between successive gaps, where $v_e = \omega/k$ is the wave velocity. Equation (6.9) can be written as

$$H_t(\omega) = E(\omega) \sum_{n=1}^{N} v_n \exp(-\mathrm{j}n\omega\tau_s). \tag{6.12}$$

This shows that the transducer behaves essentially as a transversal filter, a conceptual device shown in Fig. 6.3. Here an ideal non-dispersive delay line with regularly spaced taps produces delayed replicas of an input waveform. The replicas, with delays $n\tau_s$, are weighted using real amplitude coefficients v_n and then summed to give the output waveform. Since the frequency response of an ideal delay line with delay τ is $\exp(-\mathrm{j}\omega\tau)$, the frequency response of the

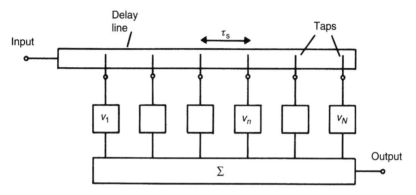

Figure 6.3 Transversal filter.

transversal filter is

$$H_s(\omega) = \sum_{n=1}^{N} v_n \exp(-jn\omega\tau_s). \tag{6.13}$$

This is the same as the transducer response, eq. (6.12), except for omission of the element factor $E(\omega)$. Consequently, transversal filter design techniques can be applied to surface-wave filters.

The transversal filter concept, introduced by Kallmann [2], is a versatile technique for obtaining accurate frequency responses. Apart from surface-wave devices, it is also applied to other technologies such as digital finite-impulse-response (FIR) filters [3–5] and charge-coupled devices (CCDs). Although these devices all use the same concept, there are some important distinctions. Surface-wave devices give zero response at zero frequency, so they cannot be used for low-pass filtering. Moreover, most surface-wave devices use two transducers, so they are effectively two transversal filters combined. We assume here that the samples are uniformly spaced, though non-uniform spacing is also possible as in pulse compression filters (Chapter 7).

Sampling and surface-wave transducers

Transforming eq. (6.13) from the frequency domain to the time domain gives the impulse response $h_s(t)$ of the transversal filter, which is the delta-function sequence

$$h_s(t) = \sum_{n=1}^{N} v_n \delta(t - n\tau_s). \tag{6.14}$$

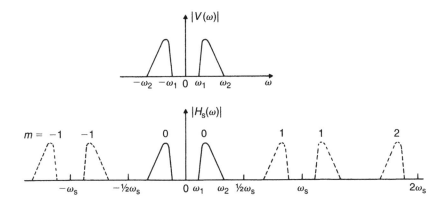

Figure 6.4 Frequency response of a transversal filter.

This equation also follows directly from Fig. 6.3. To appreciate the design principle, suppose that we can define a smooth real function $v(t)$, such that its values at times $n\tau_s$ are equal to the coefficients v_n. We also assume that $v(t) = 0$ for $t \leq 0$ and for $t > N\tau_s$. With these assumptions, eq. (6.14) may be written as

$$h_s(t) = \sum_{n=-\infty}^{\infty} v(n\tau_s)\delta(t - n\tau_s) = v(t) \sum_{n=-\infty}^{\infty} \delta(t - n\tau_s). \qquad (6.15)$$

This is a sampled version of the waveform $v(t)$. The quantity τ_s is the sampling interval, corresponding to the electrode spacing in the transducer. The corresponding frequency $\omega_s = 2\pi/\tau_s$ is the *sampling frequency*. At this frequency the electrode pitch p is equal to the surface-wave wavelength.

The frequency response of the transversal filter, $H_s(\omega)$, is the Fourier transform of $h_s(t)$. Taking $V(\omega)$ as the transform of $v(t)$, we find

$$H_s(\omega) = \frac{\omega_s}{2\pi} \sum_{m=-\infty}^{\infty} V(\omega - m\omega_s). \qquad (6.16)$$

This follows from standard theorems of Fourier analysis, given in Appendix A, eqs (A.14), (A.16) and (A.32). Figure 6.4 illustrates the relationship, showing the magnitudes of $V(\omega)$ and $H_s(\omega)$. The magnitude of $V(\omega)$ is symmetric about $\omega = 0$ because $v(t)$ is real. The figure assumes that $V(\omega)$ is a bandpass function, so that for $\omega > 0$ it is negligible except when ω is between two frequencies denoted by ω_1 and ω_2. This condition is always valid for surface-wave filter design. The additional responses for $m \neq 0$, introduced by the sampling process, are shown by broken lines and are called 'image' responses. The sampled spectrum $H_s(\omega)$ repeats with a frequency interval ω_s, and because it is zero for $\omega = 0$ it is also zero at all multiples of ω_s. In addition, $|H_s(\omega)|$ is symmetric about $\omega = n\omega_s/2$, for any n. This follows from eq. (6.16) and the fact that $|V(\omega)|$ is symmetric about $\omega = 0$. If $V(\omega)$ has a 'center' frequency ω_c and the sampling rate is a multiple of ω_c, the image responses are centered at multiples of ω_c and are therefore harmonics.

In Fig. 6.4 it has been assumed that the sampling frequency ω_s exceeds $2\omega_2$. For this case the individual terms in eq. (6.16) do not overlap, because $V(\omega)$ is negligible for $|\omega| > \omega_2$. In particular, the 'fundamental' component of $H_s(\omega)$, the term with $m = 0$, is essentially the same as $V(\omega)$. Thus,

$$H_s(\omega) = \omega_s V(\omega)/(2\pi), \quad \text{for } |\omega| < \omega_s/2, \tag{6.17}$$

provided ω_s exceeds a minimum value $2\omega_2$, which is known as the *Nyquist frequency*. This result gives an important part of the design procedure for a bandpass filter. It is assumed that the required response is specified for frequencies up to ω_2, at which point its magnitude has fallen to zero. We first generate a finite-length real waveform $v(t)$ such that its transform $V(\omega)$ is a good approximation to the required response for $|\omega| < \omega_2$, and is negligible for $\omega > \omega_2$. We then sample $v(t)$ with some sampling frequency $\omega_s \geq 2\omega_2$, giving real weighting coefficients $v_n = v(n\tau_s)$. The frequency response $H_s(\omega)$ of the transversal filter is then a good approximation to the required response, for $|\omega| \leq \omega_s/2$, apart from a multiplying constant. Since this procedure is valid for *any* specified frequency response, the transversal filter principle is very flexible. In practice, some complications arise in the calculation of $v(t)$, and the methods for this are considered later, in Section 6.3.

The unwanted image responses, shown by broken lines, are usually suppressed by low-pass filtering. Usually ω_s will be larger than $2\omega_2$ so that the filter does not need a sharp cutoff.

For a surface-wave transducer we have, from eqs (6.12) and (6.13), $H_t(\omega) = E(\omega)H_s(\omega)$. Thus, if the sampling frequency exceeds the Nyquist frequency, the transducer response is given by

$$H_t(\omega) = \omega_s E(\omega)V(\omega)/(2\pi), \quad \text{for } |\omega| < \omega_s/2. \tag{6.18}$$

The design procedure is therefore the same as for a transversal filter, except that the required frequency response is first divided by the element factor $E(\omega)$ in order to obtain $V(\omega)$. Zeros of $E(\omega)$ do not cause problems because $H_s(\omega)$ has zeros at the same frequencies.

The transversal filter has an impulse response with the simple form of eq. (6.14). In contrast, the transducer has no simple form for its impulse response because of the distortion caused by the element factor $E(\omega)$, as shown by eq. (6.12). However, for the fundamental passband the frequency response is given by eq. (6.18) and the slowly-varying factor $E(\omega)$ generally causes little distortion. Hence, the frequency response is approximately proportional to $V(\omega)$, and $v(t)$ can be regarded approximately as the transducer impulse response.

Note also that $V(\omega)$ was taken as a bandpass function. This is not actually valid because Fourier analysis shows that the finite length of $v(t)$ implies that $V(\omega)$ must have infinite extent in the ω-domain. However, it is shown in Section 6.3 that $V(\omega)$ can be designed such that its amplitude outside a specified band is very small. The above analysis is then valid to a good approximation.

Examples of particular cases

Here we give examples illustrating the relationship between the impulse response $v(t)$ and the transducer geometry. Since $V(\omega)$ is a bandpass function, the waveform $v(t)$ will be oscillatory and it may be written in the form

$$v(t) = \hat{a}(t) \cos \left[\omega_r t + \hat{\theta}(t) \right], \tag{6.19}$$

where $\hat{a}(t)$ is the envelope and ω_r is a reference frequency between ω_1 and ω_2. If the phase $\hat{\theta}(t)$ is a non-linear function of t the waveform is phase modulated. It will also be amplitude modulated if $\hat{a}(t)$ varies with t. The term 'amplitude-modulated waveform' is used here if there is no phase modulation, so that $\hat{\theta}(t)$ is a linear function of t. For this case ω_r may be chosen such that $\hat{\theta}(t)$ is a constant, and ω_r is then the carrier frequency, denoted by ω_c.

A simple example of an amplitude-modulated waveform is a pulse of carrier with a rectangular envelope, as shown in Fig. 6.5a. Here the samples are taken at a sampling frequency $\omega_s = 4\omega_c$, so that there are four samples per period. The weights are $v_n = 0, 1, 0, -1, 0, 1, \ldots$. When these are interpreted as interelectrode gaps $v_n = (u_{n+1} - u_n)/W$, the transducer geometry is seen to be the conventional double-electrode type, which is unapodized. A few zero-valued samples can be added at the ends to minimize end effects. Figure 6.5b shows the same case but with the waveform displaced slightly. In this case the transducer is apodized and has a quite different design, even though the frequency response is essentially the same. The only difference is that there is a small change of delay. Similar observations apply if there are three samples per wavelength, so that $\omega_s = 3\omega_c$. A conventional unapodized transducer (as shown in Fig. 5.11) results if the samples are taken where the phase is a multiple of $\pi/3$. The transducer response is essentially the same whatever value of ω_s is used, provided $\omega_s > 2\omega_2$.

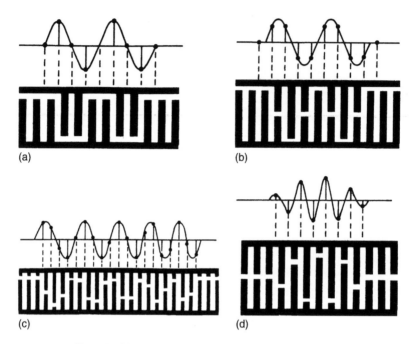

(a)

(b)

(c)

(d)

Figure 6.5 Examples of transducer design using regular sampling.

However, apodized designs are normally avoided since they are more prone to diffraction errors.

Figure 6.5c shows a design obtained by sampling a phase-modulated waveform, taking a chirp waveform as an example. The sampling rate is constant. Because of the phase modulation the resulting design is apodized. For a chirp waveform, it is usual to have non-uniform samples, with locations corresponding to the waveform phase as considered in Chapter 7. A design with constant spacing produces essentially the same frequency response, but its apodization makes it less attractive.

Figure 6.5d shows an amplitude-modulated waveform sampled with two samples per period, so that the sampling frequency is $\omega_s = 2\omega_c$. This gives a single-electrode apodized transducer. If the envelope is flat a uniform single-electrode transducer is produced. Generally, the sampling theory is *invalid* here because the sampling frequency $\omega_s = 2\omega_c$ is less than the Nyquist frequency $2\omega_2$. However, it is shown in Appendix A, Section A.5 that, if v(t) is an amplitude-modulated waveform, the sampling procedure is valid if the particular sampling frequency $\omega_s = 2\omega_c$ is chosen. This gives a single-electrode design. However, it is not usually acceptable because it makes $|V(\omega)|$ symmetrical about ω_c; some asymmetry is usually needed in order to compensate for the element factor and for second-order effects.

Waveform characteristics

Some relationships between a bandpass waveform $v(t)$ and its transform $V(\omega)$ are given in Appendix A, Section A.5. Bandpass filters are often required to have a frequency response whose phase is linear with frequency. This condition is satisfied if $V(\omega)$ has linear phase. This requires the envelope $\hat{a}(t)$ in eq. (6.19) to be a symmetric or antisymmetric function. The phase $\hat{\theta}(t)$ needs to be a constant plus an antisymmetric function; it may be linear with t. Several other cases are considered in Section A.5 of Appendix A.

6.3 DESIGN OF TRANSVERSAL FILTERS

Since an apodized transducer is closely related to a transversal filter, as shown above, established methods for transversal filter design can be applied. In principle the required frequency response can simply be transformed to the time domain and then sampled to obtain the weights. However, this is far from optimal. In particular, it is often necessary to minimize the length while taking account of tolerances which vary with frequency.

6.3.1 Use of window functions

Suppose that the required transducer response is a function $H_0(\omega)$. Since the actual response $H_t(\omega)$ is proportional to $E(\omega)V(\omega)$, it appears that we can divide $H_0(\omega)$ by $E(\omega)$ to obtain $V(\omega)$ and then take the inverse Fourier transform to obtain $v(t)$. This can then be sampled to give the weights for the transducer. However, this is not acceptable in practice because it gives a waveform $v(t)$ of infinite length. A possible solution to this problem is to simply truncate $v(t)$ to a finite length, accepting some loss of accuracy. Although this method is approximately valid, it is not generally accurate enough.

Define a function $V_0(\omega) = H_0(\omega)/E(\omega)$ with inverse transform $v_0(t)$, which will have infinite duration. The problem is to find a function $v(t)$, of finite length, such that its transform $V(\omega)$ is a good approximation to $V_0(\omega)$. The transducer response $H_t(\omega)$ will then be a good approximation to $H_0(\omega)$. The function $v(t)$ is sampled to give the transducer geometry, as explained above. $V_0(\omega)$ is required to be a bandpass function, but is otherwise arbitrary.

The truncation process can be expressed in terms of a *window function* $W(t)$, such that

$$v(t) = W(t)v_0(t). \tag{6.20}$$

This equation truncates $v_0(t)$ if $W(t)$ is unity for some finite time interval, and zero elsewhere. In the frequency domain, the multiplication in eq. (6.20) becomes a convolution of $\overline{W}(t)$, the Fourier transform of $W(t)$, with $V_0(\omega)$, so that

$$V(\omega) = \frac{1}{2\pi} \int_{-\infty}^{\infty} V_0(\omega')\overline{W}(\omega - \omega')d\omega'. \tag{6.21}$$

Equation (6.20) truncates the waveform $v_0(t)$ to a length T if the window is taken as

$$W(t) = \mathrm{rect}(t/T), \tag{6.22}$$

where $\mathrm{rect}(x) = 1$ for $|x| < 1/2$ and is zero for other x. Strictly speaking, we should ensure that $v(t)$ is zero for $t < 0$ in order to satisfy causality. However, this can be satisfied by simply delaying $v(t)$, and this makes no essential difference to the main argument. The Fourier transform of eq. (6.22) is

$$\overline{W}(\omega) = T \, \mathrm{sinc}(\omega T/2), \tag{6.23}$$

where $\mathrm{sinc}(x) \equiv (\sin x)/x$.

As an example, consider the rectangular frequency response $V_0(\omega) = 1$ for $|\omega - \omega_c| = \pi B$, with $V_0(\omega) = 0$ for other ω, so that the bandwidth is $\Delta\omega = 2\pi B$. The inverse transform $v_0(t)$ of this function is an amplitude-modulated waveform with carrier frequency ω_c and envelope proportional to $\mathrm{sinc}(\pi Bt)$, which has infinite length. Figure 6.6a shows $V_0(\omega)$ and Fig. 6.6b shows the transform $\overline{W}(\omega)$ of the window function, eq. (6.23). These are convolved in accordance with eq. (6.21) to give $V(\omega)$, the transform of the finite-length waveform $v(t)$. This is shown in Fig. 6.6c for TB = 6 and in Fig. 6.6d for TB = 16. In comparison with the ideal response $V_0(\omega)$, the actual response $V(\omega)$ exhibits ripples in the passband and sidelobes in the stop bands. These are due to the sidelobes of the function $\overline{W}(\omega)$. At any ω, some of the sidelobes of $\overline{W}(\omega - \omega')$ are in the band occupied by $V(\omega')$ and contribute to the integral of eq. (6.21). As ω changes, sidelobes enter at one side and leave at the other side, giving an oscillatory contribution. In addition, the transitions at the band edges are no longer sharp, due to the finite width of the function $\overline{W}(\omega)$. Although this refers to a particular type of frequency response, these distortions are typical of filters with finite-length impulse responses. The passband ripples, stop-band sidelobes and finite skirt widths are associated with the finite length of the impulse response.

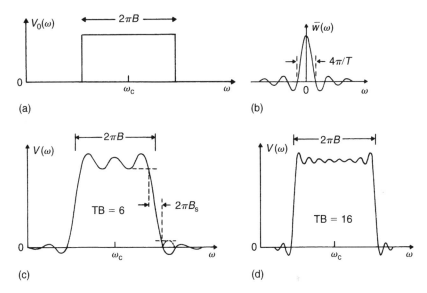

Figure 6.6 Effect of time-domain truncation for a filter with an ideally rectangular frequency response.

The distortion due to truncation can be reduced by increasing the length of the impulse response T. This can be seen in Figs. 6.6c and 6.6d, referring to TB = 6 and TB = 16. A larger value of T reduces the skirt width and the ripple at the band center. However, the ripple near the band edge, and the sidelobes near the band edge, have magnitudes almost independent of T. Thus, simply increasing the length T is not generally an adequate design solution. For large TB, the largest sidelobe is 21 dB below the passband level, and this is usually unacceptable.

The solution to this problem is to choose a more suitable window function $W(t)$. The requirement is for a finite-length function whose Fourier transform has sidelobes substantially smaller than those of the sinc function. This requirement arises in several other fields, for example digital FIR filters [5] and antenna arrays. A suitable window is the Kaiser function [3] defined by

$$W_K(t) = \begin{cases} \dfrac{I_0[\alpha\sqrt{1 - 4t^2/T^2}]}{I_0(\alpha)}, & \text{for } |t| \le T/2 \\ 0 & \text{for } |t| > T/2, \end{cases} \tag{6.24}$$

where $I_0(x)$ is the modified Bessel function of the first kind, with order zero. This can be calculated using a simple power series [6]. The parameter α is chosen to suit the application. Figure 6.7 shows $W_K(t)$ for $\alpha = 6$ and its transform $\overline{W}_K(\omega)$, which is given by an algebraic formula [7]. The figure includes for comparison the function $\text{sinc}(\omega T/2)$, which is the transform of the rectangular

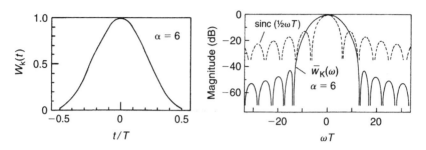

Figure 6.7 Kaiser window function $W_K(t)$ and its Fourier transform.

window function, eq. (6.23). The frequency-domain functions are expressed in decibels. The Kaiser function gives much smaller sidelobes than the sinc function, though its main peak is wider. Thus, when used for filter design the Kaiser window gives smaller ripples and sidelobes but, for a given value of T, the skirt width is larger. The skirt width can be reduced by increasing T.

The parameter α in eq. (6.24) affects the sidelobe levels of $\overline{W}_K(\omega)$ and the width of the main peak. By varying α, these factors can be traded off against each other. For an ideally rectangular response, Tancrell [8] gives some numerical examples. If $\alpha = 6$, the largest stop-band sidelobe is -62 dB below the passband, the ripple at the band edge is 0.009 dB peak to peak and the skirt width is $B_s = 4.0/T$, where B_s is in Hz and is defined as in Fig. 6.6c. Similar design data are given by Rabiner and Gold [5]. A variety of other window functions have been considered [7, 8], though none are appreciably better than the Kaiser function. They include the Hamming and Taylor functions described in Chapter 7.

Having designed $v(t)$ with the aid of a window function, the final stage of the design process is to sample to obtain the weights $v_n = v(n\tau_s)$, as in Section 6.2. A complication arises here because it was assumed previously that $V(\omega)$ is a bandpass waveform. This is not quite true because $v(t)$ has finite length, hence the presence of sidelobes of $V(\omega)$ extending indefinitely on either side of the passband. Thus, some aliasing occurs when $v(t)$ is sampled, so that the image responses generated by the sampling contribute sidelobes in the required passband. However, the distortion caused by this is usually small and acceptable. This is the case if the sidelobes of $V(\omega)$ are small and they decrease in amplitude for frequencies more remote from the passband. This applies for most window functions, including the Kaiser window.

The above discussion has assumed that the ideal response $V_0(\omega)$ is rectangular. However, the method is quite general and $V_0(\omega)$ may be any passband waveform.

This includes dispersive filters, in which the phase $\phi(\omega)$ of $V_0(\omega)$ is non-linear and the group delay $\tau = -d\phi/d\omega$ varies with frequency.

6.3.2 Optimized design: the Remez algorithm

Although the use of window functions described above is an effective design technique, it does not give optimized designs. For example, the relative magnitude of the passband ripple and the stop-band sidelobes are determined by the weighting function, so that they cannot be controlled separately. Thus, optimized design techniques are desirable, taking account of specified tolerances. Moreover, the physical length of the filter is often important, so that design techniques need to satisfy the tolerances using the minimum possible length of transducer. The optimized method described here is not expressed in terms of generating a impulse response $v(t)$ and then sampling it. Hence, the sampling process described above is not involved in this method.

The Remez exchange algorithm [4, 5] is an optimized method widely used for design of digital FIR filters and surface-wave filters. We consider the design of a transversal filter required to have linear phase in the frequency domain, so that its frequency response has the form

$$H_s(\omega) = A(\omega)\exp[j(c - \omega t_0)], \tag{6.25}$$

for $\omega > 0$, where $A(\omega)$ is the amplitude and t_0 is a constant determined by the length of the impulse response. The constant c is either 0 or $\pi/2$. This frequency response is to be designed assuming that the impulse response is a regular sequence of delta functions, as in eq. (6.14) above. The element factor is assumed to have been allowed for, so that the problem is that of transversal filter design. The Remez algorithm designs the time-domain weights v_n such that the amplitude $A(\omega)$ of the frequency response is a good approximation to a required amplitude $A_0(\omega)$. Specifically, the algorithm minimizes an error function E_A defined by

$$E_A = \text{Max}\{e(\omega)|A(\omega) - A_0(\omega)|\}, \quad \text{for } 0 < \omega < \omega_s/2, \tag{6.26}$$

where $e(\omega)$ is an error weighting function chosen arbitrarily by the designer. The frequency range for eq. (6.26) is chosen because the response in this range determines the response at all other ω, as follows from eq. (6.16). It can be shown that there is an optimal solution to this mathematical problem, and that the Remez algorithm is guaranteed to find the optimum after a finite number of iterations. The scheme gives a very high degree of flexibility. The target response $A_0(\omega)$ is quite arbitrary and the passband ripple can be specified independently of the stop-band sidelobes. A convenient way of expressing the requirement is to define two positive functions $A_1(\omega)$ and $A_2(\omega)$, such that the magnitude $|A(\omega)|$ is required to be between $A_1(\omega)$ and $A_2(\omega)$. We can then set $A_0(\omega) = [A_1(\omega) + A_2(\omega)]/2$ and $e(\omega) = |A_1(\omega) - A_2(\omega)|$. The functions $A_1(\omega)$ and $A_2(\omega)$ are usually expressed in decibels. Plots of filter responses often include lines specifying these limits, though they refer to the device response rather than a transducer response considered here. The method is extremely flexible. An example of a filter response with a very complex specification is shown in Fig. 1.12.

The method of the Remez algorithm is explained elsewhere, and a full Fortran subroutine is publicly available [5]. The subroutine uses a set of frequency bands, within each of which the response and the errors are taken to be uniform. The number of bands can be taken to be equal to the number of points used, so that $A_1(\omega)$ and $A_2(\omega)$ are specified at each data point and they may vary from point to point. The method minimizes E_A, producing the best possible solution for a given number of time-domain samples. The resulting error will generally be smaller if the number of samples is increased. If this number is suitable, the response will just meet the specification lines $A_1(\omega)$ and $A_2(\omega)$ at the most critical points, which are usually points where the response has a local maximum or minimum; at other points the response is within the lines. The required length T of the transversal filter depends on several factors, including the passband ripple and the stop-band rejection [1, p. 198]. Typically, $B_s T$ needs to be 3 to 4, where B_s is the skirt width in Hertz.

Figure 6.8 shows the response of a transversal filter designed in this way, including the specification lines. The sampling frequency was 100 MHz and the filter had 180 samples.

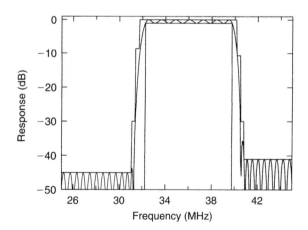

Figure 6.8 Response of a linear-phase transversal filter designed using the Remez algorithm.

On a modern computer, the Remez algorithm operates very rapidly. However, the number of samples is limited to several hundred by the numerical round-off error. Larger numbers are sometimes needed for narrow-band filters. For this, one can design a low-frequency filter, with relatively few electrodes, and then convert up in frequency. This is done by transforming to the frequency domain (using the discrete Fourier transform), shifting the response up in frequency and then transforming back to the time domain. This involves increasing the frequency window, thus reducing the spacing of the time-domain points. The new design has the response the same as the original (e.g., the same bandwidth), except for being shifted up in frequency. The design is not quite optimal, of course.

The above method applies for linear-phase designs, as expressed by eq. (6.25). Sometimes the phase is required to be non-linear. For this case, we can consider an 'in-phase' response, as in eq. (6.25) with $A(\omega)$ replaced by $A(\omega)\cos \phi$ and $c = 0$. A quadrature component is given by eq. (6.25) with $A(\omega)$ replaced by $A(\omega)\sin \phi$ and $c = \pi/2$. These are designed separately [9], typically using the Remez algorithm. Adding these two responses gives $A(\omega) \exp(j\phi) \exp(-j\omega t_0)$, where the phase ϕ may be a non-linear function of ω. The two sets of time-domain samples are simply added.

Another optimal design method is linear programming, in which the weights v_n are treated as variables, subject to constraints expressed as linear functions of the v_n [10]. Another linear function of the v_n is to be minimized. Assuming that this problem is expressed such that it has an optimal solution, a numerical method called the simplex algorithm is guaranteed to find this solution [11].

6.3.3   Withdrawal weighting

For narrow-band filters, there is an alternative to apodization weighting in which sources are removed selectively from the transducer [12]. This enables the transducer to be weighted while maintaining all the sources at the same aperture, with the advantage that the device is less subject to diffraction errors. This is illustrated in Fig. 6.9. The weighting can be achieved by removing some of the electrodes, as shown in the upper part of the figure. However, this complicates the electrostatic fields so that the usual element factor cannot be applied. A better approach is to remove sources by altering the electrode polarities, as shown in Fig. 6.9b. In this case the element factor applies and the analysis is quite straightforward, as in Chapter 5, Section 5.4. Gap sources exist where electrodes of different polarity overlap, as indicated in Fig. 6.8c. For operation in the fundamental passband the sources can be thought of as being smeared

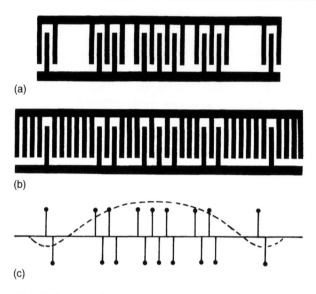

Figure 6.9 (a, b) Two types of withdrawal weighting. (c) Approximate impulse response.

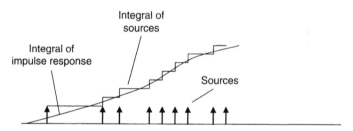

Figure 6.10 Withdrawal weighting design principle.

out. The impulse response is effectively like the broken line in the figure, being larger where the sources are more concentrated.

Figure 6.10 shows a typical design procedure, assuming that the sources are all positive. First the ideal impulse response, a smooth analog function, is calculated by methods used for apodized transducers. The aim is to make the integral of the withdrawal-weighted sources similar to the integral of the analog function, evaluating at a series of time steps. At each step, sources are added or omitted so as to minimize the difference between the two integrals. Since the resulting integrals are similar, the corresponding frequency responses are also similar. The stepped nature of the withdrawal-weighted response causes sidelobes to appear in the stop bands, at some distance from the passband, and hence the technique is best suited to narrow-band filters. The time steps can be taken as periods of the transducer. For example, the transducer might be a double-electrode type, with a positive source represented by the electrode polarities 1, 1, 0, 0 and an absence of source by 0, 0, 0, 0. A negative source would have polarities 0, 0, 1, 1. This gives the correct phase difference at the center frequency, and at other frequencies the phase is generally accurate enough because the bandwidth is small. More sophisticated methods have also been used [13]. One might hope to use standard optimization techniques such as the Remez algorithm, but these are not suitable because they assume that the variables are analog.

6.4 FILTER DESIGN AND PERFORMANCE

Filter types and design

All surface-wave bandpass filters need to have at least two transducers, and the simplest type has two transducers, one apodized and one uniform, as in Fig. 6.2. The response is ideally the product of the two transducer responses, as shown in Chapter 5, Section 5.6. The uniform transducer will be designed to have a bandwidth at least as large as the required filter bandwidth. The apodized transducer then needs to be designed to a specification given by the required filter response divided by the response of the uniform transducer. In addition, the element factors of both transducers must be allowed for, and there are several second-order effects as discussed below. Allowing for these gives a specification for the apodized transducer in terms of the equivalent transversal filter, and this can be designed using one of the methods in Section 6.3.

Many devices use two weighted transducers, giving an extra degree of flexibility. Withdrawal weighting can be used. Both transducers may be withdrawal weighted, or one withdrawal-weighted transducer may be used with an apodized transducer. Withdrawal weighting thus enables a filter to have two weighted transducers, without needing a multistrip coupler (m.s.c.). This is useful for weak-coupling substrates such as quartz, for which an m.s.c. is not feasible.

Alternatively, it is possible to use two apodized transducers. If two apodized transducers are in the same track, the response is not given by the product of transducer responses and there is no straightforward design technique. However, if the tracks are coupled using an m.s.c. the response is basically the product of the transducer responses and the m.s.c. response, as shown in Chapter 5, Section 5.6.4. For $Y–Z$ lithium niobate, the m.s.c also has the function of rejecting unwanted bulk-wave signals. The m.s.c. response varies slowly with frequency and it has little effect on the device response.

Allowing the transducer designs to be different, and assuming apodized transducers, there is a special technique which is useful for minimizing the overall length of the device [14]. We describe a transversal filter using the z-transform, where $z \equiv \exp(-j\omega\tau_s)$. The frequency response of eq. (6.13) can be written as

$$H_s(\omega) = \sum_{n=1}^{N} v_n \exp(-j\omega n\tau_s) = \sum_{n=1}^{N} v_n z^n. \tag{6.27}$$

Here the last expression is a polynomial in z so it is determined, apart from a constant, by its zeros z_n. Hence, we can write

$$H_s(\omega) = v_N \prod_{n=1}^{N} (z - z_n). \tag{6.28}$$

The zeros z_n are found by a numerical search. Zeros in the stop band occur at real ω, so that $|z| = 1$ and z is on the unit circle. It is assumed that a transversal filter has been designed, corresponding to the response of the entire surface-wave filter. The z-transform method enables us to divide this response into two, in such a way that the two new transversal filters have the same total length and the product of their responses is $H_s(\omega)$. The method is simply to allocate the zeros individually to the two transversal filters. The polynomials for each filter are then found by multiplying up terms of the form $(z - z_n)$, and the coefficients of the polynomials give the weights.

It is worth noting that all these methods can be applied for dispersive filters, in which the phase of the frequency response is non-linear. An example is a minimum-phase filter, designed such as to minimize the phase. An established method can be used to modify a given filter response such that its amplitude is unchanged while its phase is minimized, thus minimizing the delay [1, 15].

The starting point of the design process is a specification, typically giving the center frequency, bandwidth, skirt width, passband ripple, stop-band rejection and so on. Since surface-wave filters are very versatile, it is often suitable to present this data in terms of two amplitude limits specified at all frequencies up to half the sampling rate. Lines of this type are illustrated in Fig. 1.12. Similar lines can be used for the delay or phase specifications if the device is dispersive. The designer first needs to tighten this specification to allow for expected temperature shifts and fabrication tolerances. In addition, multiple-transit signals must be allowed for.

Second-order effects

Filter designs often need to compensate for second-order effects such as circuit effects and diffraction. The term 'circuit effect' refers to distortions associated with the finite source and load impedances and, if present, matching circuits. The transadmittance Y_{21} of a two-transducer device is essentially the product of the transducer responses $H_t(\omega)$ (Chapter 5, Section 5.6). However, a filter specification usually refers to the transmission coefficient S_{21}, which includes the circuit effect. S_{21} has a form similar to Y_{21} if the source and load impedances are small, but for practical cases there is usually distortion large enough to require correction. The distortion is particularly severe if the transducer is closely matched. In addition, the finite load impedances give rise to multiple-transit signals (Chapter 5, Section 5.6.1), and these generally give undesirable ripples in the response.

Appendix D gives a method for calculating S_{21} from the Y matrix. Alternatively, Section 5.6 in Chapter 5 describes a method using circuit factors. This excludes the rapidly varying multiple-transit terms, and it can be more convenient for compensating for second-order effects. Diffraction can be calculated by the method of Chapter 4, using the angular spectrum of plane waves or, if valid, the parabolic theory. Electrode resistance (Section 6.1) can also be allowed for.

Compensation of second-order effects is described by several authors [10, 16, 17]. We can represent the actual response of the filter, $H(\omega)$, in terms of an ideal response $H_0(\omega)$ and a multiplicative error function $e_m(\omega)$, so that

$$H(\omega) = e_m(\omega)H_0(\omega). \tag{6.29}$$

The ideal response $H_0(\omega)$ can include effects known initially, such as element factors and the response of an m.s.c. If the device includes an unapodized transducer, the response of this is included in $H_0(\omega)$, and it will stay constant during the error correction. The error $e_m(\omega)$, which may be complex, represents the effect of second-order effects such as the circuit effect and diffraction. Initially, a transversal filter design method, such as the Remez algorithm, is used to obtain a first-order design with response $H_0(\omega)$. The circuit effect and diffraction can be calculated for this design, and the error function $e_m(\omega)$ is deduced. To compensate for errors, the original specification for the filter is divided by $e_m(\omega)$, and the design process is repeated. The second-order effects are then recalculated, giving a new error function $e_m(\omega)$. Alternatively, it is often sufficient simply to modify an apodized transducer directly. Its frequency response is divided by $e_m(\omega)$, transformed to the time domain and truncated to the original length, giving the corrected time-domain weights. These methods do not give accurate error compensation because they assume that $e_m(\omega)$ is the same for the original and the corrected designs. In fact, $e_m(\omega)$ will have changed, but the change is small because the errors are small. Consequently, the process can be repeated a few times, and the errors become progressively smaller so that the response becomes acceptable. The method can include compensation for a non-linear phase error; in fact it can even introduce a non-linear phase, as required for many television IF filters. If an m.s.c. is present, its response does not vary much with frequency and it may be included in either $e_m(\omega)$ or $H_0(\omega)$.

The above method is not generally valid at a frequency where the ideal response $H_0(\omega)$ is zero. If the actual response $H(\omega)$ is finite at this point, then $e_m(\omega)$ defined above will be infinite. Hence, $e_m(\omega)$ is defined only in and near the passband, being set to unity elsewhere. Diffraction, for example, causes a spurious signal at frequencies in the upper stop band. For such a case we can use an alternative method, defining an additive error function $e_a(\omega)$. This is such that $H(\omega) = e_a(\omega) + H_0(\omega)$, where $e_a(\omega)$ can be calculated from the second-order effects. Compensation can then be done by a similar method, using subtraction instead of division. This method copes with diffraction. However, it is a little more complicated because the absolute phase of $e_a(\omega)$ must be known. Both methods, multiplicative or additive, can compensate for errors measured experimentally, as well as those predicted theoretically. Thus, errors of unknown origin can be corrected.

Another second-order effect is the unwanted signals due to bulk-wave excitation. Typically, these signals are at frequencies above the surface-wave passband. Signals due to the shear bulk waves are the most troublesome because their velocity is not very different from that of the surface wave. Simulations can be done using the effective permittivity considered in Chapter 3, using methods described by Milsom [18] and Hashimoto [19]. Generally, there is little that can be done about this other than choosing a suitable substrate and orientation. However, the m.s.c. can be used on strong-coupling substrates to discriminate against bulk waves. This is particularly significant on Y–Z lithium niobate, which has the advantage of being a minimal diffraction orientation. This can give excellent results, at the expense of increased substrate area.

Performance

Surface-wave bandpass are suitable for center frequencies of 20 MHz to 3 GHz, with bandwidths from about 100 kHz to 50% of the center frequency. The non-reflective type of transducer considered in this chapter is capable of excellent performance. A stop-band rejection of 60 dB is achievable. For a filter with a nominally rectangular frequency response it is common to consider the shape factor, defined as the ratio of the bandwidths at the -40 dB and -3 dB points.

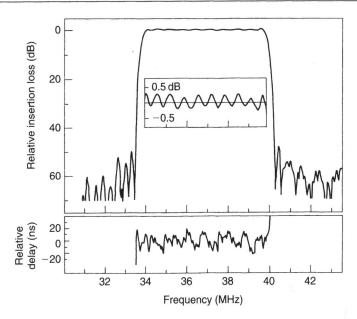

Figure 6.11 Response of a bandpass filter for television broadcast equipment (courtesy of J.M. Deacon).

Shape factors as low as 1.1 are obtainable. In the passband, the amplitude ripple may be as low as 0.1 dB and the phase can be linear to within 1°. To obtain these results it is important to minimize the triple-transit signal, which implies that quite high insertion losses must be tolerated, typically 20 dB or more. Peach [10] demonstrates a 100 MHz filter with bandwidth 1 MHz, skirt width 200 kHz, passband ripple 0.3 dB peak-to-peak and stop-band rejection 50 dB. Using an *ST–X* quartz substrate, the insertion loss is 33 dB. Another example is shown in Fig. 6.11. Here the substrate was *Y–Z* lithium niobate and an m.s.c. was used. The passband ripple was 0.4 dB peak to peak, the shape factor was 1.2 and the insertion loss was 29 dB. Another example is the television IF filter of Fig. 1.12.

Several other types of bandpass filter, using non-reflective transducers, are described elsewhere [1, p. 208]. These include two types of surface-wave filter banks, which provide multiple outputs giving responses with different center frequencies. Table 1.2 in Chapter 1 compares the performance of transversal filters with other surface-wave filters.

REFERENCES

1. D.P. Morgan. *Surface Wave Devices for Signal Processing*, Elsevier, 1991.
2. H.E. Kallmann. 'Transversal filters', *Proc. IRE*, **28**, 302–310 (1940).

3. F.F. Kuo and J.F. Kaiser. *Systems Analysis by Digital Computer*, Wiley, 1966.
4. L.R. Rabiner, J.H. McClellan and T.W. Parks. 'FIR digital filter design techniques using weighted Chebyshev approximation', *Proc. IEEE*, **63**, 595–610 (1975).
5. L.R. Rabiner and B. Gold. *Theory and Application of Digital Signal Processing*, Prentice-Hall, 1975.
6. M. Abramowitz and I. Stegun. *Handbook of Mathematical Functions*, Dover, 1968, eq. (9.6.12).
7. A.J. Slobodnik, T.L. Szabo and K.R. Laker. 'Miniature surface-acoustic-wave filters', *Proc. IEEE*, **67**, 51–83 (1979).
8. R.H. Tancrell. 'Analytic design of surface wave bandpass filters', *IEEE Trans.*, **SU-21**, 12–22 (1974).
9. P.M. Jordan and B. Lewis. 'A tolerance-related optimised synthesis scheme for the design of SAW bandpass filters with arbitrary amplitude and phase characteristics', *IEEE Ultrason. Symp.*, 1978, pp. 715–719.
10. R. Peach. 'The use of linear programming for the design of SAW filters and filterbanks', *IEEE Trans. Ultrason. Ferroelec. Freq. Contr.*, **41**, 532–541 (1994).
11. W.H. Press, B.P. Flannery, S.A. Teukolsky and W.T. Vetterling. *Numerical Recipes (Fortran version)*, Cambridge University Press, 1989, p. 312.
12. C.S. Hartmann. 'Weighting interdigital transducers by selective withdrawal of electrodes', *IEEE Ultrason. Symp.*, 1973, pp. 423–426.
13. E. Bausk, E. Kolosovsky, A. Kozlov and L. Solie. 'Optimization of broadband uniform beam profile interdigital transducers weighted by assignment of electrode polarities', *IEEE Trans. Ultrason. Ferroelec. Freq. Contr.*, **49**, 1–10 (2002).
14. C. Ruppel, E. Ehrmann-Falkenau, H.R. Stocker and W. Mader. 'A design for SAW filters with multistrip couplers', *IEEE Ultrason. Symp.*, **1**, 13–17 (1984).
15. M. Feldman and J. Henaff. 'Design of SAW filter with minimum phase response', *IEEE Ultrason. Symp.*, 1978, pp. 720–723.
16. F.Z. Bi and K. Hansen. 'Wideband diffraction compensation methods and the angular spectrum of waves (ASoW) model for TV SAW filters', *IEEE Trans. Ultrason. Ferroelec. Freq. Contr.*, **44**, 925–934 (1997).
17. G. Visintini, C. Kappacher and C.C.W. Ruppel. 'Modular two-dimensional analysis of SAW filters – Part II: Analysis and compensation results', *IEEE Trans. Ultrason. Ferroelec. Freq. Contr.*, **39**, 73–81 (1992).
18. R.F. Milsom, N.H.C. Reilly and M. Redwood. 'Analysis of generation and detection of surface and bulk acoustic waves by interdigital transducers', *IEEE Trans.*, **SU-24**, 147–166 (1977).
19. K.-Y. Hashimoto. *Surface Acoustic Wave Devices – Modelling and Simulation*, Springer, 2000, p. 107.

7

CORRELATORS FOR PULSE COMPRESSION RADAR AND COMMUNICATIONS

Pulse compression radar was one of the first applications envisaged for surface-wave devices. The function of these devices is to process a frequency-swept, or 'chirp', signal in such a way as to maximize its signal-to-noise ratio. This process is called 'correlation', and the device is known as a *correlator* or as a *matched filter*. A matched filter for a chirp waveform is essentially a type of dispersive delay line, called a *chirp filter*. The process enables the range of a radar system to be increased, for a given peak power limitation. Such devices have been widely used in radar systems. They can be implemented using interdigital transducers with varying electrode pitch, or using reflective arrays in the reflective array compressor (RAC). Some basic theory for matched filtering is given in Appendix A, Section A.3.

Correlation is also used in 'spread-spectrum' communication systems, that is, systems in which the signal bandwidth is made substantially larger than the information bandwidth. In the 'direct-sequence' spread-spectrum systems considered here, each bit of data is represented by a stream of sub-bits called 'chips'. The signal bandwidth is then approximately equal to the chip rate, which is larger than the rate of the data bits. In the communications receiver, a filter is used to strip off the chip encoding, yielding the original data. This is basically another form of correlation. Again, interdigital devices can provide this process, taking the form of tapped delay lines. Alternatively, much attention has been given to non-linear devices called convolvers, which mix the signal with a locally generated reference signal.

In this chapter, the principles of correlation in radar and communication systems are presented, and several surface-wave devices are described. These topics are described in more detail in Chapters 9 and 10 of the earlier book [1].

183

Here, Sections 7.1 to 7.5 are concerned with chirp waveforms and pulse compression. Section 7.1 describes radar principles and Section 7.2 covers chirp waveforms and correlation, emphasizing the use of the stationary-phase approximation. Section 7.3 describes interdigital chirp transducers and filters, and RACs are described in Section 7.4. Section 7.5 considers the Doppler effect and spectral analysis, topics that apply to both types of chirp filter. Section 7.6 describes spread-spectrum communications, including linear matched filters in Section 7.6.2 and non-linear convolvers in Section 7.6.3.

7.1 PULSE COMPRESSION RADAR

Radar systems using pulse compression were first envisaged during the 1940s, and a famous article by Klauder *et al.* [2] explained the principles. The technique enables the range capability of a radar system to be substantially increased, without increasing the peak transmitted power. Several surface-wave devices have been developed for this purpose.

Figure 7.1 shows a schematic of a conventional radar system compared with a pulse compression system. A conventional radar emits a short radio pulse which is reflected off a target and returns to the radar. In the receiver, a bandpass filter rejects frequency components outside the spectrum of the pulse, thus rejecting unwanted noise and interference. The range of the target is given by the delay of the received pulse, and the maximum possible range is determined by the signal-to-noise power ratio (SNR) of this pulse. In turn, this is determined by the target reflectivity and the power transmitted by the radar.

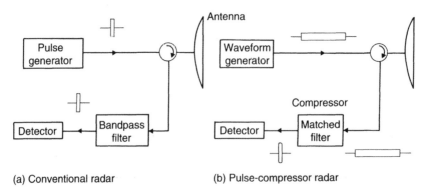

(a) Conventional radar   (b) Pulse-compressor radar

Figure 7.1 Comparison between conventional and pulse compression radar systems. Output waveforms are shown as they appear for a point target.

More accurately, it can be shown that the *energy* of the transmitted pulse is more significant than its power. The energy equals the power multiplied by the duration, so either can be increased to improve the SNR. Increasing the pulse power is often undesirable because it implies radical changes to the transmitter. Increasing the duration causes a reduction of the pulse bandwidth, so that the filter in the receiver can have smaller bandwidth. This reduces the output noise and so increases the SNR. However, this also reduces the radar resolution, that is, its ability to distinguish targets that are close together.

The pulse compression system enables the pulse length to be increased without reducing its bandwidth [2, 3]. This gives an increased range capability without compromising the radar resolution. As shown in Fig. 7.1b, this system generates a long pulse which is coded in some way. In the receiver there is a 'matched filter', a linear device which optimizes the SNR. This process, called correlation, gives a short output signal, such that the resolution of the radar has been maintained. The matched filter needs to have an impulse response with the same form as the transmitted waveform, but reversed in time.

The pulse used by a conventional radar is a simple rectangular waveform with bandwidth B and length T, and these are approximately related by $TB \approx 1$. In a pulse compression radar the duration T is increased without reducing B, giving a large time-bandwidth product so that $TB \gg 1$. This implies that some form of modulation is needed. Most commonly a *chirp* waveform is used, in which the frequency varies with time. This is illustrated in the left part of Fig. 7.2. The waveform $s(t)$ at the left is the received waveform reflected from a point target, a delayed and attenuated version of the transmitted signal. The matched filter in the receiver has an impulse response $h(t)$ which is the same as $s(t)$ except for the time reversal, so that its frequency sweeps in the opposite direction. The output waveform $g(t)$ typically has a narrow peak called the

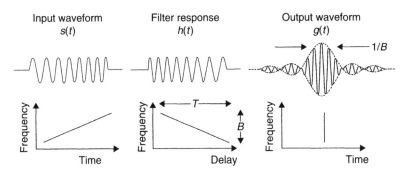

Figure 7.2 Pulse compression using linear-chirp waveforms.

correlation peak, with small time-sidelobes on either side. The width of the correlation peak is approximately $1/B$, where B is the bandwidth of the input waveform, approximately equal to the frequency sweep. Because the output pulse has length much less than the length T of the input waveform, the process is called pulse compression. The ratio of pulse lengths, called the *compression ratio*, is approximately TB. If there are two targets, they can be resolved if the delay difference exceeds the output pulse width $1/B$, so this gives the radar resolution.

The lower part of Fig. 7.2 shows a simplified interpretation. As shown, the input waveform has frequency increasing with time. The matched filter gives a delay decreasing with frequency. Thus, all frequency components arrive at the output at the same time, giving a large narrow correlation peak.

The advantage of pulse compression can be quantified by considering the SNR increase. From the analysis of Appendix A, Section A.3, the SNR at the filter output is $\text{SNR}_o = 2E_s/N_i$. Here E_s is the signal energy and N_i is the noise spectral power density, at the filter input. For a radar system the transmitted pulse has a flat envelope and its energy can be written as $E_s = P_s T$, where P_s is the average power during the pulse. Hence the SNR at the filter output is

$$\text{SNR}_o = 2P_s T/N_i. \tag{7.1}$$

This is valid for any waveform with a flat envelope, applied to its matched filter. Now consider a pulse compression radar compared with a conventional radar, assuming that both systems use an appropriate matched filter (for the conventional case, the matched filter gives an SNR similar to that given by a bandpass filter). If the pulse lengths are T for the pulse compression system, and T_c for the conventional system, the output SNRs will be in the ratio T/T_c. This assumes that P_s and N_i are the same for both. If these systems have the same bandwidth B, the systems have the same resolution and we have $B \approx 1/T_c$. The ratio of SNRs is

$$\frac{\text{SNR}_o \text{ (chirp)}}{\text{SNR}_o \text{ (conventional)}} \approx TB. \tag{7.2}$$

This ratio is called the processing gain. The use of pulse compression is equivalent to using a conventional system with the transmitted power increased by this ratio. Surface-wave devices typically have TB products of 50–500, so the improvement is very substantial. In practice, the filter design is often not quite that of a matched filter, and the processing gain is a little less than TB.

For radar systems a frequency-swept, or chirp, waveform is normally used, as explained above. Often, the frequency is a linear function of time and the waveform is called a *linear chirp*. However, non-linear chirps are also used. Amplitude weighting is often used to reduce the time-sidelobes seen in Fig. 7.2. In addition to matched filtering, a surface-wave chirp filter can also be used to generate a chirp waveform, as needed in the radar transmitter. For this purpose a short unmodulated pulse is applied to the filter, a process called 'passive generation'. However, active methods, using circuitry, are more often used for this purpose. The surface-wave devices are commonly interdigital types, described in Section 7.3, but reflective array types, Section 7.4, can also be used.

Matched filtering of coded waveforms is also done in 'spread-spectrum' communication systems, described in Section 7.6.

7.2 CHIRP WAVEFORMS

This section describes properties of chirp waveforms and filters, irrespective of the technology used to implement them. Most results are obtained using the stationary-phase approximation, which gives substantial simplification. Section 7.2.1 describes waveform characteristics and Section 7.2.2 describes weighting in order to reduce the time-sidelobes of the compressed pulse.

7.2.1 Waveform characteristics

A chirp waveform $v(t)$, representing either the transmitted waveform or the impulse response of a chirp filter, will be written in the form

$$v(t) = a(t) \cos[\theta(t)] \tag{7.3}$$

where $\theta(t)$ is the time-domain phase, a non-linear function of t. The envelope $a(t)$ is zero outside a time interval of length T, the duration of the waveform. Clearly, $\theta(t)$ must be a monotonic function of time. We define $\theta(t)$ such that it *increases* with time. This does not lose generality because a sign change in $\theta(t)$ does not affect $v(t)$. The *instantaneous frequency* $\Omega_i(t)$ is defined as the time differential of $\theta(t)$, so that

$$\Omega_i(t) \equiv \dot{\theta}(t). \tag{7.4}$$

Thus, $\Omega_i(t)$ will always be positive. It will be shown later that $\Omega_i(t)$ can be regarded approximately as the 'frequency' at time t. We also define the chirp rate $\mu(t)$ as the rate of change of $\Omega_i(t)$. It is convenient to express this in units of Hz/s, so that

$$2\pi\mu(t) = \dot{\Omega}_i(t) = \ddot{\theta}(t). \tag{7.5}$$

For the waveforms considered in this chapter, $\Omega_i(t)$ is a monotonic function, either increasing or decreasing with time. Thus, $\mu(t)$ is either positive throughout or negative throughout. Waveforms with $\mu(t) > 0$ are described as *up-chirp*, while those with $\mu(t) < 0$ are described as *down-chirp*. For a linear-chirp waveform $\mu(t)$ is a constant, so $\Omega_i(t)$ is a linear function of t and $\theta(t)$ is quadratic.

If $v(t)$ in eq. (7.3) is the transmitted radar waveform, the required impulse response of the matched filter is $h(t) = v(-t)$, apart from a constant multiplier and constant delay. The phase of $h(t)$ is taken as $-\theta(-t)$ rather than $\theta(-t)$, because it is taken to be increasing with time. The phase can also have a constant added with negligible effect.

Stationary-phase approximation

This approximation [3] gives a very convenient way of estimating the spectrum of a chirp waveform. We consider the chirp waveform of eq. (7.3), in the complex form

$$v(t) = \frac{1}{2}a(t)\{\exp[j\theta(t)] + \exp[-j\theta(t)]\}. \tag{7.6}$$

It can be assumed that $v(t)$ is a bandpass waveform, so that its spectrum is negligible for frequencies near zero. Since $\theta(t)$ increases with time, the Fourier transform of the first term in eq. (7.6) gives the positive-frequency part of the spectrum, denoted by $V_+(\omega)$. We thus have

$$V_+(\omega) = \frac{1}{2}\int_{-\infty}^{\infty} a(t)\exp[j\theta(t) - j\omega t]dt \quad \text{for } \omega > 0. \tag{7.7}$$

The negative-frequency part of the spectrum follows from this because $v(t)$ is real (Appendix A, Section A.1). Since $\theta(t)$ is non-linear, the phase $\theta(t) - \omega t$ in the above equation generally varies rapidly with t, so that the exponential gives rapid oscillations which contribute little to the integral. However, the phase varies slowly for times close to the point where its differential is zero, that is, where $\dot{\theta}(t) = \omega$. The main contribution to the integral therefore arises from times near this point. The time satisfying $\dot{\theta}(t) = \omega$ is called the *stationary-phase point*, denoted by $T_s(\omega)$, so that

$$\dot{\theta}[T_s(\omega)] = \omega. \tag{7.8}$$

For any given ω, this equation has only one solution for $T_s(\omega)$ because we assume that $\dot{\theta}(t)$ is a monotonic function.

Since the main contribution to the integral in eq. (7.7) arises from times near $T_s(\omega)$, it is a good approximation to replace $\theta(t)$ by its Taylor expansion about this point. Terms with cubic and higher orders are neglected. It is also assumed that the envelope $a(t)$ varies slowly with t, and may therefore be approximated by its value $a[T_s(\omega)]$ at the stationary-phase point. Using also eq. (7.8), we find

$$V_+(\omega) \approx \frac{1}{2} a(T_s) \exp[-j\omega T_s + j\theta(T_s)] \int_{-\infty}^{\infty} \exp\left[\frac{1}{2} j(t - T_s)^2 \ddot{\theta}(T_s)\right] dt \quad (7.9)$$

where the frequency argument of $T_s(\omega)$ is omitted for brevity. Here the integral is a form of the standard integral [4]:

$$\int_{-\infty}^{\infty} \exp(jKt^2) dt = \sqrt{\frac{\pi}{|K|}} \exp(\pm j\pi/4) \quad (7.10)$$

where $K \neq 0$ is a real constant and the sign in the exponential at the right is the same as the sign of K. We define $A(\omega)$ and $\phi(\omega)$ as the amplitude and phase of the spectrum $V_+(\omega)$, so that

$$V_+(\omega) = A(\omega) \exp[j\phi(\omega)]. \quad (7.11)$$

Using eq. (7.10), eq. (7.9) gives

$$A(\omega) \approx \frac{a(T_s)}{2} \sqrt{\frac{2\pi}{|\ddot{\theta}(T_s)|}} = \frac{1}{2} \frac{a(T_s)}{\sqrt{|\mu(T_s)|}} \quad (7.12a)$$

and

$$\phi(\omega) \approx \theta(T_s) - \omega T_s \pm \pi/4. \quad (7.12b)$$

These results assume that $\ddot{\theta}(T_s) \neq 0$, which is valid here because $\dot{\theta}(t)$ is monotonic. The sign in eq. (7.12b) is the same as the sign of $\mu(T_s)$. Note that $a(t)$ is zero outside an interval of length T, so $A(\omega)$ is zero outside a corresponding frequency band. If $v(t)$ is the impulse response of a filter the group delay is, from eqs (7.12b) and (7.8),

$$\tau_g(\omega) \equiv -d\phi(\omega)/d\omega \approx T_s(\omega). \quad (7.13)$$

The stationary-phase approximation shows that, for frequency ω, the spectrum is determined mainly by the part of the waveform near the stationary-phase point $T_s(\omega)$. Conversely, eq. (7.12) show that the waveform at time t is determined mainly by the part of the spectrum near the frequency $\omega = \dot{\theta}(t)$, which equals $\Omega_i(t)$ by definition. Thus, the instantaneous frequency $\Omega_i(t)$ can be regarded as the time-domain 'frequency' of the waveform. This gives formal meaning to the statement that the frequency varies with time. This interpretation is valid only if the stationary-phase approximation is valid, that is, for large TB. In general the term 'frequency' cannot be applied to a time-domain waveform, except in a special sense such as carrier frequency.

Linear-chirp waveforms

For a linear chirp the instantaneous frequency varies linearly with time, so that the chirp rate $\mu(t)$ is a constant. The phase $\theta(t)$ is quadratic, and the waveform can be written as

$$v(t) = a(t)\cos[\theta(t)] = a(t)\cos(\omega_c t + \pi\mu t^2 + \phi_0) \qquad (7.14)$$

where ϕ_0 is an arbitrary constant. The waveform is taken to be centered at $t = 0$, so that $a(t) = 0$ for $|t| > T/2$, where T is the length. Thus, ω_c is the center frequency. If the envelope is flat, as in Fig. 7.2, then $a(t)$ will be constant for $|t| < T/2$. The instantaneous frequency is

$$\Omega_i(t) = \omega_c + 2\pi\mu t$$

and the stationary-phase point is, from eq. (7.8),

$$T_s(\omega) = (\omega - \omega_c)/(2\pi\mu). \qquad (7.15)$$

The number of cycles in the waveform is $[\theta(T/2) - \theta(-T/2)]/(2\pi) = \omega_c T/(2\pi)$, which is independent of the chirp rate μ.

In the stationary-phase approximation the spectrum of the waveform is, from eq. (7.12),

$$A(\omega) \approx \frac{a[T_s(\omega)]}{2\sqrt{|\mu|}} \qquad (7.16a)$$

with

$$\phi(\omega) \approx -\frac{(\omega - \omega_c)^2}{4\pi\mu} + \phi_0 \pm \pi/4 \qquad (7.16b)$$

where the sign is the same as the sign of μ. Since $T_s(\omega)$ is a linear function of ω, the spectral amplitude $A(\omega)$ has the same shape as the time-domain envelope $a(t)$. However, this is not true for a non-linear chirp waveform.

Since $a(t)$ falls to zero at the ends of the waveform, where $t = \pm T/2$, the amplitude $A(\omega)$ falls to zero at the corresponding frequencies $\dot{\theta}(\pm T/2) = \omega_c \pm \pi\mu T$. The bandwidth B, in Hz, is normally taken as the range of these frequencies, so that

$$B = |\mu|T \qquad (7.17)$$

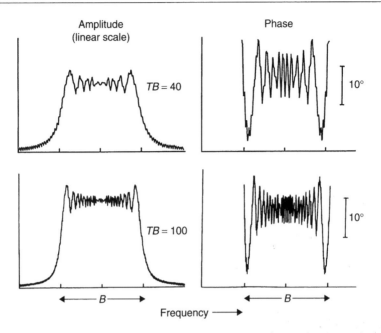

Figure 7.3 Spectra of flat-envelope linear-chirp waveforms.

and the band edges are at $\omega = \omega_c \pm \pi B$. Note that the 3 dB width of the spectrum can be quite different from this because it depends on the form of the envelope $a(t)$.

Linear chirp with flat envelope: correlation

In a radar system the transmitted signal has a flat envelope, as illustrated in Fig. 7.2. Here $a(t)$ is a constant for $|t| < T/2$ and zero elsewhere. According to the stationary-phase approximation, the spectral amplitude $A(\omega)$ is flat between the points $\omega = \omega_c \pm \pi B$, and zero elsewhere.

Some accurately calculated spectra of linear-chirp waveforms are shown in Fig. 7.3, for $TB = 40$ and 100. For the phase curves, the quadratic term predicted by eq. (7.16b) was subtracted. In contrast to the stationary-phase results of eqs (7.16), there are substantial ripples in both amplitude and phase. However, the ripples become less severe as the TB product is increased. For most purposes the stationary-phase approximation is adequate if $TB \geq 100$. It can also be seen that B is approximately the 6 dB bandwidth, for a linear chirp.

Suppose now that a flat-envelope linear-chirp waveform is applied to its matched filter, as in Fig. 7.2. The input waveform $s(t)$ is equal to $v(t)$, eq. (7.14), with

| Input waveform
$s(t)$ | Filter response
$h(t)$ | Output waveform
$g(t)$ |

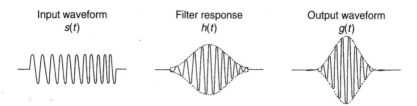

Figure 7.4 Pulse compression using linear chirps with amplitude weighting.

$a(t) = 1$ for $|t| < T/2$. Taking the filter impulse response to be the time reverse of this, the spectrum of the output waveform is $G(\omega) = |S(\omega)|^2$, where $S(\omega)$ is the spectrum of $s(t)$. This follows from eq. (A.54) of Appendix A. For positive frequencies $|S(\omega)|$ is equal to $A(\omega)$, which is given by eq. (7.16a) in the stationary-phase approximation. Hence $G(\omega)$ is a rectangular function. Using eq. (A.26) to transform to the time domain, the output waveform is found to be

$$g(t) \approx \frac{1}{2} T \cos(\omega_c t) \, \mathrm{sinc}(\pi B t) \qquad (7.18)$$

where $\mathrm{sinc}(x) \equiv (\sin x)/x$. This is illustrated in Fig. 7.2. The envelope has its maximum at $t = 0$, and the zeros nearest to this are at $t = \pm 1/B$.

A precise analysis [3, p. 133] shows that eq. (7.18) is a good approximation for times close to the main peak. The main difference is that the precise expression has finite length $2T$.

7.2.2 Weighting of linear-chirp filters

When a flat-envelope linear-chirp waveform is applied to its matched filter, the output waveform has an envelope of the form $\mathrm{sinc}(\pi B t)$ as shown by eq. (7.18). This is usually unacceptable because of the presence of the sidelobes on either side of the main peak. The largest sidelobes are only 13 dB below the peak, and a radar system could falsely interpret them as additional targets.

A common method of reducing the sidelobes is to apply amplitude weighting in the filter [2, 3], as shown in Fig. 7.4. The filter impulse response is weighted to give a maximum amplitude at the center, reducing toward the ends. In a radar system, this reduces the output SNR slightly because the filter is not exactly matched to the signal. There is also a small increase of the output pulse width, though this can be compensated by increasing the bandwidth B.

The input waveform $s(t)$, applied to the filter, has a flat envelope and can be written as

$$s(t) = \cos(\omega_c t - \pi \mu t^2 - \phi_0) \quad \text{for } |t| < T/2 \qquad (7.19)$$

where $s(t) = 0$ for other t, and ϕ_0 is a constant. The problem is to design a receiver filter which is close to being matched to $s(t)$, but also gives an output waveform with well-suppressed time-sidelobes. The filter impulse response $h(t)$ is taken to be

$$h(t) = a(t)s(-t) = a(t)\cos(\omega_c t + \pi\mu t^2 + \phi_0) \quad \text{for } |t| < T/2 \qquad (7.20)$$

with $h(t) = 0$ for other t. The waveform $h(t)$ would be matched to $s(t)$ if $a(t) = 1$, but here the amplitude is allowed to vary. The analysis remains valid if ϕ_0 is replaced by a different constant in eq. (7.20). The Fourier transforms of $s(t)$ and $h(t)$ are denoted by $S(\omega)$ and $H(\omega)$, respectively, and the spectrum of the output waveform is $G(\omega) = S(\omega)H(\omega)$. The waveforms $s(t)$ and $h(t)$ have the form of eq. (7.14), so their transforms follow from eq. (7.16) in the stationary-phase approximation. The output spectrum, for $\omega > 0$, is thus found to be the real function

$$G(\omega) \approx \frac{a[T_s(\omega)]}{4|\mu|} \quad \text{for } |\omega - \omega_c| \leq \pi B \qquad (7.21)$$

with $G(\omega) = 0$ for other ω. Here $T_s(\omega) = (\omega - \omega_c)/(2\pi\mu)$ is the stationary-phase point of the filter.

This result can be expressed in terms of a real frequency-domain *weighting function* $\overline{W}(\omega)$, which is considered to be centered at zero frequency and to be zero for $|\omega| > \pi B$. The output spectrum is proportional to $\overline{W}(\omega - \omega_c)$ for positive ω, so that

$$G(\omega) = K\overline{W}(\omega - \omega_c) \quad \text{for } \omega > 0 \qquad (7.22)$$

where K is an arbitrary real constant. If $W(t)$ is the inverse Fourier transform of $\overline{W}(\omega)$, the output waveform is given by the shifting theorem, eq. (A.6), so that

$$g(t) = 2KW(t)\cos(\omega_c t). \qquad (7.23)$$

The envelope of the output waveform is thus proportional to $W(t)$. We also have, from eqs (7.21) and (7.22),

$$a(t) = 4K|\mu|\overline{W}(2\pi\mu t). \qquad (7.24)$$

Thus the time-domain weighting function $a(t)$ is, with a change of scale, proportional to the frequency-domain weighting function. Both sides of eq. (7.24) are zero for $|t| > T/2$. In effect, the system has performed the Fourier transform of the function $a(t)$, a topic considered further in Section 7.5.

Types of weighting function

To design the filter impulse response, we need to find a weighting function $\overline{W}(\omega)$ which is zero for $\omega > \pi B$, and is such that its transform $W(t)$ has well-suppressed time-sidelobes. A similar problem was encountered in the design of bandpass filters in Chapter 6, Section 6.3.1. These problems are mathematically the same, though the bandpass filter function has finite length in the time domain, rather than the frequency domain. Functions suitable for bandpass filters can therefore be applied to pulse compression as well.

For radar applications it is usual to use either Hamming or Taylor weighting functions [1–3, 5]. The Hamming function $\overline{W}_H(\omega)$ is defined by

$$\overline{W}_H(\omega) = 0.54 + 0.46\cos(\omega/B) \quad \text{for } |\omega| < \pi B \qquad (7.25)$$

and $\overline{W}_H(\omega) = 0$ for other ω. This function, with its transform $W_H(t)$, is shown in Fig. 7.5. The output waveform has an envelope $W_H(t)$ with a series of sidelobes, and the largest of these is the fourth one, at $Bt = 4.5$. This sidelobe is 42.8 dB below the main peak. The width of the main peak is about 50% larger than in the unweighted case, for the same B. This weighting is also illustrated in Fig. 7.4, and it can be seen that the sidelobes are well suppressed in comparison with the unweighted case of Fig. 7.2.

The Taylor weighting function $\overline{W}_T(\omega)$ enables the time-sidelobe level to be set by the designer [2, 3, 5]. It is given by

$$\overline{W}_T(\omega) = 1 + 2\sum_{m=1}^{\tilde{n}-1} F_m\cos(m\omega/B) \quad \text{for } |\omega| < \pi B \qquad (7.26)$$

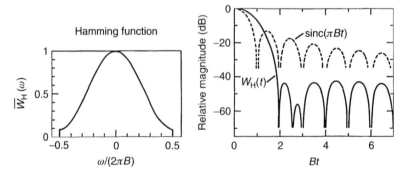

Figure 7.5 Hamming weighting function (left) and its transform. The transform is symmetric about $t = 0$.

with $\overline{W}_T(\omega) = 0$ for other ω. For this function, the largest sidelobe is the one adjacent to the main peak. The coefficients F_m are determined by a formula dependent on $\tilde{n}$ and the required sidelobe level. For a given sidelobe level, increasing $\tilde{n}$ reduces the width of the main peak. For large $\tilde{n}$ the peak width approaches that given by the Dolph–Chebyshev function, which gives the minimum possible peak width for a given sidelobe level. Typically, $\tilde{n} = 5$ for 30 dB sidelobes, or $\tilde{n} = 10$ for 50 dB sidelobes. Generally, $\overline{W}_T(\omega)$ has an appearance similar to $\overline{W}_H(\omega)$, shown in Fig. 7.5.

As well as reducing the sidelobe levels, both of these types of weighting cause a reduction of the processing gain known as the 'mismatch loss'. This arises because the filter is not exactly matched to its input signal. It can be evaluated by calculating the output SNR, using eqs (A.43–A.45) in Appendix A. Hamming weighting gives a mismatch loss of 1.4 dB. For Taylor weighting the mismatch loss is typically 0.7–1.5 dB for sidelobe levels of 30–50 dB below the main peak. Both types also increase the width of the main peak, as compared with an ideal matched filter whose output waveform is given by eq. (7.18).

Limitations of the stationary-phase approach: reciprocal ripple design

In the above description the stationary-phase approach was used in the design of the filter and in the analysis of the output waveform. This approach is usually acceptable if the time-bandwidth product, TB, is 100 or more. The filter output waveform has an envelope proportional to $W(t)$, as shown by eq. (7.24).

Comparison with accurate calculations shows that, for $TB > 100$, the stationary-phase method is valid for times close to the main peak [3]. However, it fails to predict some raised sidelobe levels which occur at times $t \approx \pm T/2$ relative to the peak. These 'gating sidelobes' arise because of the abrupt truncation of $s(t)$ and $h(t)$ at the ends. They are usually reduced by adding short extensions at each end of the filter response $h(t)$, such that it falls to zero smoothly instead of abruptly. When this is done, the gating sidelobes are typically $20 \log(TB) + 3$ dB below the main peak.

For small TB, designs produced using the stationary-phase method give raised sidelobes which may be unacceptable. For this situation, a revised method called the reciprocal ripple method can be used [6, 7]. Since the output spectrum $G(\omega) = S(\omega)H(\omega)$ is required to be proportional to the weighting function $\overline{W}(\omega - \omega_c)$, the filter response can be taken as $H(\omega) = \overline{W}(\omega - \omega_c)/S(\omega)$. This gives a design similar to the stationary-phase method if TB is large. However, it also cancels the ripples in $S(\omega)$, which can be seen in Fig. 7.3.

This enables low time-sidelobes to be obtained when TB is small. For example [6], a sidelobe rejection of 34 dB was obtained with $TB = 8$. However, the filter is not exactly matched to $s(t)$, and the method increases the system sensitivity to Doppler shifts. Consequently, the reciprocal ripple method is not usually applied if $TB > 100$.

Weighting using non-linear chirps

As noted above, the weighting introduced to reduce the output sidelobes causes some mismatch loss, so that the output SNR of the filter is less than that of an ideal matched filter. In principle, both $s(t)$ and $h(t)$ could have amplitude weighting and then the mismatch loss could be avoided. However this is not possible in practical radar systems, which need the signal $s(t)$ to have a flat envelope.

This problem can be solved by allowing the chirps to be non-linear, instead of the linear chirps considered above [3, 8]. In the stationary-phase approximation, eq. (7.12a) shows that the spectral amplitude $A(\omega)$ varies with ω if the chirp rate $\mu(t)$ varies, even if the time-domain envelope $a(t)$ is flat. To implement this, $s(t)$ is designed to have a flat envelope and to have a stationary-phase point $T_s(\omega)$ given by [1, 8]

$$\mathrm{d}T_s(\omega)/\mathrm{d}\omega = \pm C\overline{W}(\omega - \omega_c) \tag{7.27}$$

where C is a constant. Integrating this gives the waveform $s(t)$. The filter is matched to this so that $h(t) = s(-t)$ apart from constants, and both $h(t)$ and $s(t)$ have flat envelopes. The filter output waveform then has its spectrum proportional to $\overline{W}(\omega - \omega_c)$, as before. The method is valid for any weighting function, including the Hamming or Taylor types. Compared with linear-chirp designs, this method reduces the mismatch loss but it also increases the errors arising when there is a Doppler shift due to target motion.

7.3   INTERDIGITAL CHIRP TRANSDUCERS AND FILTERS

We now consider surface-wave interdigital chirp filters, which implement the chirp filters considered above. Transducer analysis and design are described in Sections 7.3.1 and 7.3.2, and Section 7.3.3 considers device design and performance. The devices considered here are all linear filters in the sense described in Appendix A, Section A.2, which gives some fundamental relationships.

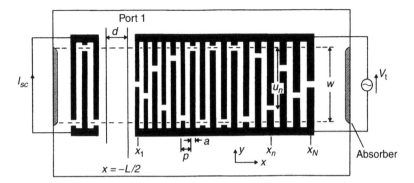

Figure 7.6 Single-dispersive interdigital chirp filter. The transducers shown are the single-electrode type ($S_e = 2$).

7.3.1 Chirp transducer analysis

A typical chirp filter is shown in Fig. 7.6. This shows a single-dispersive device, having an apodized chirp transducer and a uniform transducer. The apodized transducer has its high-frequency electrodes closest to the other transducer, so the impulse response is a down-chirp. For clarity, the figure shows single-electrode transducers, though in practice double-electrode transducers are normally used to avoid the complications of electrode reflections. Hence, the analysis is done here assuming electrode reflections to be negligible. As for bandpass filters, apodized transducers normally use dummy electrodes to improve the uniformity of the surface-wave velocity and of the electrostatic fields (Chapter 6, Section 6.2). For the basic analysis, second-order effects such as propagation loss, diffraction and waveguiding are ignored.

The analysis is based on the quasi-static method used in Chapter 5. With electrode reflections ignored, the device response is given by the product of the two-transducer responses. This is shown in Chapter 1, eq. (1.7), or more rigorously in Chapter 5, eq. (5.96). The device response is expressed as the transadmittance Y_{21}, written as

$$Y_{21} \equiv I_{sc}/V_t = H_{t1}(\omega)H_{t2}(\omega) \exp(-jkd) \tag{7.28}$$

where $H_{t1}(\omega)$ and $H_{t2}(\omega)$ are the responses of the two transducers, as in Chapter 5, Section 5.6.2. I_{sc} is the current produced by one transducer, assumed to be shorted, when a voltage V_t is applied to the other one. It is assumed that one transducer is unapodized and that its active width covers all active sources in the other transducer. For each transducer, port 1 is a reference point near the end facing the other transducer. The distance d is the distance between the transducers, referred to these ports.

The response $H_t(\omega)$ of a transducer is defined in terms of the amplitude of the wave emerging at port 1 when a voltage is applied. For an apodized transducer, this relates to the response of a two-transducer device in which the other transducer is unapodized. From Section 5.6.2 in Chapter 5, the transducer response is

$$H_t(\omega) = -j\langle\phi_s\rangle\sqrt{\omega W/\Gamma_s}/V_t \qquad (7.29)$$

where $\langle\phi_s\rangle$ is the average potential of the surface wave emerging at port 1 when a voltage V_t is applied to the transducer. The average is taken over the aperture W of the *unapodized* transducer. As in Chapter 5, Section 5.1.1, we have $\Gamma_s \equiv (\Delta v/v)/\varepsilon_\infty$.

The response of the chirp transducer is found by a simple generalization of the method in Chapter 5, Section 5.4. Consider first that a voltage V_t is applied to electrode n only, with all other electrodes grounded. This is illustrated in Fig. 5.7 assuming the electrodes to be regular. At frequency ω, the potential of the wave emerging at port 1 is, from eq. (5.57),

$$\phi_{1n} = j\Gamma_s V_t \overline{\rho}_f(kp) \exp[-jk(x_n + L/2)] \qquad (7.30)$$

where L is the transducer length between the ports and $x_n + L/2$ is the center location of electrode n, relative to port 1. The wavenumber is $k = \omega/v$, where the velocity v is close to the free-surface velocity v_f (Chapter 8, Section 8.1.4). The function $\overline{\rho}_f(kp)$ is the transform of the elemental charge density $\rho_f(x)$, which is the electrostatic charge density for unit voltage on one electrode. The transform is given by eq. (5.59). This analysis assumes that the electrodes are regular, so that the pitch p and width a are the same everywhere. The transform is a function of the metallization ratio a/p and the variable kp, and the notation has been changed slightly to show the latter. To apply eq. (7.30) to an electrode in a chirp transducer, we must allow for variation of the pitch. It is assumed that the pitch varies slowly and that the metallization ratio is constant. The electrostatic field near one electrode is not much affected by the slow variation of p, so the function $\overline{\rho}_f(kp)$ is still valid to a good approximation. However, it is necessary to specify the pitch to be used, so for electrode n we replace p by p_n and $\overline{\rho}_f(kp)$ is replaced by $\overline{\rho}_f(kp_n)$, where p_n is the pitch in the region of electrode n.

The wave generated by electrode n has a beam width corresponding to the region where the live electrode is present, extending a distance u_n from the upper limit of the active region in Fig. 7.6. To average ϕ_{1n} over the width W of the unapodized transducer, we need to multiply eq. (7.30) by u_n/W. The average wave potential $\langle\phi_s\rangle$ generated by whole transducer is obtained by adding the contributions due to individual electrodes, using superposition. Thus, $\langle \phi_s \rangle = \sum_n (u_n/W)\phi_{1n}$. Using eqs (7.30) and (7.29), the response is

$$H_t(\omega) = (\omega W \Gamma_s)^{1/2} \sum_{n=1}^{N} \frac{u_n}{W} \overline{\rho}_f(kp_n) \exp[-jk(x_n + L/2)] \qquad (7.31)$$

where N is the number of electrodes.

It is assumed here that the chirp transducer is designed by synchronous sampling, as is usually the case. Assuming that the required impulse response has the form $a(t)\cos[\theta(t)]$, synchronous sampling means that the electrode locations correspond to equal increments of $\theta(t)$. Considering a small region with electrode pitch p_n, the electrodes generate surface waves most effectively at a frequency such that the transducer period $p_n S_e$ equals the wavelength, that is, when $kp_n = 2\pi/S_e$. The function $\overline{\rho}_f(kp)$ varies slowly with kp, so $\overline{\rho}_f(kp_n)$ may be replaced by $\overline{\rho}_f(2\pi/S_e)$, which is independent of n. Equation (7.31) simplifies to

$$H_t(\omega) = (\omega W \Gamma_s)^{1/2} \overline{\rho}_f(2\pi/S_e) \sum_{n=1}^{N} \frac{u_n}{W} \exp[-jk(x_n + L/2)]. \qquad (7.32)$$

Here S_e is the number of electrodes per period, as in Chapter 5, Section 5.5, so that $S_e = 2$ for a single-electrode transducer and $S_e = 4$ for a double-electrode transducer. Equation (7.32) applies for the fundamental response. For the response at harmonic M, the term $2\pi/S_e$ is replaced by $2\pi M/S_e$. The response of a two-transducer device is given by eq. (7.28) above.

We can also write the response in terms of gap elements, following the method of Section 5.6.2 in Chapter 5. This gives

$$H_t(\omega) = (\omega W \Gamma_s)^{1/2} \overline{\rho}_g(2\pi/S_e) \sum_{n=1}^{N-1} \frac{u_{n+1} - u_n}{W} \exp(-jkx_n') \qquad (7.33)$$

where $x_n' = x_n + p_n/2 + L/2$ is the location of gap n relative to port 1, which is at $x = -L/2$. This equation has the form of the delta-function model in Chapter 1. The function $\overline{\rho}_g(kp)$ is given by eq. (6.11).

Admittance in the stationary-phase approximation

Here we consider the transducer conductance $G_a(\omega)$. Consider first an unapodized uniform transducer, with N_{pu} periods, operating at its fundamental frequency. The center-frequency conductance $G_{au}(\omega_c)$ was deduced in Chapter 5, Section 5.5.3, eq. (5.82), giving

$$G_{au}(\omega_c) = \omega_c \varepsilon_\infty^2 N_{pu}^2 W \Gamma_s \widetilde{G}_{a1} \qquad (7.34)$$

where $\tilde{G}_{a1}$ is a normalized conductance depending on S_e and a/p, shown in Table 5.1. When a voltage V_t is applied, the amplitude ϕ_s of the wave generated in either direction is given by $|\phi_s(\omega)|^2 = V_t^2 \Gamma_s G_a/(\omega W)$, from eqs (5.32) and (5.42). Using eq. (7.34), a uniform transducer gives

$$|\phi_s(\omega_c)| = V_t N_{pu} \Gamma_s \varepsilon_\infty \sqrt{\tilde{G}_{a1}}. \tag{7.35}$$

Now consider an *unapodized* chirp transducer. We regard the chirp transducer as a series of uniform transducers, each one period long ($N_{pu} = 1$), using eq. (7.35) for each period. This is valid despite the pitch variation, because the response of a one-period transducer has a large bandwidth. The total potential of the wave leaving at port 1 is

$$\phi_s(\omega) = V_t \Gamma_s \varepsilon_\infty \sqrt{\tilde{G}_{a1}} \exp(-jkL/2) \sum_{m=1}^{N_p} \exp(-jkx_m) \tag{7.36}$$

where N_p is the number of periods and x_m are the period locations, defined such that the wave potential generated by period m is real when referred to x_m. The summation in this equation is now approximated by an integral, assuming that the transducer was designed by synchronous sampling. The positions x_m correspond to times t_m obtained by setting $\theta(t_m) = 2m\pi$, where $\theta(t)$ is the time-domain phase. Thus, $kx_m = \omega t_m$. In the summation we can replace kx_m by $\omega t_m - \theta(t_m)$. The points are spaced by $t_{m+1} - t_m \approx 2\pi/\dot{\theta}(t_m)$. The summation becomes

$$S \equiv \sum_{m=1}^{N_p} \exp(-jkx_m) \approx \frac{1}{2\pi} \int_{t_1}^{t_2} \dot{\theta}(t) \exp[j\theta(t) - j\omega t] dt \tag{7.37}$$

where the limits correspond to the ends of the transducer. Using the stationary-phase approximation, comparison with eqs (7.7) and (7.12a) gives

$$|S| \approx \frac{\omega}{2\pi \sqrt{|\mu(T_s)|}}. \tag{7.38}$$

From eq. (7.36), the surface-wave amplitude generated in each direction is $|\phi_s| = V_t \Gamma_s \varepsilon_\infty |S| \sqrt{\tilde{G}_{a1}}$, and this is related to the conductance G_a by $|\phi_s|^2 = V_t^2 \Gamma_s G_a/(\omega W)$, as before. This gives the conductance as

$$G_a(\omega) \approx \frac{\omega^3 W \varepsilon_\infty^2 \tilde{G}_{a1} \Gamma_s}{4\pi^2 |\mu(T_s)|}. \tag{7.39}$$

For an *apodized* transducer we take the electrode overlaps to be given by a function $U(t)$, such that $U(t_n) = |u_{n+1} - u_n|$. At frequency ω the wave is excited mainly in the region corresponding to the stationary-phase point $T_s(\omega)$. Assuming that $U(t)$ varies slowly, $G_a(\omega)$ is given by eq. (7.39) with the aperture taken as $U[T_s(\omega)]$, so that

$$G_a(\omega) \approx \frac{\omega^3 \varepsilon_\infty^2 \widetilde{G}_{a1} \Gamma_s U(T_s)}{4\pi^2 |\mu(T_s)|}. \tag{7.40}$$

The susceptance $B_a(\omega)$ is the Hilbert transform of $G_a(\omega)$. It is often the case that $G_a(\omega)$ varies quite slowly with ω, and hence $B_a(\omega)$ is small and can be neglected.

An *effective number of periods*, $N_{\text{eff}}(\omega)$, can be defined as the number of periods in a uniform transducer, with center frequency ω, such that its conductance equals that of the chirp transducer at frequency ω. Both transducers are taken to have the same aperture. We set eqs (7.34) and (7.40) equal, and in eq. (7.34) put $\omega_c = \omega$, $N_{\text{pu}} = N_{\text{eff}}$ and $W = U(T_s)$. This gives

$$N_{\text{eff}}(\omega) = \frac{\omega}{2\pi \sqrt{|\mu(T_s)|}}. \tag{7.41}$$

Using this expression, it is easy to estimate the conductance of a chirp transducer. We can interpret N_{eff} as roughly the number of periods in the chirp transducer that are actively generating surface waves at frequency ω. For a linear chirp $|\mu|$ is equal to B/T and N_{eff} is related to the total number of periods by $N_{\text{eff}}/N_p = \omega/(\omega_c\sqrt{TB})$, where ω_c is the center frequency of the chirp. For example, if $TB = 100$, about 10% of the electrodes are active at any one frequency.

The capacitance of a chirp transducer can be derived starting from that of a uniform transducer, which is $C_{\text{tu}} = WN_{\text{pu}}\varepsilon_\infty\widetilde{C}_t$, from eq. (5.81). The constant $\widetilde{C}_t$ is given in Table 5.1. The chirp transducer is regarded as a series of one-period uniform transducers, each with capacitance C_{tu}, and the aperture $U(t_m)$ is assumed to vary slowly. Thus the capacitance is

$$C_t \approx \varepsilon_\infty \widetilde{C}_t \sum_{m=1}^{N_p} U(t_m). \tag{7.42}$$

For an unapodized chirp transducer, C_t is the same as for a uniform transducer with the same number of periods.

Some interesting results are obtained if we compare a uniform transducer with an unapodized linear-chirp transducer, with the same bandwidth and aperture. The chirp transducer has conductance G_a given by eq. (7.39) with $\mu = B/T$. The uniform transducer has conductance given by eq. (7.34), where we take $N_{\text{pu}} = \omega_c/(2\pi B)$ so that it has about the same bandwidth as the chirp transducer.

The center-frequency conductances are in the ratio $G_a(\omega_c)/G_{au}(\omega_c) \approx TB$. The capacitances are in the ratio N_p/N_{pu}, since they must be the same if $N_p = N_{pu}$. Using $N_p = T\omega_c/(2\pi)$ and $N_{pu} = \omega_c/(2\pi B)$ we have $C_t/C_{tu} = N_p/N_{pu} = TB$. Thus, the capacitance ratio is the same as the conductance ratio, and the Q-factors of the two transducers are the same.

7.3.2 Transducer design

It is assumed that the transducer is to be designed such that its impulse response approximates a chirp waveform

$$v(t) = a(t)\cos[\theta(t)], \qquad (7.43)$$

where $\theta(t)$ is a non-linear function of time. The waveform will have finite length T, so that $a(t) = 0$ for $|t| > T/2$. If the required response is specified in the frequency domain, $a(t)$ and $\theta(t)$ can be found by Fourier transformation into the time domain. It may then be necessary to truncate to a finite length, and short extensions are usually added at the ends to avoid abrupt discontinuities. Owing to the sampled nature of the transducer, the frequency response will include a fundamental and a series of harmonics. It is assumed that the fundamental is to give the response of eq. (7.43). In practical cases, the harmonics are usually irrelevant.

The design method is based on some analysis of non-uniform sampling, derived in Appendix A, Section A.4. Suppose that a waveform $v(t)$ is sampled at times t_n, such that

$$\theta(t_n) = 2\pi n/S_e + \theta_0 \qquad (7.44)$$

where θ_0 is a constant and n is an integer. Thus, synchronous sampling is assumed and the integer S_e is the number of samples per period. The sampled waveform is

$$v_s(t) = \sum_{n=1}^{N} v(t_n)\delta(t - t_n) \qquad (7.45)$$

where N is the number of samples. It is assumed that θ_0 and N are chosen such that $v_s(t) = 0$ for $t < t_1$ and for $t > t_N$. In Appendix A, Section A.4, it is shown that the *fundamental* component of $v_s(t)$ is

$$\tilde{v}_s(t) = Ca(t)\dot{\theta}(t)\cos[\theta(t)] \qquad (7.46)$$

where C is a constant. This is valid for $S_e > 2$, and for $S_e = 2$ it is valid provided θ_0 is a multiple of π.

Now consider the design of a transversal filter to give an impulse response $v(t)$. The impulse response of the transversal filter has the form

$$h_s(t) = \sum_{n=1}^{N} h_n \delta(t - t_n) \tag{7.47}$$

with t_n given by eq. (7.44). The transversal filter has the form shown in Fig. 6.3, though here the tap delays t_n are unequally spaced. Comparing with eqs (7.45) and (7.46), it can be seen that if we take $h_n = v(t_n)$ the fundamental component of $h_s(t)$ will be proportional to $v(t)$ except for distortion caused by the term $\dot{\theta}(t)$. To compensate for this, the tap weights must be given by

$$h_n = v(t_n)/\dot{\theta}(t_n). \tag{7.48}$$

The fundamental component of $h_s(t)$ will then be proportional to $v(t)$, eq. (7.43).

For a *transducer* to give an impulse response $v(t)$, we consider the design in terms of gap elements. The frequency response is given by eq. (7.33), which has the form

$$H_t(\omega) = K\omega^{1/2} \sum_{n=1}^{N} (u_{n+1} - u_n) \exp(-jkx'_n) \tag{7.49}$$

where K is a constant and x'_n is the location of gap n relative to port 1. Here N is the number of gaps and u_n are the locations of the electrode breaks, as in Fig. 7.6. The transversal filter has its impulse response given by eq. (7.47), so its frequency response is $H_s(\omega) = \sum_n h_n \exp(-j\omega t_n)$. Clearly, the gap locations in the transducer need to be given by $x'_n = vt_n$, where v is the surface-wave velocity. The transducer response in eq. (7.49) has a $\omega^{1/2}$ term. This can be compensated by noting that the response at frequency ω is mainly determined by the electrodes in the vicinity of gap n, such that $\dot{\theta}(t_n) = \omega$. Hence, the gap strength needs to be proportional to $h_n/[\dot{\theta}(t_n)]^{1/2}$, with h_n given by eq. (7.48), so that

$$u_{n+1} - u_n \propto v(t_n)/[\dot{\theta}(t_n)]^{3/2}. \tag{7.50}$$

As an example, consider the linear-chirp waveform

$$v(t) = a(t)\cos(\omega_1 + \pi\mu t^2) \quad \text{for } 0 < t < T \tag{7.51}$$

with $v(t) = 0$ for other t. This has instantaneous frequency $\dot{\theta}(t) = \omega_1 + 2\pi\mu t$. From eq. (7.44) the sampling times are given by

$$\mu t_n = [f_1^2 + \mu\theta_0/\pi + 2\mu n/S_e]^{1/2} - f_1 \tag{7.52}$$

with $f_1 = \omega_1/(2\pi)$. The gap strengths, eq. (7.50), are

$$u_{n+1} - u_n \propto \frac{a(t_n)\cos(\omega_1 + \pi\mu t_n^2)}{(\omega_1 + 2\pi\mu t_n)^{3/2}}. \tag{7.53}$$

7.3.3 Filter design and performance

Having described the design of chirp transducers in Section 7.3.2, this section now considers the design of chirp filters. The starting point is often the requirements of the radar system, for example resolution (determined by the bandwidth), processing gain, time-sidelobe rejection and temperature stability. Another consideration is Doppler shift, discussed later in Section 7.5. The first stage is therefore to design the waveforms, using criteria described in Section 7.1. This leads to a specification for the chirp filter. The devices considered here may use any of the types of chirp waveform discussed in Sections 7.1 and 7.2.

Temperature stability was discussed in Chapter 4, Section 4.3, eq. (4.32). For a linear chirp, the main consequence is that the chirp rate is changed from its ideal value μ to a modified value μ', with $\mu'/\mu = 1/(1+\varepsilon)^2 \approx 1 - 2\varepsilon$. Here ε is the fractional change of the delay τ between two points in the surface, so that $\varepsilon = \Delta\tau/\tau$. Typically [1, p. 251], one needs $|(\mu' - \mu)/\mu| < 3/(TB)$. For most cases, operating over a range of 100°C or more, a quartz substrate is needed for adequate temperature stability, but lithium niobate is sometimes applicable if TB is small. For a non-linear chirp eq. (4.32) still applies, and the non-linearity makes the temperature effects more critical. The same analysis applies for velocity errors which might arise, for example, from misorientation. These criteria depend only on the overall response of the device, not on the internal arrangement such as single or double dispersive.

The simplest form of device consists of two transducers, one chirped and the other uniform, as shown in Fig. 7.6. This is called a single-dispersive device. The first step is to design the uniform transducer, making it short enough to have adequate bandwidth. Since the device response is the product of the transducer responses (eq. (7.28)), the required device response is divided by the response of the uniform transducer to obtain the required response of the chirp transducer. This transducer can then be designed by the method of Section 7.3.2. Second-order effects that need to be compensated for include diffraction, which can be calculated as described in Chapter 4, Section 4.1. Another perturbation is the circuit effect, discussed for bandpass filters in Chapter 6, section 6.3.4. These effects can be evaluated only after an initial design has been done. It is usually adequate to compensate for them using the stationary-phase approximation, that is, an error at a particular frequency ω is corrected by adjusting the amplitude and phase at the corresponding position, given by the stationary-phase point $T_s(\omega)$. This requires modification of the electrode overlaps and positions, respectively.

Figure 7.6 is drawn assuming that the transducers are the single-electrode type. Generally, this gives rise to unacceptable distortions due to electrode reflections, as discussed earlier (in Chapter 1, Section 1.2). As for other devices, it is common practice to use double-electrode designs to avoid this problem. However, the reflections can sometimes be acceptable if the chirp rate μ is large because in this case there are relatively few electrodes active at any one frequency (eq. (7.41)).

Another problem associated with single-electrode transducers concerns scattering into bulk waves. This can occur when a monochromatic surface wave travels in a periodic structure with Bragg frequency below the surface-wave frequency. The problem occurs in up-chirp transducers with bandwidths exceeding typically 20%. In such transducers an incident wave matched to the high-frequency (long delay) end first passes through the low-frequency (small delay) end, where it can be converted into bulk waves. This causes attenuation in the high-frequency part of the response. The problem does not occur for down-chirp devices, such as that of Fig. 7.6. For these, the bulk-wave coupling occurs at a location beyond the region where the transducer responds to the surface wave. The problem can also be avoided by the use of double-electrode transducers, because the Bragg frequency is associated with the electrode pitch rather than the transducer pitch.

The triple-transit signal is not usually a problem for chirp devices. This is partly because the devices usually have quite large insertion losses, typically 30–50 dB. An additional reason is that, in the time domain, the triple-transit signal has the form of a chirp waveform with duration three times that of the main response. When the device is used as a matched filter, the triple-transit response is not matched to the input waveform and this factor gives extra suppression.

In some cases a matched pair of devices is required, one to serve as an expander in the radar transmitter and the other as a matched filter in the receiver. The expander generates a chirp waveform when a short pulse is applied. The radar system requires the expanded waveform to have a flat envelope and the expander is designed to produce this, compensating for the spectrum of the input pulse. The expansion process reduces the power level, incurring an 'expansion loss' of typically $20 \log(TB)$ dB. This is in addition to the frequency-domain insertion loss. For large TB the loss can be prohibitive, and then active generation must be used. In addition, active generation is often preferred because of its flexibility. However, a matched pair of devices has the advantage that errors due to temperature changes or fabrication conditions tend to cancel.

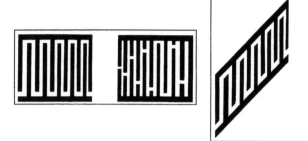

(a) Double-dispersive　　　　　　　　　(b) Slanted double-dispersive

Figure 7.7　Double-dispersive interdigital chirp filters.

Double-dispersive devices

On a weakly piezoelectric substrate such as quartz, the single-dispersive device is generally suitable only for relatively narrow bandwidths. This is due to the difficulty of matching uniform transducers over a wide bandwidth. In this situation, a double-dispersive design is often preferable. As shown in Fig. 7.7a, this device has two dispersive transducers. Each transducer contributes about half of the required dispersion, and one transducer is unapodized so that the device response is the product of the transducer responses, as in eq. (7.28). Many of the remarks on single-dispersive devices, including temperature effects, problems with single-electrode designs and diffraction, apply also to the double-dispersive type.

To design the double-dispersive device, the unapodized transducer is designed first, giving a dispersion about half of that required for the whole device. The required response of the apodized transducer is obtained by dividing the required device response by the response of the unapodized transducer. This compensates for ripples in the response of the latter.

Figure 7.7b shows a *slanted* double-dispersive device. Here, waves generated at a point in one transducer travel through relatively few electrodes before emerging on the free surface. Consequently, second-order effects are reduced. For an up-chirp device, the bulk-wave scattering problem mentioned above is ameliorated, so that single-electrode transducers may be usable. Jen [9] demonstrated a 1.2 GHz device with 200 MHz bandwidth and $TB = 100$. However, the slanted device makes less effective use of the substrate area. Another variant, applicable to either device in Fig. 7.7, is to reverse one of the transducers. If the frequency–time curves for the two transducers are the same, this has the effect of canceling the dispersion, so that a linear-phase device is produced. This scheme has been used for high-frequency delay lines and filters [9, 10].

Expander

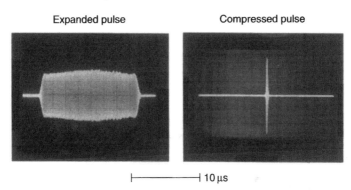

Compressor

Expanded pulse · · · · · · · · · · · · · · · Compressed pulse

⊢————————⊣ 10 μs

Figure 7.8 Performance of a matched pair of interdigital chirp filters (courtesy of Plessey Research).

Performance

Interdigital devices typically have bandwidth B up to 500 MHz and dispersion T up to 50 μs, the latter being limited by the length of substrate required. For large TB products the insertion loss can be excessive and second-order effects can be unacceptable, and consequently the TB product is usually less than 1000. Insertion losses are typically 30–50 dB. The accuracy achievable is illustrated by the fact that the time-sidelobes of the compressed waveform can be 45 dB below the main peak. This implies that the frequency response has phase errors of only a few degrees, while the phase itself typically varies by several thousand degrees over the device bandwidth. Moreover, sidelobe levels can be very close to the ideal values predicted by the analysis of Section 7.2 above. However, sidelobe levels are not always minimized because they incur a trade-off with the processing gain and the compressed pulse width.

Figure 7.8 shows a typical matched pair of devices, using linear chirps with 11 MHz bandwidth and 14 μs duration. The upper part shows the two devices in one package, and the lower figures show the expanded waveform and the

Table 7.1 Performance data for chirp filters.

Device type[a]	ID	ID	ID[c]	ID	RAC	RAC
Center frequency (MHz)	70	300	60	750	400	40
Bandwidth, B (MHz)	11	100	4	500	180	6.9
Duration, T (μs)	14	5	12	0.5	90	120
TB product	153	500	49	250	16 200	830
Time-sidelobe rejection (dB)	35	32	45	13	–	40
Substrate[b]	Q	LN	Q	LN	LN	Q
Reference	see text	[7]	[11]	[12]	[13]	[14]

[a] ID = interdigital and RAC = grooved reflective array compressor.
[b] Q = ST–X quartz, LN = Y–Z lithium niobate.
[c] Non-linear chirp.

compressor output when this waveform is applied to it. The compressor has amplitude weighting to reduce the time-sidelobes, which are 35 dB below the main peak, though they are not visible on the figure.

Table 7.1 gives the performance of a variety of devices, including reflecting array compressors (RACs) which are described in Section 7.4.

7.4 REFLECTIVE ARRAY COMPRESSORS

The reflective array compressor (RAC) is an alternative type of pulse compression filter introduced by Williamson and Smith [15, 16]. As shown in Fig. 7.9, this device relies on two arrays of angled grooves, each reflecting the wave through 90°. The wave is introduced by a uniform transducer at one end, and after reflection by the grooves it is received by another uniform transducer at the same end. The grooves are quite shallow, with depth typically 1% of the wavelength, and they reflect quite weakly. However, at a particular frequency the total reflected wave arises mainly from the region where individual reflections have similar phases, and this region contains many grooves. The periodicity of the grooves is graded along the length of the device so that different frequencies are reflected at different locations, giving a dispersive frequency response. The groove inclination depends on the substrate anisotropy, so it is not normally the 45° angle that would be suitable for an isotropic material.

The RAC can give very large time-bandwidth products, up to 16 000 for experimental devices (Table 7.1). This is possible because many second-order effects, such as multiple reflections between adjacent grooves, can be minimized simply by reducing the groove depth. The RAC also avoids a limitation of the interdigital device. The latter is limited by the fact that for one frequency most

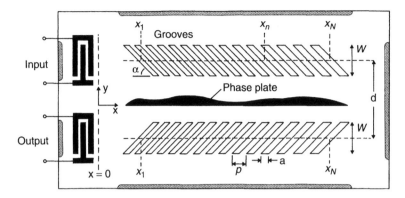

Figure 7.9 Reflective array compressor.

of the electrodes do not contribute to the response – they simply add stray capacitance which affects the transducer impedance and tends to increase the insertion loss. The RAC has no such limitation. Extra grooves can be added, increasing the *TB* product, with little effect on the insertion loss. Another advantage is that, for a given T, the RAC is shorter because the surface waves traverse the length twice. The substrate material for the RAC is often Y–Z lithium niobate. This choice, giving strong piezoelectricity, enables the input and output transducers to have relatively low conversion losses, particularly when large bandwidths are needed. It also minimizes diffraction effects as noted in Chapter 4, Section 4.1.3, an important consideration for long devices. Second-order effects in RACs are discussed in Chapter 9, Section 9.6 of the previous book [1].

As for interdigital devices, amplitude weighting is usually needed in order to suppress the time-sidelobes in the compressed waveform, and to compensate for distortions arising from causes such as the varied groove spacing. In the RAC, weighting can be implemented by varying the groove depth along the length of the device, though this complicates the fabrication procedure. To define the groove geometry, a metal film is deposited first and metal is removed at the required groove locations by conventional photolithography. The grooves are then made by exposure to an ion beam, which cuts into the surface in the unmetallized regions. To vary the depth, the ion beam is confined by means of a narrow aperture and the substrate is drawn past at a varying rate, so that different regions are subject to different exposure times.

Between the two groove arrays the RAC can have a 'phase plate' consisting of a thin aluminum film with varied width, as shown in Fig. 7.9. This introduces a phase change dependent on the plate width. In the device response, the phase change at a particular frequency corresponds to the plate width at the corresponding location. Hence the plate enables an arbitrary form of phase change to be introduced. This can be used to compensate for phase errors. It is even possible to correct devices individually, compensating for unrepeatable errors due to small changes in fabrication conditions.

The grooved RAC described above is time consuming to fabricate, and therefore expensive. A more economical and practical alternative is to use metal strip arrays as reflectors, in place of grooves. In this way Judd and Thoss [17] produced a device on a quartz substrate with 50 MHz center frequency, 10 MHz bandwidth and 50 μs dispersion, giving 30 dB time-sidelobe suppression in the compressed pulse. Weighting was realized by varying the lengths of the reflecting strips. A strongly piezoelectric substrate such as lithium niobate cannot be used in this way because the strip reflectivity is too strong. However, for this case it has been shown that the device becomes feasible if the metal strips are replaced by a sequence of disconnected segments known as 'dots'. The density of dots can be varied in order to implement amplitude weighting. A device with $TB = 1000$ was demonstrated [18].

7.5 DOPPLER EFFECTS AND SPECTRAL ANALYSIS

In addition to radar pulse compression, a variety of other applications have been developed for chirp filters, especially for RACs. Many of these have features related to the effect of a Doppler shift on a radar system, so here we consider this topic first.

If a target has a radial component of velocity relative to the radar, the familiar Doppler effect occurs and the received signal is shifted in frequency. Its spectrum is modified from an ideal form $S(\omega)$ to a new form $S'(\omega) = S(\omega - \omega_D)$, where ω_D is the Doppler shift. The band edge frequencies of $S'(\omega)$ are shifted slightly, but these shifts can usually be ignored because ω_D is small. For a *linear* chirp, the signal is still approximately matched to the compressor filter, but the frequency displacement causes the main output peak to occur at a slightly different time, with a delay change $\Delta\tau = -\omega_D/(2\pi\mu)$, where μ is the chirp slope of $S(\omega)$. Hence the position of the peak varies linearly with the Doppler frequency.

This effect can be exploited for frequency measurement using a system called the *compressive receiver* [2, 19], shown in Fig. 7.10. A short pulse is applied to a chirp expander, producing a linear-chirp waveform. This is mixed with a continuous-wave (CW) signal at frequency Ω, giving a signal in a band suitable for a matching chirp filter called the compressor. The expander has a bandwidth

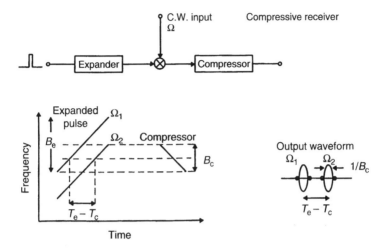

Figure 7.10 Compressive receiver. Subscripts e and c refer to the expander and compressor, respectively.

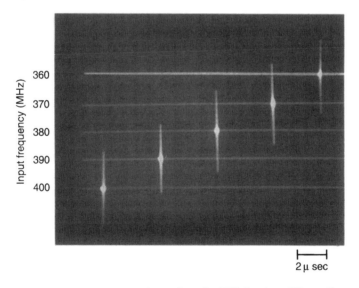

Figure 7.11 Performance of a compressive receiver using RACs (courtesy of Plessey Research).

larger than that of the compressor, so that Ω can be varied appreciably while still giving a compressed pulse with a sharp peak. The delay of the output peak is a linear function of Ω, and the compressor may have amplitude weighting in order to minimize the associated sidelobes. The system provides a vary rapid means of frequency measurement. Practical examples are described by

Williamson *et al.* [20] and Moule *et al.* [12]. Figure 7.11 shows the output of a compressive receiver for several input frequencies, with the traces displaced vertically for clarity. This system used RACs with bandwidths of 40 and 80 MHz and dispersions of 17 and 34 μs. The time-sidelobe suppression was 34 dB, and the resolution was 17 kHz.

If the CW input is replaced by a more general waveform, the compressive receiver gives the amplitude spectrum of this waveform. The output of the system almost corresponds to the Fourier transform of the input signal. The correspondence is not complete, because the compressive receiver does not give the phase of the transform correctly. However, it can be shown that the correct phase is obtained if the output waveform is multiplied by another linear-chirp waveform, generated by impulsing another chirp filter. The output waveform then has amplitude and phase corresponding to the Fourier transform of the input waveform, with Fourier frequencies linearly related to the output timing [1].

A wide variety of other signal processing functions can be obtained using chirp filters, including variable delay, chirp filtering with variable chirp rate, bandpass filtering with variable bandwidth and time scaling [1]. A very complex system for variable chirp rate matched filtering was demonstrated by Gerard *et al.* [13].

7.6   CORRELATION IN SPREAD-SPECTRUM COMMUNICATIONS

Surface-wave filters can serve various functions in communications systems, including bandpass filtering using a variety of devices described in Chapters 6, 9–11. A different class of devices are applicable to correlation in spread-spectrum systems. Here, Section 7.6.1 describes the system principles and Section 7.6.2 covers linear filters for correlation. In Section 7.6.3 we consider non-linear convolvers.

7.6.1   Principles of spread-spectrum systems

Spread-spectrum systems can take a variety of forms [21, 22], and here we summarize the basic principles of a common type known as direct-sequence spread spectrum. A basic feature is that the bandwidth of the transmitted signal is increased, so that it is greater than the bandwidth of the data. In the receiver a reverse process is employed, so that the bandwidth is reduced and the data is recovered. Figure 7.12 shows the principles of a transmitter. This system uses *phase shift keyed* (PSK) waveforms, consisting of a continuous sequence of sinusoidal segments with relative phase 0 or π. The data to be transmitted is

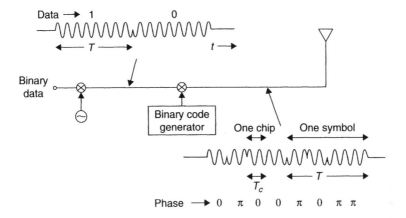

Figure 7.12 Spread-spectrum transmitter using PSK waveforms.

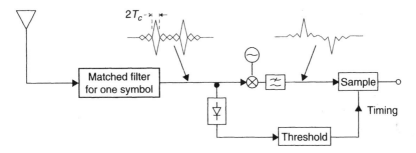

Figure 7.13 Spread-spectrum receiver using a matched filter.

encoded onto a carrier, such that phase 0 represents a binary '0' and phase π represents binary '1'. Each of these data bits is further encoded using multiplication by a binary code, giving a phase change of 0 or π corresponding to the code bits. Thus, each data bit becomes a PSK sequence, and the segments of this are called 'chips'. The chips have length T_c which is much less than the data bit length T, and consequently the signal bandwidth is much greater than the data bandwidth.

A receiver using a matched filter is shown in Fig. 7.13. The received signal is applied to a filter which is matched to the transmitter code. As for pulse compression radar, this requires the filter impulse response to be the time reverse of the code. The filter output gives a correlation peak corresponding to each bit of the data, and this waveform is converted down to baseband frequencies and then sampled at the times of the correlation peaks. The resulting samples correspond to the original data, with bandwidth approximately $1/T$. Advantages of this system are that it gives increased security and the wide-band signal can be more difficult to detect by an unauthorized receiver. In addition, the receiver responds preferentially to the waveform that it is matched to, so it is relatively insensitive to interference. The narrow correlation peak enables the system to distinguish multipath signals with different delays, due to reflections from buildings, for example.

An alternative, and more common, type of receiver uses an 'active correlator'. This simply multiplies the received signal by a binary code the same as that used in the transmitter. This strips off chip coding, since a phase of 0 is left unaltered while a phase of π is changed by π. The resulting PSK waveform has phase changes corresponding to the original data. This system is simpler than the matched filter approach, but the receiver code generator needs to be accurately timed to correspond with the signal timing, which may be unknown initially. The timing can be established by a search procedure, using trials with different receiver timings. However, this process can be time consuming, in which case a matched filter receiver can be preferable.

Much attention has been given to the question of code design, particularly in order to minimize the time-sidelobes of the correlated waveform. The pseudo-noise codes [1] can be generated by means of a recursive shift register. A group known as Barker codes have the property that the sidelobes all have amplitude $1/N$ relative to the main peak, where N is the number of chips. The longest Barker code has $N = 13$ chips, with coding 0101001100000. Detailed studies show that well-suppressed sidelobes can be obtained if amplitude weighting is included [23]. Other code forms include [1] minimum-shift keying (MSK) and frequency hopping.

7.6.2   Linear matched filters for PSK

A surface-wave matched filter for PSK can be implemented by a simple tapped delay line, as in Fig. 7.14. The device has a series of short transducers acting as taps, with spacing corresponding to the chip length T_c. Some of the taps are inverted in accordance with the code required. A uniform input transducer

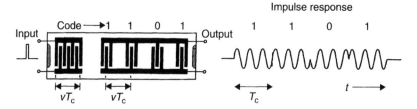

Figure 7.14 Fixed-coded matched filter for PSK.

has length similar to the tap spacing. When this device impulsed, it gives a continuous PSK waveform in which the phases correspond to the tap connections, as shown in the figure. These devices have been developed quite extensively [24] and systems with chip rate up to 200 MHz have been developed [25].

For spread-spectrum systems it is sometimes desirable to be able to change the coding at will, and this requires a *programmable* PSK filter. For this purpose, each tap can be connected to a switch which changes the phase by 0 or π according to the state of a static electrical control signal. A 31-tap system described by Hickernell *et al.* [26] used surface-wave propagation on a silicon substrate. The wave was detected by MOSFET taps connected to phase-switching circuitry in the same silicon substrate. To generate the wave, a piezoelectric zinc oxide film was deposited in the region required for the uniform transducer, and the transducer was fabricated on top of this film. Another approach to programmability is the use of gallium arsenide, which is both semiconducting and piezoelectric. Fixed-coded filters have been demonstrated [27], and the semiconducting property indicates the possibility of incorporating switching circuitry on the same substrate. A zinc oxide film can also be added on gallium arsenide, giving stronger piezoelectric coupling (Table 4.3).

7.6.3 Non-linear convolvers

Non-linear convolvers offer an alternative approach for programmable correlation of PSK and other signals, as described by Luukkala and Kino [28, 29] and in Kino's substantial review [30]. These devices differ fundamentally from all other devices in this book in that they use non-linear effects. In contrast, other devices are linear filters which conform to basic relations given in Appendix A, Section A.2.

A simple convolver structure, using a Y–Z lithium niobate substrate, is shown in Fig. 7.15. The device has two input transducers, one at each end, and in

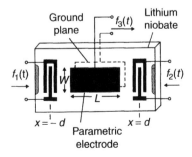

Figure 7.15 Surface-wave convolver.

between there is a uniform metal film called the parametric electrode. Signals are introduced at both ends, generating surface waves traveling toward the center of the device. The substrate has a weak non-linearity, so that the fields in the device correspond not only to the input waves, but also to extra terms including harmonics and a product of the two wave amplitudes. The product term is sensed by the parametric electrode, acting in conjunction with a ground electrode on the reverse side of the substrate.

Suppose first that sinusoidal signals with the same frequency ω are applied to both inputs. At any point within the device the two surface waves have the form $\cos(\omega t \pm kx)$. The product of these is $\cos(2\omega t) + \cos(2kx)$. Here the second term is a zero-frequency component which is eliminated by bandpass filtering at the output. The first term, with frequency 2ω, is independent of x. The field due to this is selectively sensed by the parametric electrode, which is spatially invariant. Hence the $\cos(2\omega t)$ term is selected by two mechanisms – spatial filtering by the parametric electrode and frequency filtering to select the 2ω component. Owing to this selectivity, the convolver can give an adequate output signal even though the non-linearity is in fact very weak. In particular, the selectivity helps to reject unwanted signals at the input frequency ω, which might be due to feed through, for example. In addition, the individual waves have second-harmonic signals at frequency 2ω, but these are spatially sinusoidal so the parametric electrode suppresses them.

The non-linearity in lithium niobate is very weak, and for most surface-wave devices it is of little consequence provided the power level is not too large (Chapter 4, Section 4.3). In the convolver the situation is different because the device is designed to select the required product signal and to reject other terms.

The signal processing action of the convolver becomes clear if we consider general input signals $f_1(t)$ and $f_2(t)$ applied at the left and right, respectively, as in Fig. 7.15. These are assumed to have relatively small bandwidths centered on frequency ω, so that the above arguments regarding selectivity still apply. The wave amplitudes within the device have the form $f_1(t - x/v)$ and $f_2(t + x/v)$, where v is the surface-wave velocity. A delay related to the transducer separation is ignored here. The non-linearity gives rise to a field proportional to the product, and the parametric electrode gives an output $f_3(t)$ proportional to the average of this. Thus, the output can be written as

$$f_3(t) \propto \int_{-\infty}^{\infty} f_1(t - x/v) f_2(t + x/v) \, dx. \qquad (7.54)$$

In general the integration is over the length L of the parametric electrode, but here it is assumed that the input waveforms have length less than the delay L/v along this electrode. It is also assumed that the waveforms are timed such that the waves overlap completely in the parametric region. The limits of the integral can then be taken as $\pm\infty$. With the substitution $\tau = t - x/v$, eq. (7.54) becomes

$$f_3(t) \propto v \int_{-\infty}^{\infty} f_1(\tau) f_2(2t - \tau) d\tau. \tag{7.55}$$

This equation is the *convolution* of $f_1(t)$ and $f_2(t)$, apart from the factor of 2 which causes a contraction of the time scale. The time contraction arises because of the relative motion of the two surface waves. Convolution is also associated with *linear* filtering, since the output of a linear filter is the convolution of the input waveform with the device impulse response (Appendix A, Section A.2). Hence the convolver behaves mathematically like a *linear* filter. One input waveform, called the 'reference' waveform, takes the role of the impulse response. This 'impulse response' is quite arbitrary, apart from constraints on bandwidth and duration, so the convolver is an extremely flexible filter with a very high degree of programmability. In particular, if the reference waveform is the time reverse of a coded signal waveform, the convolver can be used to correlate this signal. One limitation is that the convolver can correlate the signal only if the reference waveform is present at the same time. However, this can be overcome by applying the reference repeatedly. It has been shown that this can give the complete output waveform for arbitrary timing of the input signal, as given by a linear filter, apart from a time-domain segmentation associated with the time contraction [31].

Figure 7.16 shows some output waveforms, for a convolver such as that of Fig. 7.15. On the left is the output produced when the same rectangular pulse, of length $T = 4.6\,\mu\text{s}$, is applied to both inputs. The convolution is ideally a triangular pulse of length $2T$, and the convolver gives a pulse of length T because of the time contraction. On the right is the correlation of a PSK waveform using the 13-chip Barker code. Ideally, this gives a main peak with equal-amplitude sidelobes on each side, with a peak-to-sidelobe amplitude ratio of 13. The upper part of the figure shows the baseband signals used to produce the phase modulation of the input waveforms. The convolver is particularly suited to correlation of waveforms such as PSK, because these can be easily generated.

A non-degenerate device uses a similar principle but is designed for input waveforms with different frequencies [1, 30]. In this case the parametric electrode becomes an interdigital structure, with period corresponding to the frequency difference of the inputs.

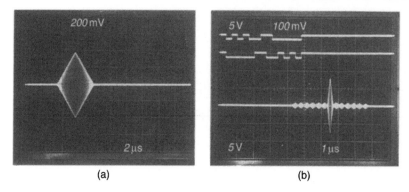

(a)   (b)

Figure 7.16 Output waveforms produced by a convolver similar to that of Fig. 7.15, for input waveforms that are (a) identical rectangular pulses and (b) PSK waveforms coded according to the 13-chip Barker code and its time reverse. In both the cases the input waveforms are 4.6 μs long. (courtesy of J.H. Collins, University of Edinburgh).

Efficiency: dual-channel convolvers

Owing to the weakness of the non-linear effect, the convolver of Fig. 7.15 gives a very weak output signal. To quantify the efficiency we consider sinusoidal input waveforms. Ideally the device is *bilinear*, that is, the output power P_o is proportional to the product of the input powers P_1 and P_2. The bilinearity factor, in decibels, is defined as

$$C = 10 \log_{10} \left(\frac{P_o}{P_1 P_2} \right) \text{(dBm)}. \tag{7.56}$$

The powers are usually in units of mW, in which case C has units dBm. Since P_o is proportional to $P_1 P_2$, this equation gives the same value of C for any input power levels. Note that there is no output if either input is zero. For example, if there is no input at port 1 we have $P_1 = 0$ and this gives $P_o = 0$ since C is finite. The convolver of Fig. 7.15 typically gives $C = -95$ dBm. For illustration, the input powers might be 20 dBm (100 mW), and the output power is then -55 dBm. This limits the applications and is inconvenient in view of the amplification needed. One factor affecting the value of C is the choice of substrate. Measurements have been made on many materials and most have shown weaker non-linearity than lithium niobate. However, potassium niobate ($KNbO_3$) has been found to be stronger, increasing the bilinearity factor by 25 dB [32].

The efficiency of a lithium niobate convolver can be improved by reducing the width of the surface-wave beam, thus increasing its energy density. This necessitates the use of a surface-wave *waveguide*, so that the energy is prevented from spreading outwards due to diffraction. Waveguides will be considered

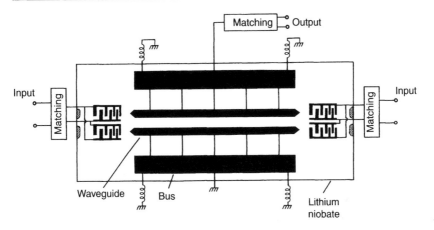

Figure 7.17 Waveguide convolver using chirp transducers.

later, in Chapter 10, where it will be seen that a simple metallic strip can have this function. Thus, the parametric electrode becomes a thin metallic strip, typically three wavelengths wide, which can be fabricated in the same step as the transducers. An additional consideration is the output capacitance. To obtain a reasonable value, it is best to use a ground electrode on the surface instead of the rear of the substrate. In this way, Defranould and Maerfeld [33] obtained a much improved bilinearity factor of $C = -71$ dBm.

This device was refined further by several authors, leading to the designs such as that of Fig. 7.17. Here there are two waveguides. One of the four input transducers is reversed so that the non-linear fields on the two waveguides have opposite sign, and these signals are added by connecting the waveguides across the output live and ground terminals. This dual-channel approach has the advantage of reducing unwanted signals due to reflections from the transducers, as well as other effects. The waveguides are connected periodically to wide bus bars to reduce loss due to resistance. The inductors shown are included to reduce errors due to electromagnetic propagation along the structure, which behaves like a transmission line in this respect. Alternatively, the transmission-line effect can be reduced by making connections to a network of transmission lines, equalizing the electromagnetic transit times [34].

An additional complication is the need to generate narrow beams efficiently. A conventional transducer with small aperture has a high impedance, making it difficult to match effectively. Figure 7.17 shows the use of chirp transducers [35], which give a reduced impedance for a given aperture as noted in Section 7.3.1

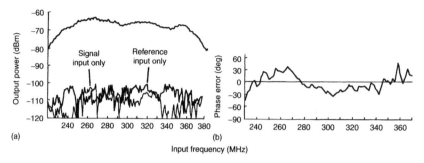

Figure 7.18 Performance of a waveguide convolver for CW input signals with the same frequency. (a) The upper curve shows the output power level for 0 dBm input levels, giving the bilinearity factor. (b) Phase error of output waveform (courtesy of Plessey Research).

above. Other effective methods are the use of beam-compressing waveguide horns [36] or multistrip beam compressors [37].

Figure 7.18 shows the measured efficiency for a convolver such as that of Fig. 7.17. On the left, the upper curve is the bilinearity factor C, in dBm. The lower curves show the output level when only one input is connected, which should ideally give no output. On the right is the phase error. This device had a center frequency of 300 MHz, a bandwidth of 120 MHz and a parametric length of 16 μs, so that it was capable of correlating signals with time-bandwidth products up to 1920.

The measurement in Fig. 7.18 does not completely characterize the device. For more information, a two-dimensional frequency response $H(\omega_1, \omega_2)$ can be used. This is defined by considering CW input signals at frequencies ω_1 and ω_2 and taking the output signal, at frequency $\omega_1 + \omega_2$, to have amplitude and phase given by $H(\omega_1, \omega_2)$. This function fully characterizes any bilinear device, that is, a device whose output amplitude is proportional to the product of two input amplitudes. Experimentally, one can measure $H(\omega_1, \omega_2)$, regard it as a function of the difference frequency $\Delta\omega = \omega_1 - \omega_2$, and then Fourier transform from the $\Delta\omega$ domain to the time domain. This gives the response as a function of position along the parametric region [1], giving excellent resolution because $\Delta\omega$ can be varied widely. The result shows, for example, spatial variations due to the transmission-line effect.

Semiconductor devices

In order to further improve the efficiency, several types of semiconductor devices have been investigated. An air-gap convolver shown in Fig. 7.19a uses surface waves on a Y–Z lithium niobate substrate, coupled to a silicon sheet through a

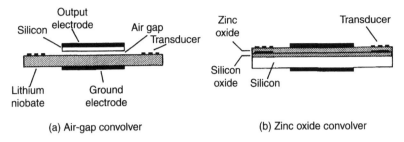

Figure 7.19 Semiconductor convolvers.

narrow air gap [30]. A non-linear interaction takes place in the silicon. To avoid unwanted attenuation the silicon is separated from the substrate by a small gap, but this gap must be less than one wavelength in order to obtain adequate coupling. The gap is obtained by supporting the silicon on an array of 'posts' fabricated by etching the substrate, with the posts arranged sparsely so that they do not cause significant attenuation. Cafarella *et al.* [38] describes a device with 100 MHz input bandwidth and 10 µs interaction length, with bilinearity factor $C = -60$ dBm.

Another semiconductor device is the zinc oxide convolver of Fig. 7.19b. A piezoelectric zinc oxide film is deposited on a (non-piezoelectric) silicon substrate, and the fields accompanying the waves penetrate into the silicon where the non-linear interaction takes place [30]. The zinc oxide also enables the waves to be generated using conventional interdigital transducers. This technology has the advantage of avoiding the need for a small air gap, though care is needed in fabrication of the zinc oxide film. An experimental device [39] gave $C = -44$ dBm, with 5 MHz input bandwidth and 3.5 µs interaction length. A variant of this is the 'strip-coupled convolver', in which a semiconductor is deposited on a lithium niobate substrate at the side of the surface-wave track, and coupled into it by means of metal strips normal to the propagation direction [30]. Using a GaAs semiconductor with embedded diodes, a bilinearity factor of -16 dB has been obtained [40]. Similar structures have been used to amplify surface waves [30].

Semiconductors have been used in a considerable variety of surface-wave devices, as described by Kino [30]. These include a variety of convolvers and other devices not considered here. A storage convolver due to Ingebrigtsen [41] is illustrated in Fig. 7.20. Here an air-gap device has an array of Schottky diodes fabricated in the silicon surface, used to store signals in the form of charges. When a negative voltage is applied to the upper electrode the diodes are forward biassed, with a time constant of typically 0.5 ns. For a negative voltage the time

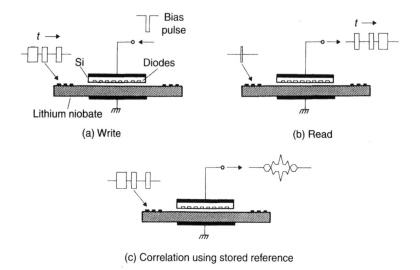

(a) Write   (b) Read

(c) Correlation using stored reference

Figure 7.20 Air-gap storage convolver.

constant is $1-100\,\mu s$. A finite length coded waveform is applied to the transducer at the left, and when the surface-wave packet is within the interaction region a short bias pulse is applied to the top electrode. This causes the waveform to be stored as a pattern of charges on the diodes. At a later time, a short pulse can be applied at the left and a time-reversed version of the stored waveform appears at the top electrode, as in Fig. 7.20b. If the original coded waveform is applied again at the left, the output from the top electrode gives the correlation function of this waveform. In contrast to the basic convolver, the correlation is obtained without the need for time reversal, and it is obtained for any timing so long as the original waveform is stored in the device. Similar functions have been obtained using zinc oxide convolvers [42].

REFERENCES

1. D.P. Morgan. *Surface-Wave Devices for Signal Processing*, Elsevier, 1991.
2. J.R. Klauder, A.C. Price, S. Darlington and W.J. Albersheim. 'The theory and design of chirp radars', *Bell Syst. Tech. J.*, **39**, 745–808 (1960).
3. C.E. Cook and M. Bernfeld. *Radar Signals*, Academic Press, 1967.
4. M. Abramowitz and I.A. Stegun. *Handbook of Mathematical Functions*, Dover, 1968, p. 301.
5. E.C. Farnett, T.B. Howards and G.H. Stevens. 'Pulse compression radar', in M.I. Skolnik (ed.), *Radar Handbook*, McGraw-Hill, 1970, pp. 20.1–20.37.
6. G.W. Judd. 'Technique for realising low time sidelobe levels in small compression ratio chirp waveforms', *IEEE Ultrason. Symp.*, 1973, pp. 478–481.

7. G.A. Armstrong. 'The design of SAW dispersive filters using interdigital transducers', *Wave Electron.*, **2**, 155–176 (1976).

8. C.O. Newton. 'Non-linear chirp radar signal waveforms for surface acoustic wave pulse compression filters', *Wave Electron.*, **1**, 387–401 (1976).

9. S. Jen. 'Synthesis and performance of precision wideband slanted array compressors', *IEEE Ultrason. Symp.*, **1**, 37–49 (1991).

10. C.C.W. Ruppel, L. Reindl, S. Berek, U. Knauer, P. Heide and M. Vossiek. 'Design, fabrication and performance of precise delay lines at 2.45 GHz', *IEEE Ultrason. Symp.*, **1**, 261–265 (1996).

11. M.B.N. Butler. 'Radar applications of SAW dispersive filters', *Proc. IEE*, **127F**, 118–124 (1980).

12. G.L. Moule, R.A. Bale and T.I. Browning. 'A 1 GHz bandwidth SAW compressive receiver', *IEEE Ultrason. Symp.*, **1**, 216–219 (1980).

13. H.M. Gerard, P.S. Yao and O.W. Otto. 'Performance of a programmable radar pulse compression filter based on a chirp transformation', *IEEE Ultrason. Symp.*, 1977, pp. 947–951.

14. A. Rønnekleiv. 'Amplitude and phase compensation of RAC-type chirp lines on quartz', *IEEE Ultrason. Symp.*, **1**, 169–173 (1988).

15. R.C. Williamson and H.I. Smith. 'The use of surface elastic wave reflection gratings in large time-bandwidth pulse compression filters', *IEEE Trans.*, **MTT-21**, 195–205 (1973).

16. R.C. Williamson. 'Properties and applications of reflective array devices', *Proc. IEEE*, **64**, 702–710 (1976).

17. G.W. Judd and J.L. Thoss. 'Use of apodised metal gratings in fabricating low cost quartz RAC filters', *IEEE Ultrason. Symp.*, **1**, 343–347 (1980).

18. F. Huang and E.G.S. Paige. 'The design of SAW RACs using arrays of thin metal dots', *IEEE Trans. Ultrason. Ferroelec. Freq. Contr.*, **41**, 236–244 (1994).

19. S. Darlington. 'Demodulation of wide band low-power FM signals', *Bell. Syst. Tech. J.*, **43**, 339–374 (1964).

20. R.C. Williamson, V.S. Dolat, R.R. Rhodes and D.M. Boroson. 'A satellite-borne SAW chirp-transform system for uplink demodulation of FSK communication signals', *IEEE Ultrason. Symp.*, 1979, pp. 741–747.

21. R.C. Dixon. *Spread Spectrum Systems*, Wiley, 1976.

22. J.K. Holmes. *Coherent Spread Spectrum Systems*, Wiley, 1982.

23. O. Hikino, M. Belkerdid and D.C. Malocha. 'Code optimisation for direct sequence spread spectrum and SAW matched filter implementation', *IEEE Trans. Ultrason. Ferroelec. Freq. Contr.*, **47**, 974–983 (2000).

24. D.T. Bell and L.T. Claiborne. 'Phase code generators and correlators', in H. Matthews (ed.), *Surface Wave Filters*, Wiley, 1977, pp. 307–346.

25. G. Kipens, H. Leib, J. Saw, J. Nisbet and J.-D. Dai. 'A SAW-based commutation signalling modem for broadband indoor wireless communication', *IEEE Trans. Ultrason. Ferroelec. Freq. Contr.*, **45**, 634–649 (1988).

26. F.S. Hickernell, M.D. Adamo, R.V. DeLong, J.G. Hinsdale and H.J. Bush. 'SAW programmable matched filter signal processor', *IEEE Ultrason. Symp.*, **1**, 104–108 (1980).

27. F. Moeller, J. Enderlein, M.A. Belkerdid, D.C. Malocha and W. Buff. 'Direct sequence spread spectrum differential phase shift keying SAW correlator on GaAs', *IEEE Trans. Ultrason. Ferroelec. Freq. Contr.*, **46**, 842–848 (1999).

28. M. Luukkala and G.S. Kino. 'Convolution and time inversion using parametric interactions of acoustic surface waves', *Appl. Phys. Lett.*, **18**, 393–394 (1971).

29. G.S. Kino, S. Ludvik, H.J. Shaw, W.R. Shreve, J.M. White and D.K. Winslow. 'Signal processing by parametric interactions in delay-line devices', *IEEE Trans.*, **MTT-21**, 244–263 (1973).

30. G.S. Kino. 'Acoustoelectric interactions in acoustic-surface-wave devices', *Proc. IEEE*, **64**, 724–748 (1976).
31. D.P. Morgan, J.H. Collins and J.G. Sutherland. 'Asynchronous operation of a surface acoustic wave convolver', *IEEE Ultrason. Symp.*, 1972, pp. 296–299.
32. Y. Cho, N. Oota, K. Morozumi, H. Odagawa and K. Yamanouchi. 'Quantitative study on the non-linear piezoelectric effect of $KNbO_3$ single crystal for super highly efficient SAW elastic convolver', *IEEE Ultrason. Symp.*, **1**, 289–292 (1998).
33. P. Defranould and C. Maerfeld. 'A SAW planar piezoelectric convolver', *Proc. IEEE*, **64**, 748–751 (1976).
34. I. Yao. 'High performance elastic convolver with extended time-bandwidth product', *IEEE Ultrason. Symp.*, **1**, 181–185 (1981).
35. D.P. Morgan, D.H. Warne and D.R. Selviah. 'Narrow-aperture chirp transducers for SAW convolvers', *Electron. Lett.*, **18**, 80–81 (1982).
36. B.J. Darby, D.J. Gunton, M.F. Lewis and C.O. Newton. 'Efficient miniature SAW convolver', *Electron. Lett.*, **16**, 726–728 (1980).
37. H. Gautier and C. Maerfeld. 'Wideband elastic convolver', *IEEE Ultrason. Symp.*, **1**, 30–36 (1980).
38. J.H. Cafarella, W.M. Brown, E. Stern and J.A. Alusow. 'Acoustoelectric convolvers for programmable matched filtering in spread-spectrum systems', *Proc. IEEE*, **64**, 756–759 (1976).
39. J.B. Green and B.T. Khuri-Yakub. 'A $100\,\mu m$ beam width ZnO on Si convolver', *IEEE Ultrason. Symp.*, 1979, pp. 911–914.
40. K. Hohkawa, T. Suda, Y. Aoki, C. Hong, C. Kaneshiro and K. Koh. 'Design of a semiconductor-coupled SAW convolver', *IEEE Trans. Ultrason. Ferroelec. Freq. Contr.*, **49**, 466–474 (2002).
41. K.A. Ingebrigtsen. 'The Schottky diode acoustoelectric memory and correlator – a novel programmable signal processor', *Proc. IEEE*, **64**, 764–769 (1976).
42. H.C. Tuan, J.E. Bowers and G.S. Kino. 'Theoretical and experimental results for monolithic SAW memory correlators', *IEEE Trans.*, **SU-27**, 360–369 (1980).

8

REFLECTIVE GRATINGS AND TRANSDUCERS

In Chapters 5–7 we have considered non-reflective transducers and their application to transversal filters. Here we consider reflective transducers, that is, transducers with internal reflections due to the electrodes. Reflective gratings are conveniently included here. The analysis for a grating is considered to be the same as that of a shorted single-electrode transducer, on the assumption that the bus bars have no effect other than providing lossless connections. Gratings are essential elements in surface acoustic wave resonators and in many low-loss bandpass filters. In addition, many such filters make use of single-electrode transducers. These transducers are also suited to high-frequency devices because of the use of single electrodes.

Much of this chapter is concerned with regular electrodes, which have constant width a and pitch p. Strong reflections occur at the Bragg frequency, at which the wavelength λ is equal to $2p$. Two theoretical approaches are used – the reflective array model (RAM) and the coupling-of-modes (COM) model. These models have their individual advantages, and a comparison is made in Section 8.2.2. Neither of them is very accurate, but they both provide fairly straightforward ways of handling the complex behavior of reflective components. They are often adequate in practice. Bulk-wave excitation is excluded, so while the models can apply for piezoelectric Rayleigh waves they are not strictly correct for leaky waves. However, they can be applied approximately to leaky waves provided the attenuation is small.

In this chapter, it is assumed that there is no diffraction or waveguiding. The resistivity of the electrodes and bus bars is ignored.

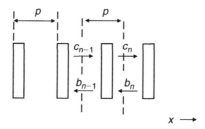

Figure 8.1 RAM analysis of an infinite array.

8.1 REFLECTIVE ARRAY METHOD FOR GRATINGS AND TRANSDUCERS

8.1.1 Infinite-length grating

We consider an infinite array of regular strips, with pitch p, as in Fig. 8.1. The physical nature of the strips is not specified here, but usually they will be shorted metal electrodes. Alternatively, grooves are often used in resonators, and open-circuit (disconnected) electrodes are also possible. The physical nature of the waves is not specified at this point, but it is assumed that the wavefronts are parallel to the strips and that only one wave mode is involved. Initially, we assume power conservation, but losses will be considered later.

The RAM approach [1, 2] starts by considering the scattering by individual strips. Waves traveling to the right and left have amplitudes denoted by c_n and b_n, respectively. These are specified at the port positions, as shown by the broken lines which are midway between the strips. Each strip has port 1 on the left and port 2 on the right. The strips are identical, and they have a transmission coefficient t_s. The reflection coefficient is r_{s1} for waves incident from the left, and r_{s2} for waves incident from the right. These coefficients refer to the center of the strip. In terms of the amplitudes at the ports, the strip has a P-matrix defined as in Appendix D, and denoted by p_{ij}. The components are

$$p_{11} = r_{s1} \exp(-jk_e p),$$
$$p_{12} = p_{21} = t_s \exp(-jk_e p),$$
$$p_{22} = r_{s2} \exp(-jk_e p). \tag{8.1}$$

Here k_e is the wavenumber, which is real because we assume no losses.

From reciprocity considerations in Appendix D we have $p_{21} = p_{12}$, so that the transmission is the same in both directions. We can take t_s to be real, because a non-zero phase in t_s can be compensated by changing the value of k_e. It is assumed that the value of k_e allows for any change of velocity due to the strips, so that in the absence of reflections the wave has an effective phase velocity $v_e = \omega/k_e$. The evaluation of this is considered later (Section 8.14).

We take t_s to be positive, and since there is no loss we have

$$t_s = \sqrt{1 - |r_{s1}|^2} = \sqrt{1 - |r_{s2}|^2} > 0 \tag{8.2}$$

and so $|r_{s1}| = |r_{s2}|$. Substituting eq. (8.1) into eq. (D.4), power conservation gives

$$r_{s2} = -r_{s1}^*. \tag{8.3}$$

Generally we are concerned with strips that have a symmetrical geometry, and this usually implies that they behave symmetrically. In this case $r_{s2} = r_{s1}$, and eq. (8.3) shows that they are both imaginary. However, here we allow r_{s2} to be different from $rs1$ because this allows for analysis of the Natural SPUDT effect, considered later in Section 8.2.3.

The scattering by one strip can be written as

$$b_{n-1} = (r_{s1}c_{n-1} + t_s b_n) \exp(-jk_e p);$$

$$c_n = (r_{s2}b_n + t_s c_{n-1}) \exp(-jk_e p). \tag{8.4}$$

Expressing waves on the right in terms of waves on the left, this gives

$$c_n = (1/t_s)c_{n-1} \exp(-jk_e p) + (r_{s2}/t_s)b_{n-1},$$

$$b_n = -(r_{s1}/t_s)c_{n-1} + (1/t_s)b_{n-1} \exp(jk_e p). \tag{8.5}$$

This makes use of the relation $r_{s1}r_{s2} = t_s^2 - 1$, which follows from eqs (8.2) and (8.3). The waves c_n and b_n are constituents of the overall wave motion. We now look for a *grating-mode* solution, which is a combination of the constituents such that

$$c_n = c_{n-1} \exp(-j\gamma p); \quad b_n = b_{n-1} \exp(-j\gamma p). \tag{8.6}$$

Substitution into eq. (8.5) is found to give

$$\cos(\gamma p) = \frac{\cos(k_e p)}{t_s}. \tag{8.7}$$

Assuming no loss, k_e is real so $\cos(\gamma p)$ is real. At most frequencies γ is real, and there is a solution with $\gamma \approx k_e$. However, because $t_s < 1$ the right side of eq. (8.7) has magnitude greater than unity when $k_e p$ is near $M\pi$, that is, when $p \approx M\lambda/2$. Then γ has an imaginary part, giving a stop band. The real and

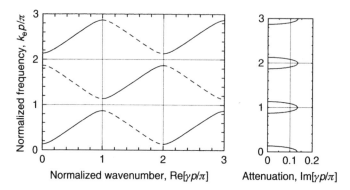

Figure 8.2 Dispersion relation for grating mode.

imaginary parts of γ are plotted in Fig. 8.2. For illustration, $|r_{s1}|$ and $|r_{s2}|$ have been given a large value independent of frequency, though this is unrealistic in practice. Assuming that v_e is independent of frequency the stop bands all have the same width, given by

$$\Delta f/f_{c1} = (2/\pi)\sin^{-1}|r_{s1}| \approx 2|r_{s1}|/\pi, \tag{8.8}$$

where $f_{c1} = v_e/(2p)$ is the center frequency of the first stop band ($M = 1$). The approximation is valid when $|r_{s1}|$ is small. In each stop band, γ can be written as $\gamma = M\pi/p + j\alpha$, where α is the attenuation coefficient. The latter is maximized at the band center, where $\cosh(\alpha p) = 1/t_s$. For small $|r_{s1}|$ this gives $\alpha \approx |r_{s1}|/p$.

The grating mode given by eqs (8.6) and (8.7) is also called a Bloch wave. At any frequency, eq. (8.7) shows that there is an infinite number of solutions for γ. The solutions repeat with an interval $\Delta\gamma = 2\pi/p$. Moreover, there are two solutions in each of these intervals. Solutions with γ differing by $2\pi/p$ are really the same solution, because the waves have been specified at the ports, separated by a distance p. Consequently it is sufficient to consider the solution in the region $-1 < \gamma p/\pi < 1$, which is the first Brillouin zone. The nth Brillouin zone occupies the regions $n - 1 < |\gamma p/\pi| < n$. In the passbands, the group velocity is given by the slope of the curves for $\mathrm{Re}[\gamma p/\pi]$. At any one frequency, the curves shown as solid lines in Fig. 8.2 all have the same group velocity because they are all the same solution. Curves shown by broken lines are the second solution, propagating in the opposite direction. This also relates to Floquet's theorem, which states that wave fields in a periodic material have the form

$$\psi(x) = \sum_{n=-\infty}^{\infty} a_n \exp[-j(k + 2\pi n/p)]$$

where $k + 2\pi n/p$ corresponds to one of the γ values. The coefficients a_n are determined by the boundary conditions. If $2\pi/p$ is added to k, the series is exactly the same – all that has happened is that a_n has been replaced by a_{n+1}.

Physically, the assumption of localized reflection coefficients may not be quite correct. For example, the behavior of a metal electrode in a surface-wave grating depends on the configuration of several neighboring electrodes because of coupling due to the electric fields. For open-circuit electrodes, a wave incident on one electrode induces fields extending over several neighbors, so the reflection mechanism is not restricted to the locality of one electrode. However, it is assumed that an array of strips can be modeled by using effective reflection coefficients r_{s1}, r_{s2}, with the values adjusted if necessary so as to give the correct behavior of the array.

8.1.2 Finite-length grating

This is the practical case, of course. The grating is characterized by calculating its P-matrix, defined in Chapter 5, Section 5.3 and Appendix D. We consider an array with N strips, which might represent a grating or a shorted transducer. The waves on the left have amplitudes c_0 and b_0, and those on the right have amplitudes c_N and b_N. With $b_N = 0$, we have $P_{11} = b_0/c_0$ and $P_{21} = c_N/c_0$. With $c_0 = 0$ we have $P_{22} = c_N/b_N$ and $P_{12} = b_0/b_N$. The P-matrix is found by diagonalizing the P-matrix of one strip, giving the result of Appendix D, eq. (D.26). In terms of the notation here, the matrix is

$$P_{11} = \frac{p_{11} \sin N\gamma p}{\sin N\gamma p - p_{12} \sin (N-1)\gamma p}, \tag{8.9a}$$

$$P_{12} = P_{21} = \frac{p_{12} \sin \gamma p}{\sin N\gamma p - p_{12} \sin (N-1)\gamma p}, \tag{8.9b}$$

$$P_{22} = \frac{p_{22} \sin N\gamma p}{\sin N\gamma p - p_{12} \sin (N-1)\gamma p}, \tag{8.9c}$$

with the p_{ij} given by eqs (8.1). These parameters are relative to the ports, which are outside the structure, at a distance $p/2$ from the centers of the end strips. The strips are assumed to behave identically, so end effects have been excluded. Using power conservation, the magnitude of P_{11} can be shown to be given by

$$\frac{1}{|P_{11}|^2} = 1 + \left|\frac{t_s}{r_{s1}}\right|^2 \frac{\sin^2 \gamma p}{\sin^2 N\gamma p}, \tag{8.10}$$

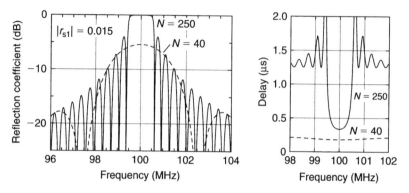

Figure 8.3 Reflection coefficient and delay of a periodic grating with center frequency 100 MHz.

which shows that $|P_{11}|$ cannot exceed unity. At the center of the first stop band, where $p = \lambda/2$, the reflection coefficient is

$$|P_{11}| = \tanh(N\alpha p) \approx \tanh(N|r_{s1}|). \tag{8.11}$$

This follows from eq. (8.9a), after some manipulation. At this frequency, the phase of P_{11} differs from the phase of r_{s1} by π. If P_{11} is referred to the center of the first strip, its phase is the same as that of r_{s1}.

For a weakly reflecting grating, with $N|r_{s1}| \ll 1$, eq. (8.10) gives

$$|P_{11}|^2 \approx |r_{s1}|^2 [\sin(N\,k_e p)/\sin(k_e\,p)]^2.$$

This is just the form expected when multiple reflections are insignificant.

Figure 8.3 shows examples of the reflection coefficient P_{11} for two grating lengths. For a strongly reflecting grating, $|P_{11}|$ is close to unity when the frequency is in the stop band given by eq. (8.8). It becomes unity in this band when $N \rightarrow \infty$. The delay is $\tau_g = -d\phi/d\omega$, where ϕ is the phase of P_{11}. This quantity is important for the performance of resonators. It can be shown from eq. (8.9a) that $\tau_g = \Delta t/|r_{s1}|$ for a semi-infinite grating ($N \rightarrow \infty$) at the center of the stop band. Here $\Delta t = p/v_e$ is the strip spacing in time units. Gratings are also described, using the COM method, in Section 8.2 of Chapter 8 and Section 11.1 of Chapter 11, the latter giving a derivation of τ_g.

Allowance for propagation loss

The above description has assumed that there is no loss. However, loss can be included in the model by assuming that it occurs only in the propagation between strips. The strips themselves can be taken to have no loss, so that

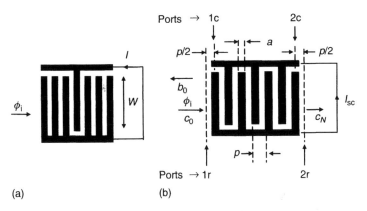

Figure 8.4 RAM analysis of single-electrode transducer for an incident wave with potential ϕ_i.
(a) Current due to one live electrode in an infinite array. (b) Current for general electrode polarities.
Ports 1r and 2r are used for RAM analysis. Ports 1c and 2c, at the centers of the end electrodes, are
used for COM analysis. For RAM analysis, the electrode connections do not need to alternate.

eqs (8.2) and (8.3) still apply. With lossy propagation, k_e becomes complex
but eqs (8.4)–(8.7) and (8.9) are still valid. The effective velocity is taken as
$v_e = \omega/\mathrm{Re}\{k_e\}$. The constant γ is complex at all frequencies, so there is no clear
distinction between the stop bands and passbands.

8.1.3 Transducer with regular electrodes

Assuming that there is no resistance in the bus bars, a shorted transducer behaves
like a shorted grating. This is described by the acoustic scattering parameters
P_{11}, P_{12} and P_{22}, given by eq. (8.9). For other parameters we consider the current
produced when the transducer is shorted, as in Fig. 8.4b. Here the electrodes are
taken to be regular and they are connected alternately to the bus bars, forming
a uniform single-electrode transducer. However, the analysis applies for an
arbitrary pattern of connections.

For the parameters P_{11}, P_{12} and P_{22} the wave amplitudes can be expressed
in relative terms, because these parameters are ratios of the wave amplitudes.
However, for transducer analysis the wave amplitudes c_n and b_n need to be
expressed in absolute terms. The amplitude $c(x)$ will now be defined such that
$|c(x)|^2/2$ equals the wave power, and $c(x)$ has the same phase as the surface
potential $\phi(x)$ of the wave. As shown by eq. (5.53), this requires that

$$c(x) = \phi(x)[\omega W/(2\Gamma_s)]^{1/2} \qquad (8.12)$$

and a corresponding relation applies for $b(x)$. Here $\Gamma_s = (\Delta v/v)/\varepsilon_\infty$ as in
eq. (3.31).

The derivation starts from a quasi-static formula for the current produced by a shorted transducer with one live electrode, as in Fig. 8.4a. If a wave with potential ϕ_i is incident on this transducer, the current flowing into the live electrode is given by

$$I = -j\omega W \phi_i \overline{\rho}_f(k_e).$$

This follows from the quasi-static theory, eq. (5.58), assuming that the electrodes do not reflect the waves. W is the transducer aperture, and ϕ_i is the surface potential of the incident wave, measured at the center of the live electrode. The function $\overline{\rho}_f(k_e)$ is given by eq. (5.59). The live electrode is taken to be in an infinite array, so that it is not affected by end effects, and all electrodes are identical.

To apply this to a reflective transducer, as in Fig. 8.4b, it is assumed that the reflections are weak. The current produced by one electrode depends on electrostatic fields in and near that electrode. These fields will be affected by the reflectivity. However, if the reflectivity is weak the wave amplitudes will not vary much over a distance of a few electrodes. Assuming that this is so, the current is related to the wave amplitudes as for a non-reflecting transducer. The main difference is that we now have to include waves in both directions, so that the current flowing into electrode n depends on the incident wave amplitudes c_{n-1} and b_n. These need to be multiplied by $\exp(-jk_e p/2)$ to refer to the center of the electrode. The current can also be found from the waves c_n and b_{n-1} leaving the electrode, giving slightly different results because the amplitudes change as the waves pass through the electrode. For better accuracy we take the average of currents due to waves incident on and leaving the electrode. Equation (8.12) is also needed, relating the wave amplitudes to the potentials. The current due to electrode n thus becomes

$$I_n = -j\overline{\rho}_f(k_e)\sqrt{\omega W \Gamma_s/2}[(c_{n-1} + b_n)e^{-jk_e p/2} + (c_n + b_{n-1})e^{jk_e p/2}]. \quad (8.13)$$

Since the transducer is shorted, the wave amplitudes are given by eqs (8.5). Starting at port 1, we can set $c_0 = 0$, so that there is no input wave here, and eqs (8.5) can be used repeatedly to obtain all the c_n and b_n. Equation (8.13) is used for each live electrode, and the currents I_n are summed to give the transducer current I_{sc}. The only input wave is b_N, and the ratio I_{sc}/b_N is by definition P_{32}. Cascading in the opposite direction gives P_{31}, and by reciprocity we have $P_{13} = -P_{31}/2$ and $P_{23} = -P_{32}/2$. If there is no power loss (k_e is real), this procedure is found to obey the conservation equations (D.6) and (D.7). The resulting P-matrix is referred to ports 1r and 2r which are just outside the transducer, as shown in Fig. 8.4b.

The above theory gives all the P-matrix components apart from the admittance P_{33}. This is generally more difficult to obtain. However, if we assume power conservation the conductance G_a is given by eq. (D.5), so that $G_a = |P_{13}|^2 + |P_{23}|^2$. The susceptance B_a is the Hilbert transform of this (Appendix A, Section A.6), and the capacitance C_t can be calculated as for a non-reflective transducer (Chapter 5, Section 5.4). This gives the admittance $P_{33} = G_a + j(B_a + \omega C_t)$. We now have all components of the P-matrix. The method applies for arbitrary electrode connections, so it is not restricted to the uniform single-electrode transducer of Fig. 8.4b.

Although we have only considered unapodized transducers, apodization can be included by dividing the device into channels, as described in Chapter 5, Section 5.6.2. Alternatively, an approximate analysis for a device can be obtained by first analyzing the response assuming no electrode reflections, and then mutiplying by a distortion function. This function depends only on r_{s1} and the number of electrodes [1].

8.1.4   Reflectivity and velocity for single-electrode transducers

This section considers the values of k_e and r_{s1}, which occur in the above equations. This is relevant to both gratings and transducers. For a transducer the responses for launching or receiving waves are basically the same, as shown by the reciprocity relation of eq. (B.25), so the same velocity applies. For reception the transducer is considered to be shorted, so k_e and r_{s1} are the same as for a shorted grating. We also consider open-circuit (disconnected) electrodes and grooves. Datta [3–5] has given some convenient approximate expressions for these parameters, derived from Auld's perturbation theory. In this approach, the perturbations due to the electrodes are considered to arise from two factors – electrical loading and mechanical loading. In practice, these perturbations are generally small, and the effects of the two types of loading can be simply added.

Electrode reflection coefficient

The reflection coefficient r_{s1} is considered to arise from an electrical term r_{s1E} and a mechanical term r_{s1M}, with corresponding relations for r_{s2}. Thus we have

$$r_{s1} \approx r_{s1E} + r_{s1M}; \quad r_{s2} \approx r_{s2E} + r_{s2M}. \tag{8.14}$$

For r_{s1}, the two terms r_{s1E} and r_{s1M} are each evaluated as if the other were absent. The electrical term applies if the film thickness h is zero. This term, derived by Datta [3], is considered in Appendix E. It is the same for both incident wave directions, so r_{s1E} and r_{s2E} will both be represented as r_{sE}. At the center of the

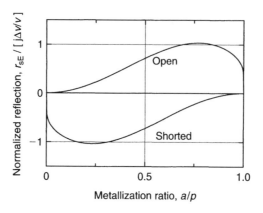

Figure 8.5 Electrode reflection coefficient due to electrical loading, at center of first stop band.

first stop band, we have

$$r_{sE} = \pm\frac{1}{2}j\pi\frac{\Delta v}{v}\left[\mp\cos\Delta + \frac{P_{1/2}(\pm\cos\Delta)}{P_{-1/2}(\pm\cos\Delta)}\right]. \tag{8.15}$$

The upper signs refer to open-circuit electrodes, and the lower signs to short-circuit electrodes. $P_\nu(x)$ is the Legendre function and $\Delta = \pi a/p$, where a and p are the width and pitch of the electrodes. The reflection coefficient varies slowly with frequency, and it is usually adequate to take it to be constant. The derivation accounts for neighboring electrodes. Figure 8.5 shows r_{sE} for the two cases, as functions of metallization ratio a/p. Apart from the coupling parameter $\Delta v/v$, these functions are independent of the choice of material. For $a/p = 1/2$, both cases give $|r_{sE}| = 0.718\Delta v/v$. The value of r_{sE}/j is negative for shorted electrodes and positive for open-circuit electrodes.

The mechanical terms r_{s1M} and r_{s2M} depend on the film thickness h and on the material and orientation, but they are independent of the electrode connection. For the center of the first stop band, Datta [5] gives r_{s1M} as

$$r_{s1M} = jR_M(h/\lambda_0)\sin(2\pi a/\lambda_0) \tag{8.16}$$

where $\lambda_0 = 2p$. From eq. (8.3) we have $r_{s2M} = -r_{s1M}^*$. The coefficient R_M is

$$R_M = -\frac{2\pi(\Delta v/v)}{\varepsilon_\infty}\left[(u_1/\phi)^2(\alpha_1 + \rho v_f^2) + (u_2/\phi)^2(\alpha_2 + \rho v_f^2) + (u_3/\phi)^2\rho v_f^2\right]. \tag{8.17}$$

Here the u_j are surface displacements for the wave on a free surface, with axes defined as in other chapters (x_1 is the propagation direction, x_3 is the surface normal pointing into the vacuum) and ϕ is the associated surface potential.

Table 8.1 Electrode reflection data for common materials, for shorted Al electrodes* or grooves, with $a/p = 1/2$. $r_{s1} = r_{s2} = jC_1 + jC_2 h/\lambda_0$

Material	C_1	C_2	Reference
Shorted metal strips			
LiNbO$_3$, Y–Z	−1.7%	−0.24	[6]
LiNbO$_3$, 128°Y–X	−2%	+0.8	[11, 9]
LiTaO$_3$, X–112°Y	−0.23%	−0.45	[12, 7]
Quartz, ST–X	−0.04%	−0.50	[6]
Grooves			
LiNbO$_3$, Y–Z	0	+0.65	[6, 13]
Quartz, ST–X	0	+0.6	[7]

*For open-circuit electrodes, the sign of C_1 is reversed.

The free-surface velocity is v_f. These quantities are given by the wave analysis method described in Chapter 2. The constants α_1, α_2 and ρ (density) are properties of the film material, assumed to be isotropic. In terms of Lamé constants, $\alpha_1 = 4\mu(\lambda + \mu)/(\lambda + 2\mu)$ and $\alpha_2 = \mu$. For aluminum, $\alpha_1 = 7.8 \times 10^{10}$ N/m^2, $\alpha_2 = 2.5 \times 10^{10}$ N/m^2 and $\rho = 2695$ kg/m^3. The $(u_j/\phi)^2$ terms are real for most common materials, so that r_{s1M} is imaginary. However they can be complex, and this gives the natural SPUDT effect, discussed later (Section 8.2.3). If r_{s1M} is imaginary, eq. (8.3) gives $r_{s2M} = r_{s1M}$.

An alternative approach is to define r_+ and r_- as reflection coefficients for an up-step (electrode front edge) and a down-step, respectively. For a wave incident from the left, these are given by

$$r_\pm = \pm(1/2)R_M h/\lambda_0.$$

Assuming that the reflections are weak and there is little velocity perturbation, this gives the electrode reflection coefficient of eq. (8.16).

A wide variety of measurements are reported by, for example, Dunnrowitz *et al.* [6], Wright [7] and Koshiba and Mitobe [8]. Most results agree well with Datta's theory. Experimentally, measurements are made on reflecting gratings, and the RAM or COM analysis can be used to deduce r_{s1}. Table 8.1 gives examples for common materials, for $a/p = 1/2$. In all these cases, r_{s1} is imaginary so $r_{s2} = r_{s1}$. For example, ST–X quartz with $a/p = 1/2$ gives $r_{s1} = r_{s2} \approx \pm 4 \times 10^{-4}$ j-0.5j h/λ_0. For shorted electrodes (lower sign), r_{s1}/j is negative. For open-circuit electrodes (upper sign), r_{s1}/j is positive for small h and negative for large h, with a zero at $h/\lambda_0 \approx 0.0008$. This agrees well with Wright's experimental results [7]. For 128°Y–X lithium niobate, C_2 is positive, an unusual case. Consequently, the short-circuit reflection coefficient is positive for small h and negative for large h, with a zero at $h/\lambda_0 \approx 0.04$. This is confirmed experimentally [9].

Table 8.2 Magnitude and phase for surface displacement and potential of Rayleigh waves*.

Material	u_1/ϕ (nm/V)	u_2/ϕ (nm/V)	u_3/ϕ (nm/V)
LiNbO$_3$, Y–Z	0.137, 174.9°	0	0.196, 85.0°
LiNbO$_3$, 128°Y–X	0.189, 90°	0.014, 180°	0.213, 0°
LiTaO$_3$, X–112°Y	0.335, 180°	0.121, 180°	0.446, 90°
Quartz, ST–X	0.405, 90°	0.059, 270°	0.615, 270°

*x_1 is the propagation direction and x_3 is the outward-directed surface normal.

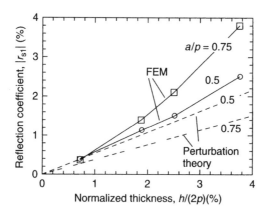

Figure 8.6 Electrode reflection coefficient for 38°Y–X quartz. Points: data from Ventura's FEM analysis [10]. Broken lines: from perturbation theory, eqs (8.15) and (8.16).

Table 8.2 has values of the u_j/ϕ for several materials. If the direction of the wave is reversed, these variables take on new values u_j', ϕ', given by [14]

$$u_1'/\phi' = -(u_1/\phi)^*, \quad u_2'/\phi' = -(u_2/\phi)^* \quad \text{and} \quad u_3'/\phi' = (u_3/\phi)^*,$$

where the asterisk indicates a complex conjugate. As usual x_1 is defined as the propagation direction, so reversing the wave direction reverses the direction of x_1 and x_2.

Figure 8.6 shows some electrode reflection data for ST–X quartz. The electrical term is small here. The perturbation theory [eqs (8.15) and (8.16)] is shown by broken lines. The points are accurate results obtained by Ventura [10] using the finite-element method (FEM), which will be described later. The perturbation theory agrees well for a metallization ratio $a/p = 1/2$. However, for wider electrodes with $a/p = 0.75$, the perturbation theory gives significant errors at the larger values of h/λ_0. This distinction is important for SPUDTs, described in Chapter 9. It appears that the perturbation theory is reasonable provided h is small. The FEM results on this figure agree very well with experimental measurements [10].

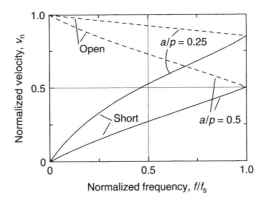

Figure 8.7 Velocity variation due to electrical loading. The normalized velocity is defined as $v_n = (v_e - v_m)/(v_f - v_m)$.

Effective wave velocity

The wavenumber k_e above is related to an effective wave velocity $v_e = \omega/k_e$. This refers to the velocity as perturbed by the strips, but not taking account of the reflections. The velocity is quite close to the free-surface velocity v_f. Generally, the changes are not very significant, though they become crucial in the case of the transversely coupled resonator filter (Chapter 10, Section 10.4). To first order, the changes due to electrical and mechanical loading, Δv_E and Δv_M are independent and can be simply added. Thus we have

$$v_e = v_f + \Delta v_E + \Delta v_M. \tag{8.18}$$

When h is small, mechanical effects are insignificant. The velocity change for this case was derived by Bløtekjaer [15], as shown in Appendix E. This is given by

$$\Delta v_E \approx \frac{v_m - v_f}{2}\left[1 \mp \frac{P_s(\pm\cos\Delta)}{P_{s-1}(\pm\cos\Delta)}\right] \tag{8.19}$$

with upper signs for open-circuit electrodes and lower signs for shorted electrodes. Here $s \equiv k_f\, p/(2\pi)$, with $k_f = \omega/v_f$, and we assume $s < 1$. Because the velocity change is small and there is little dispersion, we have $s \approx f/f_s$, where $f_s = v_e/p$ is the sampling frequency. For a single-electrode transducer, f_s is the second-harmonic frequency. Figure 8.7 shows velocities obtained using this formula. The shorted case applies for transducer analysis. For $a/p = 1/2$, the velocity for shorted electrodes varies almost linearly, from v_m at $f = 0$ to $(v_m + v_f)/2$ at $f = f_s$. Stop bands at frequencies $nf_s/2$ are not shown in Fig. 8.7 because they were excluded from the theory.

The mechanical loading depends on the film thickness h, and for many cases it is known to have the form

$$\Delta v_M/v_f \approx D_M(a/p)(h/\lambda_0) - K_V(h/\lambda_0)^2. \qquad (8.20)$$

The constant D_M is given by [4, 16–18]

$$D_M = \frac{2\pi(\Delta v/v)}{\varepsilon_\infty} \left[\left|\frac{u_1}{\phi}\right|^2 (\alpha_1 - \rho v_f^2) + \left|\frac{u_2}{\phi}\right|^2 (\alpha_2 - \rho v_f^2) - \left|\frac{u_3}{\phi}\right|^2 \rho v_f^2 \right].$$

$$(8.21)$$

For aluminum electrodes on ST–X quartz, this gives $D_M = -0.13$. The K_V term in eq. (8.20) is attributed to energy stored at the electrode edges [19]. For ST–X quartz, a typical value [20, 21] is $K_V \approx 0.8$. Experiments on grooves show a velocity shift proportional to $(h/\lambda_0)^2$, in agreement with eq. (8.20) because D_M is zero in this case.

For a continuous film the velocity shift is given by eq. (8.20), with $a/p = 1$ and $K_V = 0$.

8.2   COUPLING OF MODES (COM) EQUATIONS

This method for analyzing reflective transducers has become very widespread owing to its relative simplicity and wide applicability. Coupled-mode methods have been used for microwave devices since the 1950s and, more recently, in optics [22]. These applications involve coupling between co-directional or contra-directional waves. For surface waves, there is the added complication of distributed transduction. Suitable equations were considered by Koyamada and Yoshikawa [23], followed by many others [18, 24–26]. In the derivation, the reflection and transduction phenomena are modeled empirically. The physical evaluation of these parameters is left to later, when the solutions of the equations are examined (Section 8.2.2). The derivation here is related to the RAM of Section 8.1 above, but a more fundamental method is derived from a Floquet expansion [26].

8.2.1   Derivation of equations

We first consider reflection, taking transduction to be absent. As shown in Fig. 8.8, consider waves with amplitude $c(x)$ propagating to the right, and $b(x)$ propagating to the left. The physical meaning of these amplitudes is not considered yet, but later they will be related to the wave powers. The solid dots represent reflectors, with spacing P, assumed to be localized. The broken lines are reference points at $x = x_1 + nP$, midway between the reflectors, where

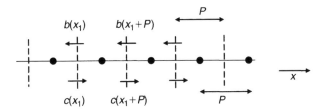

Figure 8.8 Reflection modeling for COM method.

the wave amplitudes are measured. For the wave to the right, the change of amplitude for one period is written as

$$c(x_1 + P) = c(x_1)T \exp(-jk_e P) + b(x_1 + P)R \exp(-jk_e P) \qquad (8.22)$$

where R and T are reflection and transduction parameters for one cell, respectively. Here k_e is an effective wavenumber allowing for velocity changes, as for the RAM method above. For the COM method, we define the slowly varying amplitudes $c'(x) = c(x)\exp(jk_e x)$ and $b'(x) = b(x)\exp(-jk_e x)$. Strong reflections occur for a frequency such that k_e is close to the Bragg wavenumber $k_g = \pi/P$, and also at the harmonics of this frequency. We consider harmonic number m, and we define a 'detuning parameter' δ by the equation

$$\delta = k_e - mk_g; \quad k_g = \pi/P. \qquad (8.23)$$

Thus, the reflected waves are in phase when $\delta = 0$. Making these changes, and with a little manipulation, eq. (8.22) becomes

$$c'(x_1 + P) - c'(x_1) = (T - 1)c'(x_1) + Rb'(x_1 + P)\exp(2j\delta x_1). \qquad (8.24)$$

Here a term $\exp(2jx_1 mk_g)$ has been absorbed into R. We assume that the reflections are very weak, so the amplitude $c'(x)$ varies slowly with x. The left side of eq. (8.24) has the form $P(\Delta c'/P)$, which is approximately $P(dc'/dx)$.

On the right, there are terms depending on $(T - 1)$ and R. Assuming power conservation, we have $|T|^2 = 1 - |R|^2$. We now assume that T is real. This is possible because a phase change of T can be compensated by changing the wavenumber k_e, as seen from eq. (8.22). For real T we have $T^2 = 1 - |R|^2$, and for small R this gives $T - 1 \approx -|R|^2/2$, so that $|T - 1| \ll |R|$. Hence, if $|R|$ is small enough the $(T - 1)$ term can be neglected in eq. (8.24), leading to

$$dc'(x)/dx = c_{12}b'(x)\exp(2j\delta x) \qquad (8.25)$$

where we define a reflection parameter $c_{12} = \text{Lim}\{R/P\}$ for $R \to 0$. By an analogous argument, we find

$$db'(x)/dx = c_{21}c'(x)\exp(-2j\delta x) \qquad (8.26)$$

where c_{21} is another reflection parameter.

For transduction analysis, we ignore reflections. As shown in Fig. 8.9, we consider a series of 'transducers' represented by dots, with spacing Λ. The wave to the right has amplitude given by

$$c(x_1 + \Lambda) = c(x_1) \exp(-jk_e\Lambda) + \alpha V \exp(-jk_e\Lambda/2)$$

where V is the applied voltage and α is a transduction parameter. We define a detuning parameter $\varepsilon = k_e - nk_T$, where $k_T = 2\pi/\Lambda$ and n is a harmonic number. Writing $c(x) = c'(x)\exp(-jk_ex)$ as before, we find

$$c'(x_1 + \Lambda) - c'(x_1) = \alpha V \exp(j\varepsilon x_1) \tag{8.27}$$

where a term $\exp(2jnx_1/\Lambda)$ has been absorbed into α. As before, the interaction is assumed to be weak, and the left side of eq. (8.27) approximates to $\Lambda dc'/dx$. Defining α_1 as the limit of α/Λ for $\alpha \to 0$, eq. (8.27) becomes

$$dc'(x)/dx = \alpha_1 V \exp(j\varepsilon x). \tag{8.28}$$

By a similar argument,

$$db'(x)/dx = \alpha_2 V \exp(-j\varepsilon x) \tag{8.29}$$

where α_2 is another transduction parameter.

For the transducer current, we consider a shorted transducer. The dots in Fig. 8.9 are now regarded as current sources, and $I(x_1)$ is the current flowing in the bus bar at $x = x_1$. Consider the wave to the right. The source at $x = x_1 + \Lambda/2$ gives a current $I(x_1 + \Lambda) - I(x_1) = \eta c(x_1) \exp(-jk_e\Lambda/2)$, where η is some transduction parameter. We can write this as

$$I(x_1 + \Lambda) - I(x_1) = \eta c'(x_1) \exp(-j\varepsilon x_1)$$

where some exponentials have been absorbed into η. The left side of this equation approximates to $\Lambda \, dI/dx$ and, taking the limit $\eta \to 0$, we find

$$dI(x)/dx = \eta_1 c'(x) \exp(-j\varepsilon x)$$

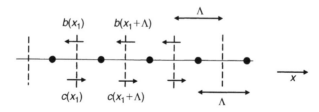

Figure 8.9 Transduction modeling for COM method.

where η_1 is the limit of η/Λ. A similar derivation gives the current due to the wave $b(x)$ directed to the left. In addition, when $V \neq 0$ there is a contribution due to the capacitance. The total current becomes

$$dI(x)/dx = \eta_1 c'(x) \exp(-j\varepsilon x) + \eta_2 b'(x) \exp(j\varepsilon x) + j\omega C_l V \qquad (8.30)$$

where η_2 is another transduction parameter and C_l is the capacitance per unit length.

For a total description of the device, it is assumed that the reflection and transduction terms can be simply added. From eqs (8.25), (8.26), (8.28)–(8.30), we have

$$dc'/dx = c_{12} b' \exp(2j\delta x) + \alpha_1 V \exp(j\varepsilon x)$$
$$db'/dx = c_{21} c' \exp(-2j\delta x) + \alpha_2 V \exp(-j\varepsilon x)$$
$$dI/dx = \eta_1 c' \exp(-j\varepsilon x) + \eta_2 b' \exp(j\varepsilon x) + j\omega C_l V \qquad (8.31)$$

with $\delta = k_e - mk_g = k_e - m\pi/P$ and $\varepsilon = k_e - nk_T = k_e - 2n\pi/\Lambda$. Here m and n are harmonic numbers, P is the reflection pitch and Λ is the transduction pitch.

In practice, the COM equations are of interest only when transduction and reflection occur simultaneously. For these cases, δ and ε are equal, and we can write

$$\delta = \varepsilon = k_e - k_c = k_e - mk_g$$

where $k_c = mk_g$ is the wavenumber at the center frequency, which may be a harmonic. For example, a single-electrode has $P = p$, where p is the electrode pitch, so $k_g = \pi/p$. Also, $\Lambda = 2p$. At the fundamental frequency, we have $m = 1$ giving $mk_g = \pi/p$, and $n = 1$ so that $nk_T = \pi/p = k_g$. Hence $\delta = \varepsilon$. For a distributed acoustic reflection transducer (DART) transducer, considered later in Chapter 9, Section 9.2, the reflection pitch P is equal to the transducer pitch Λ. This transducer operates at the fundamental transduction frequency and the second-harmonic reflection frequency. So we have $m = 2$ and $n = 1$, giving $mk_g = m\pi/P = 2\pi/\Lambda$ and $nk_T = 2n\pi/\Lambda = 2\pi/\Lambda$. Hence, again we have $\delta = \varepsilon$. When using the COM equations, the integers m and n do not need to be considered explicitly. However, the parameters in the equations, such as c_{12}, will generally depend on them.

A modified form of the COM equations is obtained by defining another pair of slowly varying amplitudes as

$$C(x) = c(x) \exp(jk_c x); \quad B(x) = b(x) \exp(-jk_c x), \qquad (8.32)$$

where $k_c = m k_g$ is the center-frequency wavenumber. Using also $\varepsilon = \delta$, the equations become

$$dC(x)/dx = -j\delta C(x) + c_{12}B(x) + \alpha_1 V$$
$$dB(x)/dx = j\delta B(x) + c_{21}C(x) + \alpha_2 V$$
$$dI(x)/dx = \eta_1 C(x) + \eta_2 B(x) + j\omega C_l V. \qquad (8.33)$$

The coefficients in these equations are related though constraints on the P-matrix (Appendix D). Considering a short transducer with length Δx operating at its center frequency ($\delta = 0$), the above equations give $P_{13} = -\alpha_2 \Delta x$, $P_{23} = \alpha_1 \Delta x$, $P_{31} = \eta_1 \Delta x$ and $P_{32} = \eta_2 \Delta x$. By reciprocity we have $P_{31} = -2P_{13}$ and $P_{32} = -2P_{23}$, giving $\eta_1 = 2\alpha_2$ and $\eta_2 = -2\alpha_1$. Power conservation gives $\text{Re}\{V^* dI/dx\} = (d/dx)[|C|^2 - |B|^2]$. For $V = 0$ this gives $c_{21} = c_{12}^*$. For $V \neq 0$ and $c_{12} = 0$ we find $\alpha_2 = \alpha_1^*$. With these relations, the COM equations finally become

$$dC(x)/dx = -j\delta C(x) + c_{12}B(x) + \alpha_1 V \qquad (8.34a)$$
$$dB(x)/dx = j\delta B(x) + c_{12}^* C(x) + \alpha_1^* V \qquad (8.34b)$$
$$dI(x)/dx = 2\alpha_1^* C(x) - 2\alpha_1 B(x) + j\omega C_l V. \qquad (8.34c)$$

These equations are derived more rigorously by Plessky and Koskela [26], and they are also given by Chen and Haus [27]. The wave amplitude $C(x)$ is defined such that the wave power is $|C(x)|^2/2$, so that $|C(x)|$ is the same as $|c(x)|$, eq. (8.12). A similar relation applies for $B(x)$. These definitions are needed to comply with the above power considerations. Hashimoto [28] quotes several other forms of the equations, including the effect of electrode resistivity. Although power conservation has been used to deduce relations between the parameters, power loss can be included in the model by allowing k_e to have an imaginary part.

8.2.2 General solution for a uniform transducer

It is assumed throughout that the parameters c_{12} and α_1 are constants, independent of x. This implies that we consider uniform (unweighted) transducers and gratings. A non-uniform transducer can be analyzed as a cascade of uniform transducers, using formulae in Appendix D. Hence it is sufficient to consider a uniform transducer here. The main aim here is to derive the P-matrix for a grating or transducer.

Grating-mode solution

The first step is to consider grating modes, that is, waves propagating in an infinite structure. We assume $V = 0$, which applies for a grating or a shorted transducer. Consider a solution in which $C(x)$ and $B(x)$ are both proportional to $\exp(-\mathrm{j}sx)$. Using $\mathrm{d}/\mathrm{d}x = -\mathrm{j}s$ in eqs (8.34a) and (8.34b) we find

$$s^2 = \delta^2 - |c_{12}|^2 \qquad (8.35)$$

and also $B(x)/C(x) = \mathrm{j}(\delta - s)/c_{12} = \mathrm{j}c_{12}^*/(\delta + s)$. This ratio is independent of x because both waves are proportional to $\exp(-\mathrm{j}sx)$. This solution is similar to that found by the RAM method, eq. (8.7). If there is no loss, k_e is real and so δ is also real, and eq. (8.35) shows that s^2 is real. At most frequencies s is real, but it becomes imaginary when $|\delta| < |c_{12}|$. This region is a stop band, where the wave is cut off. Figure 8.10 shows the dispersion relation. Unlike the RAM method, the COM gives only one stop band, and this is a consequence of the approximations used in the derivation. The stop band edges are at $\delta = \pm|c_{12}|$. For a single-electrode transducer or grating, it is shown later that $c_{12} = -r_{s1}^*/p$, where r_{s1} is the reflection coefficient of one electrode. From this, the fractional width of the stop band is $\Delta f/f_\mathrm{c} = 2|r_{s1}|/\pi$, in agreement with the RAM result of eq. (8.8). This formula assumes that the velocity $v_\mathrm{e} = \omega/k_\mathrm{e}$ is constant. The group velocity is given by the slopes of the curves in Fig. 8.10, and the branch with negative group velocity is shown as a broken line.

The full solution for $V = 0$, involving a grating mode traveling in each direction, is written as

$$C(x) = h_1\mathrm{e}^{-\mathrm{j}sx} + h_2\mathrm{e}^{\mathrm{j}sx}; \quad B(x) = h_1p_1\mathrm{e}^{-\mathrm{j}sx} + h_2p_2\mathrm{e}^{\mathrm{j}sx}$$

where

$$p_1 = \mathrm{j}(\delta - s)/c_{12}; \quad p_2 = \mathrm{j}(\delta + s)/c_{12} \qquad (8.36)$$

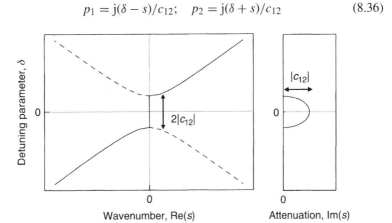

Figure 8.10 Dispersion relation for grating mode, according to COM.

and h_1, h_2 are arbitrary constants. The wavenumber s is either solution of eq. (8.35).

General solution

When V is non-zero, eqs (8.34a) and (8.34b) have a solution that is independent of x. This solution is easily found to be

$$C(x) = K_1 V; \quad B(x) = K_2 V, \tag{8.37}$$

where

$$K_1 = (\alpha_1^* c_{12} - \mathrm{j}\delta\alpha_1)/s^2; \quad K_2 = (\alpha_1 c_{12}^* + \mathrm{j}\delta\alpha_1^*)/s^2. \tag{8.38}$$

If there is no loss we find $K_2 = K_1^*$. In mathematical terminology, eq. (8.37) is a 'particular integral'. The grating-mode solution is a 'complementary function', a solution obtained when the driving term V is zero. Because the equations are linear, the complementary function can be added to the particular integral to obtain a more general solution. The general solution is therefore

$$C(x) = h_1 \exp(-\mathrm{j}sx) + h_2 \exp(\mathrm{j}sx) + K_1 V;$$

$$B(x) = h_1 p_1 \exp(-\mathrm{j}sx) + h_2 p_2 \exp(\mathrm{j}sx) + K_2 V, \tag{8.39}$$

with p_1 and p_2 given by eq. (8.36). The constants h_1 and h_2 are to be determined from the boundary conditions.

These equations are applied to the single-electrode transducer of Fig. 8.4b. As in Chapter 5, Section 5.3, the P-matrix is defined by the equation

$$[A_{t1}, A_{t2}, I]^{\mathrm{T}} = [P_{ij}][A_{i1}, A_{i2}, V]^{\mathrm{T}}$$

where superscript T indicates a transpose. A_i and A_t are amplitudes of the incident and emerging surface waves, with additional subscript 1 or 2 to show that they are measured at the acoustic ports. For the COM analysis it is usual to take the ports to be at points 1c and 2c on Fig. 8.4b, that is, at the centers of the end electrodes. These are taken to be at $x = 0$ and $x = L$, respectively. At $x = 0$ the input wave amplitude is $A_{i1} = c(0) = C(0)$ and the output wave is $A_{t1} = b(0) = B(0)$. At $x = L$ the input wave is $A_{i2} = b(L) = B(L) \exp(\mathrm{j}k_c L)$ and the output wave is $A_{t2} = c(L) = C(L) \exp(-\mathrm{j}k_c L)$. Using eqs (8.39) and (8.34c), the P-matrix is found to be as follows:

$$P_{11} = -c_{12}^* \sin(sL)/D; \tag{8.40a}$$

$$P_{12} = P_{21} = s \exp(-jk_cL)/D; \tag{8.40b}$$

$$P_{22} = c_{12} \sin(sL) \exp(-2jk_cL)/D; \tag{8.40c}$$

$$P_{31} = -2P_{13} = \{2\alpha_1^* \sin(sL) - 2sK_2[\cos(sL) - 1]\}/D; \tag{8.40d}$$

$$P_{32} = -2P_{23} = \exp(-jk_cL)\{-2\alpha_1 \sin(sL) - 2sK_1[\cos(sL) - 1]\}/D; \tag{8.40e}$$

$$P_{33} = -K_1P_{31} - K_2P_{32}\exp(jk_cL) + 2(\alpha_1^*K_1 - \alpha_1K_2)L + j\omega C_t, \tag{8.40f}$$

where $D = s\cos(sL) + j\delta\sin(sL)$. Also, $C_t = LC_l$ is the transducer capacitance. The expression for the admittance P_{33} can be obtained from the particular integral, eq. (8.37), in which $C(x)$ and $B(x)$ are constants. This solution gives the current easily, but it includes waves incident on the transducer. Terms proportional to P_{31} and P_{32} are added to cancel these waves. Equations (8.40a)–(8.40c) give the scattering by a reflecting grating, and eqs (8.40d)–(8.40f) are needed in addition for a transducer. These equations satisfy reciprocity. If δ is real, the equations also satisfy the P-matrix conditions for no loss (Appendix D). Losses can be included by allowing k_e, and therefore δ, to have an imaginary part. This assumes that the unperturbed waves decay exponentially.

The reflection coefficient P_{11} is considered further in Chapter 11, Section 11.1.

COM parameters for single-electrode transducer

For a single-electrode transducer, c_{12} and α_1 can be evaluated using the RAM analysis. Assuming that the transducer length L is small, the COM equations give $P_{11} \approx -c_{12}^*L$, $P_{22} \approx c_{12}L$ and $P_{13} \approx -\alpha_1^*L$, where P_{13} ignores the reflection term. Now take $L = p$, referring to a 'transducer' with only one electrode. The RAM gives $P_{11} = -r_{s1}$ and $P_{22} = -r_{s2}$. These are equated to the COM result. The difference in port locations (Fig. 8.4b) is allowed for by changing the signs. This gives

$$c_{12} = -r_{s1}^*/p = r_{s2}/p. \tag{8.41}$$

The COM thus gives $r_{s2} = -r_{s1}^*$, agreeing with eq. (8.3). Hence, the COM implicitly takes the strip transmission coefficient t_s to be real. This is expected, because T was assumed to be real in the derivation of Section 8.2.1.

For transduction, consider a transducer of length $2p$, with one live electrode located centrally. This corresponds to ports 1c and 2c in Fig. 8.4b.

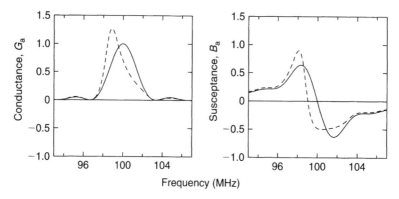

Figure 8.11 Conductance and susceptance for a uniform single-electrode transducer with 30 periods, using COM analysis with $r_{s2} = r_{s1}$. Full lines are calculated for $r_{s1} = 0$, broken lines for $r_{s1}/j = -0.017$. The curves are normalized to the maximum of G_a for $r_{s1} = 0$.

The short-circuit current in response to an incident wave is, from eq. (5.58), $I_{sc} = -j\omega W\phi_i\overline{\rho}_f(k_e)$, assuming that the reflectivity is negligible. Here ϕ_i is the potential of the incident wave, measured at the center of the live electrode. This formula refers to an electrode within an infinite array, which is appropriate for the COM theory. From this formula we obtain P_{13}, and equating this to the COM result gives

$$\alpha_1 = j\overline{\rho}_f(k_e)\sqrt{2\omega W\Gamma_s}/(2p) \qquad (8.42)$$

with $\overline{\rho}_f(k_e)$ given by eq. (5.59). For the COM analysis, c_{12} and α_1 are often taken to be independent of ω.

Figure 8.11 shows the conductance and susceptance of a uniform transducer, calculated using the COM theory. It can be seen that the reflections cause considerable distortion, reducing the frequency where G_a is maximized. Figure 8.12 shows the conductance G_a of uniform single-electrode transducers on Y–Z lithium niobate, for metallization ratios $a/p = 0.2$, 0.5 and 0.75, compared with experimental results [29]. For these transducers the mechanical loading is negligible so the reflectivity is given by the electrical term, shown in Fig. 8.5, using the value for shorted electrodes. The reflections cause G_a to be distorted. The distortion is largest for small a/p because $|r_{s1}|$ is largest for this case, as shown by Fig. 8.5. The experiments confirm the theory. They also confirm the sign of r_{s1}/j. If r_{s1}/j were positive instead of negative, the peak of G_a would shift to higher frequencies instead of lower.

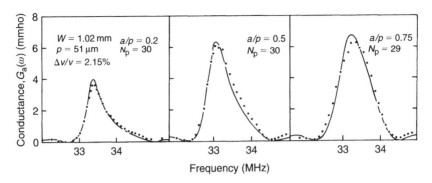

Figure 8.12 Conductance of single-electrode transducers on $Y-Z$ lithium niobate, for different electrode widths. Experimental points from Ref. [29].

The response can also be calculated using the RAM theory of Section 8.1, which allows for arbitrary sequences of electrode polarities. For a uniform transducer the results are very similar.

Comparison between COM and RAM methods

Both of these methods are quite convenient to use, though they are rather empirical. For example, end effects are ignored. The COM method has the advantage that it gives algebraic formulae for the P-matrix components of a grating and a uniform transducer, and for other properties which will be considered later. It is valid for almost any type of periodic reflective transducer, not just for the single-electrode type considered here. It can allow for propagation loss. The susceptance B_a is obtained directly, without needing a Hilbert transform. On the other hand the parameters, notably c_{12} and α_1, need to be provided from elsewhere. For example, for a single-electrode transducer the RAM results can be used as discussed above.

The RAM method is less general but it can be more informative. For example, a non-periodic transducer (with regular electrodes) can be accurately modeled because any sequence of electrode polarities is valid for eq. (8.13). The RAM also gives the reflection and transduction parameters for a single-electrode transducer, and these provide the required COM parameters as described above. A disadvantage is that G_a can be calculated only if propagation loss is ignored. However, for a shorted transducer or grating the loss can be included. The RAM described here represents the transduction quite accurately, being related to the electrostatic fields. In consequence, it correctly describes the capacitance and the harmonic responses, allowing for variation of the electrode width a. Unlike the COM, the RAM method takes account of the fact that a voltage applied to one electrode will generate charges on neighboring electrodes. Because of this, the RAM is not equivalent to use of a P-matrix for each electrode – there are capacitive terms coupling neighboring electrodes, and these cannot be represented in the conventional P-matrix approach.

The two methods use different wave amplitudes, $c(x)$ and $b(x)$ for the RAM, and $C(x)$ and $B(x)$ for the COM. The RAM dispersion relation, eq. (8.7), approximates to the COM result if the reflectivity is weak. Assume that $k_e p$ and γp are both close to π, and expand the cosines. Noting that t_s is real, we have $t_s \approx 1 - |r_{s1}|^2/2$. This leads to $(\gamma p - \pi)^2 = (k_e p - \pi)^2 + |r_{s1}|^2$, which corresponds to the COM relation of eq. (8.35).

8.2.3 The Natural SPUDT effect in single-electrode transducers

In the above analysis, the reflection coefficients for a shorted strip, r_{s1} and r_{s2} were allowed to be different. These coefficients refer to waves incident from the left and the right, respectively. For a symmetrical electrode, these coefficients will often be the same. However, they are affected by the symmetry of the substrate, as well as the electrode geometry. Some substrate materials can give $r_{s2} \neq r_{s1}$, even if the electrodes are symmetrical. If this is so, a single-electrode transducer will show directivity even if it has a symmetrical geometry. The term 'directivity' means that, when a voltage is applied, the transducer generates waves with different amplitudes in the two directions, so that $|P_{23}| \neq |P_{13}|$. This is called the 'Natural SPUDT', or 'N-SPUDT', effect. The term SPUDT refers to 'Single-Phase UniDirectional Transducer'. As explained in Chapter 9, there are various design techniques for introducing geometrical asymmetry such that directivity is produced. In contrast, the term 'natural' indicates that directivity can occur even when the geometry is symmetrical. The effect was first seen in quartz [30].

Most surface-wave materials do not show the N-SPUDT effect because their orientations are chosen to give attractive surface-wave properties. This often implies some degree of symmetry, and the symmetry can eliminate the N-SPUDT effect. However, general orientations can easily give the effect.

In a single-electrode transducer, the directivity depends on the relative phase of the reflection and transduction parameters. The latter always has the same phase, as shown by eq. (8.42). An electrode behaves symmetrically if $r_{s2} = r_{s1}$, and then r_{s1} and c_{12} are both imaginary (eqs (8.3) and (8.41)). A symmetrical single-electrode transducer then has no directivity. This can be shown from the COM results of eqs (8.40d) and (8.40e), which give $P_{23} = P_{13}$ if c_{12} is imaginary. This situation occurs for *any* material if the film thickness h is small, because in this case r_{s1} is dominated by the electrical term, which is imaginary by eq. (8.15). In addition, the electrical Green's function $G(x)$ applies when h is small, and it gives $G(-x) = G(x)$, as shown in Appendix B. A geometrically symmetric transducer governed by a symmetric Green's function must behave symmetrically, so there is no directivity.

Equations (8.16) and (8.17) show that the mechanical term r_{sM} is imaginary if the factors u_j/ϕ are real or imaginary. This is often the case for surface-wave materials, as for some examples in Table 8.2. In general, however, the phases of the u_j/ϕ depend on the substrate orientation. If the phases of these terms are not multiples of $\pi/2$, the N-SPUDT effect will be seen if h is large enough for the mechanical term to be significant. For ST-cut quartz, the effect is maximized for propagation at about $25°$ to the X-axis [16, 31]. The orientation is $34°Y–X + 25°$. At typical values of h the mechanical term dominates and r_{s1} becomes real, as confirmed experimentally [31].

Another example is langasite, $La_3Ga_5SiO_{14}$, with Euler angles λ, μ, $\theta = 0°$, $138.5°$, $26.6°$. Figure 8.13 shows the calculated amplitude and phase of r_{s1} as functions of normalized film thickness $h/(2p)$, for $a/p = 1/2$ [32]. Here the dots refer to the perturbation method, eqs (8.15)–(8.17). The continuous lines show a more accurate result obtained by the FEM, described later. The calculations were done using Hashimoto's program FEMSDA [33]. The perturbation method gives good accuracy for small film thickness, with $h/(2p) < 2\%$. The data also show that r_{s1} is imaginary when h is small, so that there is no N-SPUDT effect, as discussed above. At larger thicknesses the mechanical term becomes significant,

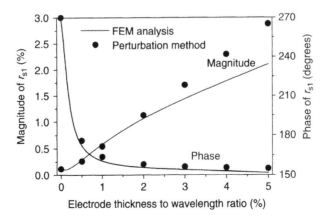

Figure 8.13 Theoretical electrode reflection coefficient on langasite N-SPUDT orientation, for aluminum electrodes with metallization ratio 0.5. FEM results from program FEMSDA [33] (courtesy of A. Shvetsov and S. Zhgoon).

and the phase of r_{s1} approaches about 150°. This shows a strong, but not ideal, N-SPUDT effect. Similar results are reported by Hasegawa *et al.* [44].

The behavior of SPUDTs will be discussed in more detail in the next chapter. It will be shown that the directivity of an SPUDT increases with the number of electrodes, and this also applies to N-SPUDT transducers. Practical cases can give quite substantial directivity, for example 6 dB or more. If the directivity is high, the transducer can be electrically matched such that its acoustic reflection coefficient is ideally zero, and also its conversion loss is small. In principle, this gives low loss and good triple-transit suppression simultaneously. However, the usefulness of an N-SPUDT transducer is limited because the substrate orientation determines the 'forward' direction of the transducer. Figure 8.14 illustrates the problem. In Fig. 8.14a, both transducers are the single-electrode type, assumed to have the 'forward' direction to the right. In principle, the left transducer can be electrically matched such that its acoustic reflectivity is zero, and most of the applied power emerges toward the right transducer. However, the directivity of the right transducer is unsuitable. For waves incident from the left, it cannot be matched for low reflection, or to give low loss. One solution to this is to use a double-electrode design for one transducer, as in Fig. 8.14b. A double-electrode transducer has negligible internal reflections, so it is not affected by the N-SPUDT effect and its behavior is bidirectional. In principle, this device can give zero triple-transit signal and an insertion loss not much more than the 3 dB bidirectional loss of the right transducer. However, the performance is not very good because reflections are suppressed only in one transducer, and only at one frequency.

In view of these comments, the N-SPUDT effect is of rather limited application. However, it is important because it exists in some materials with attractive surface-wave properties, such as the langasite case above and a low-diffraction orientation of quartz (Table 4.2). To exploit these in practical devices, the N-SPUDT effect needs to be accommodated.

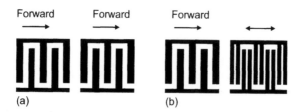

Figure 8.14 Devices using the N-SPUDT effect.

8.3 NUMERICAL EVALUATION OF COM PARAMETERS

Section 8.1.4 has given relations for the electrode reflection coefficient and the effective velocity, for a regular array of strips. Section 8.2.2 gave the corresponding COM parameters, eqs (8.41) and (8.42). These data were derived from Datta's approximate formulae, which give reasonable agreement with experiment. However, for some situations this method is not adequate. More sophisticated methods are needed for several reasons:

(a) For some transducers, each period has two or more electrodes with different geometry, for example the SPUDTs in the next chapter.
(b) Some waves, such as leaky waves and surface transverse waves, do not behave in the manner of Rayleigh waves considered above.
(c) For Rayleigh-wave devices, bulk-wave generation can be significant so that two modes are involved simultaneously.
(d) The finite thickness of the electrodes can introduce significant mechanical effects. The analysis needs to take account of the non-uniform distribution of stress and strain in each electrode. For this problem, the FEM can be used.
(e) The cross-section shape of the electrodes can vary because of the fabrication process, affecting the device performance.
(f) The presence of metal strips can substantially alter the character of the waves. In this case, it may not be adequate to analyze the wave on a free or metallized surface and then regard the strips as a perturbation.

The COM parameters can be deduced from experiments designed for the purpose [34], or from measurements of practical devices. This approach has the merit that it takes account of fabrication conditions. Small variations in these conditions can give significant changes in electrode behavior. However, experimental tests to cover all cases of practical interest would be very tedious.

8.3.1 Theoretical methods for periodic structures

Basic concepts

For accurate results numerical techniques are used, and many of these rely on special concepts applied to infinite gratings [28, 35–38]. Consider an infinite array of identical electrodes with period p. Assume initially that the voltage v_n of electrode n is given by

$$v_n = v_0 \exp(-j2\pi\gamma n), \qquad (8.43)$$

as in Fig. 8.15a. The real constant γ, which can be restricted to the range $0 < \gamma < 1$, has the role of a wavenumber. The current entering electrode n is

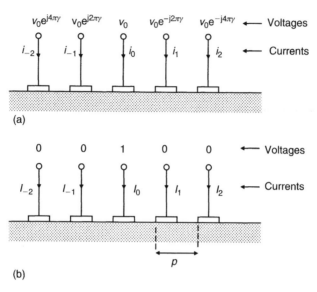

Figure 8.15 Definitions for analysis of an infinite array of electrodes. (a) Voltages for harmonic admittance. (b) Voltages for discrete Green's function, $I_n = G_d(n)$.

denoted by i_n, and the ratio i_n/v_n is independent of n. This follows because all electrodes are identical, so that we cannot identify which electrode has $n = 0$. Thus we have $i_n = i_0 \exp(-j2\pi\gamma n)$. The ratio i_n/v_n is the *harmonic admittance* $Y_h(\gamma)$, so that

$$Y_h(\gamma) = i_n/v_n. \tag{8.44}$$

This function is therefore independent of n. It depends on frequency as well as γ. Using this concept, we can calculate the response for other voltage distributions, using Fourier synthesis. For $\gamma = 1/2$ the voltages v_n alternate in sign, corresponding to a single-electrode transducer of infinite length. The solution for a shorted grating ($v_n = 0$) is given by a pole of the harmonic admittance. This function is also mentioned in Appendix E, and it has been used for numerical analysis of devices with regular electrodes [28, 35, 39].

Now consider a general set of electrode voltages V_n, with corresponding currents I_n. By superposition, the current in one electrode is linearly related to the voltages on all electrodes, so that

$$I_m = \sum_{n=-\infty}^{\infty} G_d(m-n)V_n. \tag{8.45}$$

The function $G_d(n)$ is the current in electrode n due to the voltage on electrode 0, as in Fig. 8.15b. This is called the *discrete Green's function*. If we consider the particular voltage distribution of eq. (8.43) and substitute into eq. (8.45), the corresponding current i_n is found to be

$$i_m = v_0 e^{-j2\pi\gamma m} \sum_{k=-\infty}^{\infty} G_d(k) \exp(j2\pi k\gamma) \qquad (8.46)$$

where $k = m - n$. This shows that $i_n = i_0 \exp(-j2\pi\gamma n)$, confirming the earlier conclusion that i_n has the same form as v_n in eq. (8.43). In addition, the harmonic admittance is seen to be

$$Y_h(\gamma) \equiv i_n/v_n = \sum_{m=-\infty}^{\infty} G_d(m) \exp(j2\pi m\gamma). \qquad (8.47)$$

The function $G_d(m)$ has the symmetry $G_d(-m) = G_d(m)$. This follows by considering a device in which electrodes 0 and m are taken as input and output terminals, and applying the reciprocity condition $Y_{21} = Y_{12}$. It follows that $Y_h(-\gamma) = Y_h(\gamma)$. We also have $Y_h(\gamma + 1) = Y_h(\gamma)$ from eq. (8.47), and it follows that $Y_h(\gamma)$ is symmetric about $\gamma = 1/2$.

Equation (8.47) can be inverted to obtain the $G_d(n)$ from $Y_h(\gamma)$. Consider voltages of the form

$$V_n = \int_0^1 \exp(-j2\pi n\gamma) d\gamma. \qquad (8.48)$$

This integral gives $V_0 = 1$ and $V_n = 0$ for $n \neq 0$ as in Fig. 8.15b. For each γ, the voltages are $\exp(-j2\pi\gamma n)$ and the currents are $I_n = Y_h \exp(-j\gamma 2\pi n)$, from eqs (8.43) and (8.45). Integrating the currents, we find

$$I_n = \int_0^1 Y_h(\gamma) \exp(-j2\pi n\gamma) d\gamma \qquad (8.49)$$

using eq. (8.47) in eq. (8.49) gives $I_n = G_d(n)$. Thus, $G_d(n)$ are simply the Fourier components of the repetitive function $Y_h(\gamma)$.

Equation (8.49) shows that $G_d(n)$ can be obtained from $Y_h(\gamma)$. Since $G_d(n)$ can be used to obtain the electrode currents for any voltage distribution, $Y_h(\gamma)$ also gives this information. This is subject to the constraint that the electrodes form an infinite regular array. This constraint simplifies the numerical computation somewhat. The method is particularly suited to analysis of reflection and transduction in periodic arrays.

The function $G_d(n)$ can be used to analyze an unapodized two-transducer device, provided the transducers have the same electrode pitch. We simply imagine

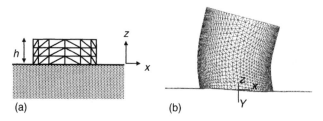

Figure 8.16 (a) Division of one electrode into elements for FEM analysis, schematic. (b) Motion of electrode for propagation of longitudinal leaky wave on lithium niobate. From Solal *et al.* [40], copyright 2004 IEEE, reproduced with permission.

applying a voltage to the live electrodes of one transducer, with zero voltage on the other. The currents in both transducers are then obtained from $G_d(n)$. This gives two components of the device Y-matrix. Interchanging the transducers gives the other components of Y_{ij}. This assumes that the electrode positions are periodic throughout the device. This implies that the space between the transducers is a multiple of the electrode pitch, but some simple adjustments can be made if this is not so.

Green's function analysis for finite-thickness gratings

For thin films, the above functions can be evaluated numerically using the effective permittivity described in Chapter 3. However, in general it is necessary to take account of the finite electrode thickness h. For small h the boundary conditions involve the potential $\phi(x)$ and charge density $\sigma(x)$, as in Chapter 3. For finite h it is necessary to include mechanical effects in addition. This is a substantial subject, so the following account is only a summary. For simplicity, we consider a grating in which all strips have the same geometry, with pitch p.

Each electrode, assumed to be isotropic, can be analyzed using the finite element method (FEM). The cross-section is divided into a number of elements [8, 28, 38, 37] as illustrated in Fig. 8.16a. In each element the equations of elasticity are represented in a discrete form, leading to a set of simultaneous equations. This method is used to derive a relation between the elastic stress and displacement at the lower surface of the strip, $z = 0$, which is in contact with the substrate. The three stress components T_{3i} need to be continuous across the surface $z = 0$. The stress and displacement are specified at a number of discrete points. Figure 8.16b is an example showing the calculated electrode motion [40]. In this case, the wave involved in a longitudinal leaky surface wave, which will be described in Chapter 11.

The substrate behavior is described by a boundary element method (BEM). This is expressed as a generalized form of the Green's function discussed earlier (Chapter 3). The surface displacements $\mathbf{u} = [u_1, u_2, u_3]$ and potential ϕ are related to a stress vector $\mathbf{T} = [T_{31}, T_{32}, T_{33}]$ and the charge density $\sigma(x)$. Assume initially that all components are proportional to $\exp(j\beta x)$, with this form indicated by a tilde. The relation can be written as

$$\begin{bmatrix} \widetilde{\mathbf{u}}(\beta) \\ \widetilde{\phi}(\beta) \end{bmatrix} = (1/|\beta|)\widetilde{\mathbf{G}}(\beta) \begin{bmatrix} \widetilde{\mathbf{T}}(\beta) \\ \widetilde{\sigma}(\beta) \end{bmatrix}. \tag{8.50}$$

Here $\widetilde{\mathbf{G}}(\beta)$, with dimensions 4×4, is the *dyadic Green's function*, in the wavenumber domain. It can be shown that $\widetilde{\mathbf{G}}(\beta)$ depends only on the slowness β/ω, and not on β or ω individually. Specific formulations for the $\widetilde{\mathbf{G}}(\beta)$ are given by Peach [41], Hashimoto [28] and Smith [42]. For a substrate supporting piezoelectric Rayleigh waves, all components have a pole at the slowness of a free-surface wave. For a free surface, the stresses $\mathbf{T}$ are all zero and the potential is $\widetilde{\phi}(\beta) = (1/|\beta|)\widetilde{G}_{44}(\beta)\widetilde{\sigma}(\beta)$. Hence, $\widetilde{G}_{44}(\beta)$ is the reciprocal of the effective permittivity $\varepsilon_s(\beta)$ defined in Chapter 3.

To find a solution, it is necessary to use the substrate properties, as determined by the Green's function matrix $\widetilde{\mathbf{G}}(\beta)$, allowing for the electrode behavior determined by the FEM analysis. To analyze a grating mode, all field quantities can be taken to be given by a Floquet expansion. For example, the u_j and T_{3i} are sums of terms proportional to $\exp(j\beta_n x)$, with $\beta_n = \beta_0 + 2n\pi/p$. In the substrate, eq. (8.50) applies for each Floquet component. Hashimoto [28] has given a method relating potential to charge density for each Floquet component, incorporating the motion of the electrode. This gives a relation for the ratio $\widetilde{\phi}(\beta_n)/\widetilde{\sigma}(\beta_n)$, which generalizes the effective permittivity $\varepsilon_s(\beta)$. This takes account of the mechanical effects in the electrodes. The remaining calculation can proceed as if these mechanical effects were absent, using the method following Bløtekjaer *et al.* as in Appendix E.

An alternative method [38] starts from the x-domain Green's function $\mathbf{G}(x)$, in which each component is the inverse Fourier transform of the β-domain function in eq. (8.50). $\mathbf{G}(x)$ gives the substrate response $[\mathbf{u}(x), \phi(x)]$ when the sources $[\mathbf{T}(x), \sigma(x)]$ have a delta-function form. For a general source distribution, the response is the convolution

$$\begin{bmatrix} \mathbf{u}(x) \\ \phi(x) \end{bmatrix} = \int_{-\infty}^{\infty} \mathbf{G}(x - x') \begin{bmatrix} \mathbf{T}(x') \\ \sigma(x') \end{bmatrix} dx'. \tag{8.51}$$

We assume the voltages have the form $V_n = V_0 \exp(-jn\gamma)$, so that the currents have the form $I_n = I_0 \exp(-jn\gamma)$ as discussed above. It follows that $\mathbf{u}(x + np) = \mathbf{u}(x) \exp(-j2n\pi\gamma)$, and corresponding expressions for $\phi(x)$, $\mathbf{T}(x)$

and $\sigma(x)$. The response can be written as

$$\begin{bmatrix} \mathbf{u}(x) \\ \phi(x) \end{bmatrix} = \int\limits_{-p/2}^{p/2} \mathbf{G}^{\mathrm{P}}(x - X) \begin{bmatrix} \mathbf{T}(X) \\ \sigma(X) \end{bmatrix} dX \tag{8.52}$$

where $\mathbf{G}^{\mathrm{P}}(x)$ is another 4×4 matrix called the *periodic Green's function* [43], defined by

$$\mathbf{G}^{\mathrm{P}}(x) = \sum_{n=-\infty}^{\infty} \mathbf{G}(x - np) \exp(-\mathrm{j}2n\pi\gamma). \tag{8.53}$$

Using eq. (A.33), we find that

$$\mathbf{G}^{\mathrm{P}}(x) = (1/p) \sum_{m=-\infty}^{\infty} \widetilde{\mathbf{G}}(k_m) \exp(-\mathrm{j}k_m x) \tag{8.54}$$

where $k_m = 2\pi(m + \gamma)/p$. Equation (8.52) describes the substrate behavior, involving $\mathbf{u}$ and ϕ on the left, and $\mathbf{T}$ and σ on the right. A further relation between $\mathbf{u}$ and $\mathbf{T}$ follows by FEM analysis of the electrode. This yields a relation between ϕ and σ. Integration of σ over one electrode then gives the harmonic admittance $Y_{\mathrm{h}}(\gamma)$.

In practice, there are considerable problems in numerical implementation. The charge density $\sigma(x)$ is proportional to $1/\sqrt{x}$ at the electrode edges, giving poles which need special attention. For an electrode with edges at $x = \pm 1$, $\sigma(x)$ is commonly represented as a sum of terms with the form $T_n(x)/\sqrt{(1 - x^2)}$, where $T_n(x)$ is a Chebychev polynomial. Poles due to surface-wave excitation are usually dealt with by extracting algebraic forms for which the transform is known exactly. Details are given by, for example, Ventura *et al.* [38] and Koskela *et al.* [37].

8.3.2 Coupled-mode parameters from band edge frequencies

It was shown in Section 8.2.2 that the width of the grating stop band is given by the electrode reflection coefficient. In fact, most of the parameters needed for COM analysis can be deduced from the frequencies of the band edges, provided we assume that these parameters are independent of frequency [27]. This provides a convenient way of finding the COM parameters if the dispersion relation is known from sophisticated methods such as the FEM. In this way, the accurate infinite-array methods of Section 8.3.1 can be applied to finite-length devices, though with some limitations associated with the COM analysis. It is assumed that there is no loss, because otherwise the band edges cannot be identified. This implies that the COM variable δ is real. In the following, we first consider a single-electrode transducer having regular electrodes with pitch p, and with no N-SPUDT effect. A more general case is considered later.

Regular electrodes with no N-SPUDT effect: shorted

For a shorted transducer or grating, we take $V = 0$. From eq. (8.35) the band edges are at $\delta = \pm|c_{12}|$, where $\delta = k_e - k_c = \omega/v_e - k_c$. Here v_e is the effective velocity, taken to be constant. The width of the stop band is given by

$$\Delta\omega/v_e = \Delta\delta = 2|c_{12}| = 2|r_s|/p \qquad (8.55)$$

so the bandwidth gives $|c_{12}|$. The electrode reflection coefficients r_{s1} and r_{s2} are the same so they are written as r_s, which is taken to be constant. In addition, the center of the stop band is at $\delta = 0$, or $k_e = k_c$, and this gives the velocity v_e. A grating mode with amplitude proportional to $\exp(-jsx)$ is given by eq. (8.39), with $h_2 = V = 0$ and $p_1 = j(\delta - s)/c_{12}$. At the band edges, the real and imaginary parts of s are both zero, so that $C(x)$ and $B(x)$ are constants and $\delta = \pm|c_{12}|$. We have

$$p_1 \equiv B(x)/C(x) = j\delta/c_{12} = \pm j|c_{12}|/c_{12} \quad \text{at band edges.} \qquad (8.56)$$

Hence $|p_1| = 1$, so the constituent waves $C(x)$ and $B(x)$ have the same amplitude and the solutions are standing waves. Because of the symmetry, c_{12} is imaginary. For metal electrodes, r_{s1} is usually negative imaginary and eq. (8.41) shows that c_{12} is negative imaginary, giving $c_{12} = -j|c_{12}|$. This gives $p_1 = 1$ for $\delta = -|c_{12}|$ (the lower band edge), and $p_1 = -1$ for $\delta = |c_{12}|$ (the upper band edge). The current is given by eq. (8.34c), in which α_1 is imaginary (eq. (8.42)). Setting $V = 0$ and $B(x)/C(x) = p_1$ we find

$$dI/dx = -2\alpha_1(1 + p_1)C(x) \quad \text{at band edges} \qquad (8.57)$$

Hence the mode with $p_1 = -1$ gives no current, and this mode is at the upper band edge if $r_s/j < 0$. The fact that the total current is zero implies that all the electrode currents are zero. This follows because the fields are all periodic, with period $2p$. The electrode currents alternate in sign, but because they are connected alternately to the bus bars they contribute equally to the transducer current. Hence, zero transducer current implies that all electrodes have zero current. It follows that *this solution for a shorted grating is also a solution for an open-circuit grating.*

If r_s is positive imaginary instead of negative imaginary, the zero-current mode is at the lower band edge instead of the upper band edge; again, this mode is also a band edge of the open-circuit grating.

Regular electrodes with no N-SPUDT effect: open circuit

The COM equations also give a solution for an open-circuit grating. In this case $I = 0$ and eq. (8.34c) gives $j\omega C_l V = 2\alpha_1 B(x) - 2\alpha_1^* C(x)$. Substituting into eqs (8.34a) and (8.34b) gives, after some manipulation [27],

$$dC/dx = -j\delta' C + c_{12}' B \qquad (8.58a)$$

$$dB/dx = j\delta' B + (c_{12}')^* C. \qquad (8.58b)$$

These equations correspond to the COM eqs (8.34a) and (8.34b), with δ and c_{12} replaced by the primed quantities

$$\delta' = \delta - 2|\alpha_1|^2/(\omega C_l) = k_e - k_c - 2|\alpha_1|^2/(\omega C_l) \qquad (8.59)$$

$$c_{12}' = c_{12} - 2j\alpha_1^2/\omega C_l. \qquad (8.60)$$

Equation (8.59) shows that the center frequency of the stop band has shifted, such that k_e has been changed to $k_e + \Delta k_e$, with $\Delta k_e = 2|\alpha_1|^2/\omega C_l$. If ω_c is the center frequency for shorted electrodes, the value for open-circuit electrodes is $\omega_c + \Delta\omega_c$, with

$$\Delta\omega_c/\omega_c = \Delta k_e/k_e = 2|\alpha_1|^2 p/(\pi\omega C_l) \qquad (8.61)$$

where we have used $k_c = \pi/p$. Hence, the change in ω_c gives $|\alpha_1|$, assuming that C_l is known. The stop band width is given by

$$\Delta\omega/v_e' = \Delta\delta = 2|c_{12}'| \qquad (8.62)$$

where v_e' is the new velocity, and $v_e' \approx v_e$.

The sign of the reflection coefficient is given by identifying which band edge is in common with a shorted grating. Equations (8.58) have the same form as eqs (8.34a) and (8.34b), with $V = 0$ and with δ and c_{12} replaced by δ' and c_{12}'. They therefore have a grating-mode solution as considered in Section 8.2.2. At each band edge we have $B(x)/C(x) = j\delta'/c_{12}'$ and $\delta' = \pm jc_{12}'$. As stated earlier, the voltage is given by $j\omega C_l V = 2\alpha_1 B(x) - 2\alpha_1^* C(x)$, and because of the symmetry we know that α_1 is imaginary, so this becomes $j\omega C_l V = 2\alpha_1(1 + j\delta'/c_{12}')C$. This is zero if $\delta' = jc_{12}'$, and this corresponds to one band edge of the shorted grating. If c_{12}' is positive imaginary we have $\delta' < 0$, so this refers to the lower band edge of the open-circuit grating. In addition, from eq. (8.41) we have $c_{12}' = r_s'/p$, where r_s' is the reflection coefficient for open-circuit electrodes. To clarify this, define ω_l' and ω_u' as the lower and upper band edges of the open-circuit grating, and ω_l and ω_u as the lower and upper band edges of the shorted grating. Remembering that r_s' and r_s are imaginary because of the symmetry, Table 8.3 summarizes the results.

The commonality of the band edges thus gives the signs of the electrode reflection coefficients.

Table 8.3 Band edge conditions from signs of reflection coefficients.

Signs of reflection coefficients		Band edge conditions
Shorted	Open	
$r_s/j > 0$	$r'_s/j > 0$	$\omega'_l = \omega_l$
$r_s/j > 0$	$r'_s/j < 0$	$\omega'_u = \omega_l$
$r_s/j < 0$	$r'_s/j > 0$	$\omega'_l = \omega_u$
$r_s/j < 0$	$r'_s/j < 0$	$\omega'_u = \omega_u$

General case

In general, each period of the array could have more than one electrode, with different properties, and the N-SPUDT effect could be present. If so, most of the above analysis still applies, though generally r_s is not defined and p_1 does not necessarily have values ± 1 at the band edges. Equations (8.55), (8.59) and (8.61) are still valid. The parameters c_{12} and α_1 are not necessarily imaginary. Their relative phases will depend on the geometry, and this can be deduced from the reflection coefficients. Taking the squared modulus of eq. (8.60) gives

$$|c'_{12}|^2 = |c_{12}|^2 + [2|\alpha_1|^2/(\omega C_l)]^2 - 4\mathrm{Re}\{j\alpha_1^2 c_{12}^*\}/(\omega C_l).$$

Now define ϕ_α and ϕ_c as the phases of α_1 and c_{12}, so that $\alpha_1 = |\alpha_1| \exp(j\phi_\alpha)$ and $c_{12} = |c_{12}| \exp(j\phi_c)$, and define $\psi = \phi_c - 2\phi_\alpha$. We then find

$$|c'_{12}|^2 = |c_{12}|^2 + [2|\alpha_1|^2/(\omega C_l)]^2 - 4|\alpha_1^2 c_{12}/(\omega C_l)| \sin\psi, \qquad (8.63)$$

and $|c'_{12}|$ is known from eq. (8.62). From this we can find ψ, which determines the directivity for a directional transducer such as an SPUDT (Chapter 9) or N-SPUDT.

With these relations, we now have nearly all the parameters needed for the COM analysis. The main exception is the absolute phase of α_1. However, this can be found from an electrostatic analysis, as explained for regular electrodes in Section 8.1.3. If the phase of α_1 is known, the phase of c_{12} follows by using eq. (8.63) for ψ.

Reflection relationship for regular electrodes

Equation (8.60) gives an interesting relation between the reflection coefficients for the two cases. If the electrodes are regular and there is no directivity, the reflection coefficients are given by $r_s = -pc_{12}$ for the shorted case and $r'_s = -pc'_{12}$ for the open-circuit case (for incident waves propagating in the $+x$-direction). Equation (8.60) gives

$$r'_s - r_s = -2j\alpha_1^2 p/(\omega C_l). \qquad (8.64)$$

This shows that $r'_s - r_s$ is independent of mechanical effects, and therefore independent of the film thickness. This is expected because the mechanical effects are independent of the electrode connections. It also follows that $r'_s - r_s$ is not affected by the N-SPUDT effect, and that it is independent of the direction of the incident waves.

Equation (8.64) is confirmed by Datta's relations. The electrical reflection coefficients, at the Bragg frequency, are given by eq. (8.15). Using eq. (C.6), these are found to give

$$r'_s - r_s = \frac{2j(\Delta v/v)}{P_{-1/2}(\cos \Delta)P_{-1/2}(-\cos \Delta)}. \tag{8.65}$$

For the right side of eq. (8.64), C_l is given by eq. (5.79) and α_1 is given by eq. (8.42). Using these in eq. (8.64) gives agreement with eq. (8.65).

REFERENCES

1. D.P. Morgan. *Surface-Wave Devices for Signal Processing*, Elsevier, 1991, p. 387.
2. D.P. Morgan. 'Reflective array modelling for reflective and directional SAW transducers', *IEEE Trans. Ultrason. Ferroelect. Freq. Contr.*, **45**, 152–157 (1998).
3. S. Datta and B.J. Hunsinger. 'An analytical theory for the scattering of SAW by a single electrode in a periodic array on a piezoelectric substrate', *J. Appl. Phys.*, **51**, 4817–4823 (1980).
4. S. Datta and B.J. Hunsinger. 'First-order reflection coefficient of SAW from thin-strip overlays', *J. Appl. Phys.*, **50**, 5661–5665 (1979).
5. S. Datta. *Surface Acoustic Wave Devices*, Prentice-Hall, 1986.
6. C. Dunnrowitz, F. Sandy and T. Parker. 'Reflection of surface waves from periodic discontinuities', *IEEE Ultrason. Symp.*, 386–390 (1976).
7. P.V. Wright. 'Modeling and experimental measurements of the reflection properties of SAW metallic gratings', *IEEE Ultrason. Symp.*, **1**, 54–63 (1984).
8. M. Koshiba and S. Mitobe. 'Equivalent networks for SAW gratings', *IEEE Trans. Ultrason. Ferroelect. Freq. Contr.*, **35**, 531–535 (1988).
9. S. Jen and C.S. Hartmann. 'An improved model for chirped slanted SAW devices', *IEEE Ultrason. Symp.*, **1**, 7–14 (1989).
10. P. Ventura. 'Full strip reflectivity study on quartz', *IEEE Ultrason. Symp.*, **1**, 245–248 (1994).
11. N. Mishima, M. Takase, S. Mitobe and Y. Ebata. 'IIDT type SAW filter using acoustic reflection cancel condition with solid IDT', *IEEE Ultrason. Symp.*, **1**, 31–35 (1989).
12. Y. Ebata, K. Sato and S. Morishita. 'A LiTaO₃ SAW resonator and its application to a video cassette recorder', *IEEE Ultrason. Symp.*, **1**, 111–116 (1981).
13. R.C. Williamson and H.I. Smith. 'The use of surface elastic wave reflection gratings in large time-bandwidth pulse compression filters', *IEEE Trans.*, **MTT-21**, 195–205 (1973).
14. B.A. Auld. *Acoustic Waves and Fields in Solids*, Vol. 2, Krieger, 1990, p. 188.
15. K. Bløtekjaer, K.A. Ingebrigtsen and H. Skeie. 'A method for analysing waves in structures consisting of metal strips on dispersive media', *IEEE Trans. Electron. Devices*, **ED-20**, 1133–1138 (1973).

16. T. Thorvaldsson. 'Analysis of the natural single phase unidirectional SAW transducer', *IEEE Ultrason. Symp.*, **1**, 91–96 (1989).

17. T. Suzuki, H. Shimizu, M. Takeuchi, K. Nakamura and A. Yamada. 'Some studies on SAW resonators and multiple-mode filters', *IEEE Ultrason. Symp.*, 297–302 (1976).

18. S.V. Biryukov, G. Martin, V.G. Polevoi and M. Weinacht. 'Derivation of COM equations using the surface impedance method', *IEEE Trans. Ultrason. Ferroelec. Freq. Contr.*, **42**, 602–611 (1995).

19. R.C.M. Li and J. Melngailis. 'The influence of stored energy at step discontinuities on the behaviour of surface-wave gratings', *IEEE Trans. Son. Ultrason.*, **SU-22**, 189–198 (1975).

20. T. Uno and H. Jumonji. 'Optimization of quartz SAW resonator structure with groove gratings', *IEEE Trans. Son. Ultrason.*, **SU-29**, 299–310 (1982).

21. S. Datta and B.J. Hunsinger. 'An analysis of energy storage effects on SAW propagation in periodic arrays', *IEEE Trans. Son. Ultrason.*, **SU-27**, 333–341 (1980).

22. H.A. Haus and W. Huang. 'Coupled-mode theory', *Proc. IEEE*, **79**, 1505–1518 (1991).

23. Y. Koyamada and S. Yoshikawa. 'Coupled mode analysis of a long IDT', *Rev. Electr. Commun. Labs.* (NTT), **27**, 432–444 (May/June 1979).

24. C.S. Hartmann, P.V. Wright, R.J. Kansy and E.M. Garber. 'An analysis of SAW interdigital transducers with internal reflections and the application to the design of single-phase unidirectional transducers', *IEEE Ultrason. Symp.*, **1**, 40–45 (1982).

25. P.V. Wright. 'A new generalized modeling of SAW transducers and gratings', *Proc. 43rd Ann. Freq. Contr. Symp.*, 1989, pp. 596–605.

26. V. Plessky and J. Koskela. 'Coupling-of-modes analysis of SAW devices', in C.C.W. Ruppel and T. Fjeldy (eds.), *Advances in Surface Acoustic Wave Technology, Systems and Applications*, Vol. 2, World Scientific, 2001, pp. 1–81.

27. D.P. Chen and H.A. Haus. 'Analysis of metal-strip SAW gratings and transducers', *IEEE Trans. Son. Ultrason.*, **SU-32**, 395–408 (1985).

28. K.-Y. Hashimoto. *Surface Acoustic Wave Devices in Telecommunications*, Springer, 2000.

29. M.R. Daniel and P.R. Emtage. 'Distortion of the central resonance in long interdigital transducers', *Appl. Phys. Lett.*, **20**, 320–322 (1972).

30. P.V. Wright. 'The natural single-phase unidirectional transducer: a new low-loss SAW transducer', *IEEE Ultrason. Symp.*, **1**, 58–63 (1985).

31. T. Thorvaldsson and B.P. Abbott. 'Low loss SAW filters utilizing the natural single phase unidirectional transducer (NSPUDT)', *IEEE Ultrason. Symp.*, **1**, 43–48 (1990).

32. D.P. Morgan, S. Zhgoon, A. Shvetsov, E. Semenova and V. Semenov. 'One-port SAW resonators using Natural SPUDT substrates', *IEEE Ultrason. Symp.*, **1**, 446–449 (2005).

33. K.-Y. Hashimoto and Y. Yamaguchi. 'Free software products for simulation and design of surface acoustic wave and surface transverse wave devices', *IEEE Intl. Freq. Control Symp.*, 1996, pp. 300–307.

34. C.S. Hartmann and B.P. Abbott. 'Experimentally determining the transduction magnitude and phase and the reflection magnitude and phase of SAW SPUDT structures', *IEEE Ultrason. Symp.*, **1**, 37–42 (1990).

35. Y. Zhang, J. Desbois and L. Boyer. 'Characteristic parameters of surface acoustic waves in a periodic metal grating on a piezoelectric substrate', *IEEE Trans. Ultrason. Ferroelect. Freq. Contr.*, **40**, 183–192 (1993).

36. E. Danicki. 'Spectral theory for IDTs', *IEEE Ultrason. Symp.*, **1**, 213–222 (1994).

37. J. Koskela, V.P. Plessky and M. Salomaa. 'Suppression of the leaky SAW attenuation with heavy mechanical loading', *IEEE Trans. Ultrason. Ferroelect. Freq. Contr.*, **45**, 439–449 (1998).

38. P. Ventura, J.M. Hodé, J. Desbois and M. Solal. 'Combined FEM and Green's function analysis of periodic SAW structures, application to the calculation of reflection and scattering parameters', *IEEE Trans. Ultrason. Ferroelect. Freq. Contr.*, **48**, 1259–1274 (2001).
39. P. Ventura, J.M. Hodé, M. Solal, J. Desbois and J. Ribbe. 'Numerical methods for SAW propagation characterization', *IEEE Ultrason. Symp.*, **1**, 175–186 (1998).
40. M. Solal, R. Lardat, T. Pastureaud, W. Steichen, V. Plessky, T. Makkonen and M.M. Salomaa. 'Existence of harmonic metal thickness mode propagation for longitudinal leaky waves', *IEEE Ultrason. Symp.*, **2**, 1207–1212 (2004).
41. R.C. Peach. 'A general Green function analysis for SAW devices', *IEEE Ultrason. Symp.*, **1**, 221–225 (1995).
42. P.M. Smith. 'Dyadic Green's functions for multilayer SAW substrates', *IEEE Trans. Ultrason. Ferroelect. Freq. Contr.*, **48**, 171–179 (2001).
43. V.P. Plessky and T. Thorvaldsson. 'Periodic Green's function analysis of SAW and leaky SAW propagation in a periodic system of electrodes on a piezoelectric crystal', *IEEE Trans. Ultrason. Ferroelect. Freq. Contr.*, **42**, 280–293 (1995).
44. K. Hasegawa, K. Inagawa and M. Koshiba. 'Extraction of all coefficients of coupled-mode equations for natural single-phase unidirectional SAW transducers from dispersion characteristics computed by hybrid finite element method', *IEEE Trans. Ultrason. Ferroelect. Freq. Contr.*, **48**, 1341–1350 (2001).

9

UNIDIRECTIONAL TRANSDUCERS AND THEIR APPLICATION TO BANDPASS FILTERING

In Chapters 5–7 we considered devices using non-reflective transducers, that is, transducers that do not reflect surface waves when they are shorted. The performance given by these devices can be of very high quality except for one major factor, which is the presence of unwanted multiple-transit signals. These arise because the transducers reflect the waves when they are connected to finite electrical impedances. This feature is associated with the bidirectional nature of the transducers, as shown in Appendix D, Section D.1. To obtain adequate suppression of these signals it is necessary to adjust the electrical matching, and this has the consequence that the insertion loss is large, typically 15–25 dB. There are several special techniques for overcoming this problem. In this chapter we consider *unidirectional transducers* (UDTs), which can be designed to give both low reflectivity and low loss. Other techniques will be described in Chapters 10 and 11.

A unidirectional transducer (UDT) has directivity, that is, it responds to an applied voltage by generating waves with different amplitudes in the two directions. Directivity is needed if the acoustic reflections in a two-transducer device are to be reduced. This is expected because, if the directivity is strong, the transducer becomes essentially a two-port device, with one acoustic port and one electrical. Ideally, a two-port device gives no reflectivity if both ports are matched. Directional behavior can be obtained by applying voltages with different phases to a regular set of electrodes, and several types of UDT were devised on this principle. However, these *multi-phase transducers* have the inconvenience that three or more bus bars are needed and the fabrication is clumsy. Later, a variety of *single-phase unidirectional transducers*, or SPUDTs, were developed, using only two bus bars. Since about 1980, the single-phase types have ousted the multi-phase types almost entirely, so the latter are now of little practical interest. However, a brief mention of multi-phase types is included at the end of this chapter.

To behave as a SPUDT, a transducer needs to have internal reflectivity, and the structure must have some asymmetry. In most cases, this involves a sequence of cells, each with electrodes of different width. A prime example is the distributed acoustic reflection transducer (DART), considered in some detail here. Such transducers have been widely used for low-loss bandpass filtering at intermediate frequencies in cellular phone systems (handsets and basestations). In Section 9.1 we describe some basic considerations concerning SPUDTs, followed by description of DARTs and filters using them in Sections 9.2 and 9.3. Later sections consider other types of SPUDT and, in Section 9.5, some other techniques for obtaining low insertion losses.

The analysis here assumes that there is no propagation loss, waveguiding or diffraction, and that the only acoustic mode present is a piezoelectric Rayleigh wave. It is assumed that the substrate behaves symmetrically, so that a symmetrical transducer generates identical waves in the two directions when a voltage is applied. An absence of this symmetry is described by saying that the Natural-SPUDT (N-SPUDT) effect is occurring. The N-SPUDT effect is described in Chapter 8, Section 8.2.3. In this chapter, we assume that it is absent.

9.1 GENERAL CONSIDERATIONS

Assuming a single type of wave with no power loss or waveguiding, a variety of basic requirements can be deduced using the P-matrix of Appendix D. These topics are also discussed by Hodé et al. [1]. Reciprocity is assumed. The P-matrix is defined as in eq. (D.1) in terms of incident wave amplitudes A_i and transmitted wave amplitudes A_t, with additional subscripts 1 or 2 to indicate the acoustic ports. The transducer voltage and current are V and I. These quantities are related by the equation

$$[A_{t1}, A_{t2}, I]^T = [P_{ij}][A_{i1}, A_{i2}, V]^T, \tag{9.1}$$

where superscript T indicates a transpose.

We first show that if a transducer is to have directivity it must have internal reflections, that is P_{11} and P_{22} must be non-zero. This follows from the equation

$$P_{11}P_{13}^* + P_{12}P_{23}^* + P_{13} = 0, \tag{9.2}$$

which is obtained from power conservation as in Appendix D Section D.1. If $P_{11} = 0$ we have $|P_{12}| = 1$ from eq. (D.3). It then follows that $|P_{23}| = |P_{13}|$, so that there is no directivity. Hence, internal reflectivity is needed if the transducer is to have directivity.

We define S_{11} as the acoustic reflection coefficient at port 1 when the transducer is connected to a load with admittance Y_L. To find S_{11} we set $A_{i2} = 0$, so that there are no waves incident on port 2, and we use $I = -Y_L V$. Using reciprocity in the form $P_{31} = -2P_{13}$, we find from eq. (9.1)

$$S_{11} \equiv \frac{A_{t1}}{A_{i1}} = P_{11} + \frac{2P_{13}^2}{P_{33} + Y_L}. \tag{9.3}$$

Thus, for no reflectivity ($S_{11} = 0$), the required load admittance is

$$Y_L = -P_{33} - 2P_{13}^2/P_{11}. \tag{9.4}$$

Since $P_{13} \neq 0$, this shows that P_{11} needs to be non-zero because otherwise Y_L would be infinite. Thus the transducer needs to have internal reflections. The required load admittance has a real part

$$G_L \equiv \mathrm{Re}\{Y_L\} = -G_a - 2\,\mathrm{Re}\{P_{13}^2/P_{11}\}, \tag{9.5}$$

since the transducer conductance G_a equals the real part of P_{33}.

The directivity is defined as the ratio $D = |P_{13}/P_{23}|$, and from eq. (9.2) we find

$$\frac{1}{D^2} = \left| \frac{P_{23}}{P_{13}} \right|^2 = \frac{1 + |P_{11}|^2}{|P_{12}|^2} + \frac{X}{|P_{12}P_{13}|^2}, \tag{9.6}$$

with $X \equiv 2\,\mathrm{Re}\{P_{11}(P_{13}^*)^2\}$. When there is no internal reflectivity we have $P_{11} = 0$ and $|P_{12}| = 1$, and eq. (9.6) gives $D = 1$. Hence there is no directivity for this case. We also have $G_a = |P_{13}|^2 + |P_{23}|^2$, and from eq. (9.2) we find $G_a = (X + 2|P_{13}|^2)/|P_{12}|^2$. In addition, $2\,\mathrm{Re}\{P_{13}^2/P_{11}\} = X/|P_{11}|^2$. Using eq. (9.5), the required G_L value is

$$G_L = -[X + 2|P_{11}P_{13}|^2]/|P_{11}P_{12}|^2. \tag{9.7}$$

Assuming that the phase of P_{13} can be determined arbitrarily, the directivity D has its maximum value D_{max} when X has its minimum value $X_{min} = -2|P_{11}(P_{13}^*)^2|$. This gives

$$D_{max} = \frac{|P_{12}|}{1 - |P_{11}|} = \frac{1 + |P_{11}|}{|P_{12}|}. \tag{9.8}$$

Since $|P_{12}|$ can be deduced from $|P_{11}|$, eq. (9.8) gives the maximum directivity if $|P_{11}|$ is specified. D_{max} increases with $|P_{11}|$, so for strong directivity we need $|P_{11}|$ to be close to unity. With $X = X_{min}$, eq. (9.7) gives

$$G_L = \frac{2|P_{13}|^2(1 - |P_{11}|)}{|P_{11}P_{12}^2|} = \frac{G_a}{|P_{11}|}. \tag{9.9}$$

With $X = X_{min}$, we also find that the phase of $(P_{13}^*)^2 P_{11}$ is π, and it follows that P_{13}^2/P_{11} is real and negative. Equation (9.4) then gives $\text{Im}\{Y_L\} = -\text{Im}\{P_{33}\}$. Thus, for strong reflectivity ($|P_{11}| \to 1$) we have $Y_L \to P_{33}^*$. This approaches the condition for electrical matching. Since we also have high directivity, the power conversion loss will be small. The maximum value of X is $X_{max} = 2|P_{11}(P_{13}^*)^2|$, and this value gives the minimum directivity as $D_{min} = 1/D_{max}$. This maximizes the directivity in the opposite direction, of course. For this condition, P_{13}^2/P_{11} is positive real.

If the transducer is bidirectional we have $D = 1$, and eq. (9.6) shows that $X = -2|P_{11}P_{13}|^2$. It follows that $G_L = 0$. Although this cancels the acoustic reflections ($S_{11} = 0$), it also gives no power transfer so it is of little practical interest. For the general case with $D > 1$ we have $X < -2|P_{11}P_{13}|^2$, and eq. (9.7) gives a positive value for G_L. Hence the transducer can be matched for zero reflectivity for any $D > 1$.

9.2 DART MECHANISM AND ANALYSIS

Several types of SPUDT have been developed, and here we discuss the principles of one of the commonest. This is the distributed acoustic reflection transducer (DART), introduced by Kodama *et al.* [2]. As shown in Fig. 9.1a, the DART consists of a sequence of identical cells with length equal to λ_0, the center-frequency wavelength. Each cell has two electrodes of width $\lambda_0/8$ and one electrode of width $\lambda_0/4$, and all the interelectrode spaces are $\lambda_0/8$.

The transducer can be regarded as a sequence of localized reflectors and transducers, taken to be located at reflection and transduction centers. The reflection and transduction processes are imagined to be happening independently. To define a reflection center we imagine the transducer to be shorted. The reflection center refers to some point such that the reflection coefficient of incident waves is the same for waves incident in both directions. The waves can be regarded as reflected at a point localized at this reflection center. If the transducer geometry is symmetrical about some point, this point will be a reflection center. For the DART transducer, the center of the wide electrode satisfies this criterion, so this will be a reflection center, as shown by the notation 'RC' on

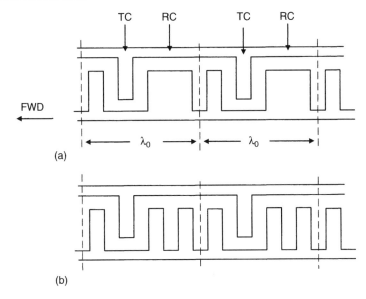

Figure 9.1 DART-type SPUDT. (a) Normal DART structure. (b) Structure with no reflectivity. λ_0 is the transducer pitch, equal to the center-frequency wavelength.

Fig. 9.1a. As shown in Appendix D, the reflection coefficient will be imaginary relative to this point.

To define a transduction center, we ignore the reflectivity and consider genertion of waves when a voltage is applied. The transduction center is a point relative to which the waves generated in the two directions have the same amplitude and phase. If there is a point about which the transducer is symmetrical, taking account of the applied voltage, then this point will be a transduction center. For example, in a single-electrode transducer the center of each electrode is a transduction center, by symmetry. In the DART this is not so clear. However, a good approximation is obtained by considering the structure of Fig. 9.1b, where the wide electrode has been replaced by two narrow electrodes of width $\lambda_0/8$, with a space of $\lambda_0/8$. In this modification, reflections are negligible because the electrodes are regular with a pitch of $\lambda_0/4$, as in a double-electrode transducer. This implies that the quasi-static theory of Chapter 5 can be applied, so the wave generation is governed by the electrostatic charge density. In this transducer, the 3rd and 4th electrodes in each period have the same potential, so the electric fields between them are relatively small. For this reason, the fields are similar to the fields in the DART of Fig. 9.1a. [3]. In Fig. 9.1b, we clearly have symmetry about the center of each 'live' electrode, so these are the locations of the transduction center. These are indicated as 'TC' on Fig. 9.1a. The similarity between the electric fields in Figs. 9.1a and 9.1b is confirmed by accurate electrostatic

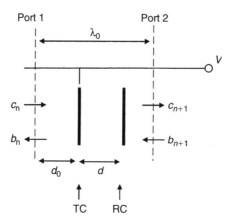

Figure 9.2 One cell of a SPUDT, represented in terms of its TC and RC. Waves to the right and left are given by c_n and b_n, respectively.

calculations [4]. The capacitance and conductance for the uniform transducer of Fig. 9.1b are given in Chapter 5, Section 5.5.3.

Figure 9.2 represents a cell of a SPUDT in terms of its transduction and reflection centers. The distance between these centers is d, which equals $3\lambda_0/8$ in the case of the DART. The vertical broken lines are reference positions with spacing λ_0. The centers can be regarded as a localized transducer and reflector. The reflector is taken to have reflection coefficient R. As discussed in Chapter 8, the symmetry implies that we can take R to be imaginary, and the transmission coefficient $T = \sqrt{(1 - |R|^2)}$ of the reflector can be taken to be positive real. These statements depend on the wavenumber k being adjusted, if necessary, to compensate for any phase change in transmission. Considering one cell, the transducer generates identical waves in the two directions. The wave to the right is reflected back by the reflector, so that the total wave emerging at the left is proportional to $1 + R\exp(-2jkd)$. At the center frequency, where $k = 2\pi/\lambda_0$, the two terms reinforce if

$$d = (2n \pm 1/2)\lambda_0/4. \tag{9.10}$$

Here the sign is the sign of R/j. With this formula, the 'forward' direction is to the left on Fig. 9.1 or 9.2. The wave emerging at the left has amplitude proportional to $1 + |R|$. The wave to the right has amplitude proportional to T, so the directivity is $(1 + |R|)/T$, in agreement with eq. (9.8). If R is negative imaginary, suitable values of d are $(2n - 1/2)\lambda_0/4$. This applies for the DART, which has $n = 1$ and $d = 3\lambda_0/8$.

Transduction is expressed by defining a parameter a_T such that the waves generated in one cell have amplitude $ja_T V$ when a voltage V is applied. For this purpose, reflectivity is ignored as in Fig. 9.1b. Here the electrodes are regular, and the quasi-static theory of Chapter 5 can be applied. The potentials $\phi_{s\pm}$ of the waves generated are the same, and they are given by $\phi_{s\pm} = j\Gamma_s V \overline{\rho}_f(k)$, where $\overline{\rho}_f(k)$ is the Fourier transform of the elemental charge density, given by eq. (5.59). The phase of $\phi_{s\pm}$ is relative to the center of the live electrode. For the parameter a_T the wave amplitudes $A_\pm$ are defined to have the same phase as $\phi_{s\pm}$, but with magnitude such that $|A_\pm|^2/2$ equals the wave power. As shown in Chapter 5, Section 5.3, the required relation is $A_\pm = \phi_{s\pm}[\omega W/(2\Gamma_s)]^{1/2}$. We thus find

$$a_T = \overline{\rho}_f(k)\sqrt{\omega W\Gamma_s/2}. \tag{9.11}$$

We now consider the P-matrix of one cell, which has length λ_0. The 'transducer' is a distance d_0 from port 1 of the cell, which is at the left. Taking $V = 0$ initially, the acoustic scattering coefficients of the cell are seen to be

$$P_{11} = R\exp[-2jk(d_0 + d)]; \quad P_{12} = T\exp(-jk\lambda_0),$$
$$P_{22} = R\exp[-2jk(\lambda_0 - d_0 - d)]. \tag{9.12}$$

At the center frequency f_0, where $k = 2\pi/\lambda_0$, we have $P_{11} = |R|\exp(-2jkd_0)$, $P_{22} = -|R|\exp(2jkd_0)$ and $P_{12} = T$. These results apply assuming that the 'forward' direction is to the left, at port 1 in Fig. 9.1. They use eq. (9.10) and the fact that R is imaginary. They apply for either sign of R/j. Also at f_0, the acoustoelectric terms are $P_{13} = ja_T(1 + |R|)\exp(-jkd_0)$ and $P_{23} = ja_T T\exp(jkd_0)$. Hence the directivity at f_0 is

$$D \equiv |P_{13}/P_{23}| = (1 + |R|)/T = (1 + |P_{11}|)/|P_{12}|, \tag{9.13}$$

in agreement with eq. (9.8). Noting that a_T is real, we also see that P_{13}^2/P_{11} is negative real. This is the condition for optimum directivity, as shown in Section 9.1.

Directivity for non-uniform SPUDTs

Here we consider a SPUDT having more than one cell, and we deduce its directivity at the center frequency f_0. Generally, a SPUDT may have different reflection and transduction coefficients in individual cells, and this is allowed for here. It is assumed that all cells have their 'forward' directions in the same direction, and that each cell has P_{13}^2/P_{11} real and negative. This will be true if all cells have transduction and reflection center at corresponding points. It remains true if a_T changes sign. The transduction center of the cells are spaced from each other by multiples of λ_0. Consequently, the phase of P_{13}^2 at f_0 is not changed when extra cells are added. Similarly, reflection center are spaced by multiples of λ_0, and adding cells does not change the phase of P_{11}. Hence,

P_{13}^2/P_{11} is negative real for the entire SPUDT. As shown in Section 9.1, this implies that the directivity is given by eq. (9.8).

We consider a SPUDT with two cells, one with P-matrix P_{ij}, and the other added on the right with P-matrix P_{ij}^a. The P-matrix of the entire SPUDT is P_{ij}', given by eq. (9.1). The directivity is $D' \equiv |P_{13}'/P_{23}'|$. At f_0, $(P_{13}')^2/P_{11}'$ is negative real, and from eq. (9.8) we have $D' = (1 + |P_{11}'|)/|P_{12}'|$. Using eq. (D.16) for P_{11}' and P_{12}', the directivity is found to be

$$D' = D \cdot D^a, \tag{9.14}$$

where D and D^a are the directivities of the individual cells. Thus, the total directivity is simply the product of the individual directivities. The same argument can be used repeatedly if further cells are added. The directivity for a SPUDT with N cells is therefore the product of the directivities of the individual cells. This applies even if the cells all have different transduction and reflection coefficients. If the cells are identical, each with directivity D, the directivity of a SPUDT with N cells is D^N.

A similar argument can be used if the individual cells have directivity in different directions. In this way, it can be shown that two cells with directivities D and D^a combine to give a SPUDT with directivity $D' = D \cdot D^a$, even if $D > 1$ and $D^a < 1$ (or if $D < 1$ and $D^a > 1$). As before, this can be extended to an arbitrary number of cells, and we conclude that for N cells with directivities D_n the whole transducer has directivity $D = D_1 \cdot D_2 \cdots D_N$. This situation occurs in weighted DART filters, discussed below. These results apply for any type of SPUDT, provided P_{13}^2/P_{11} is real for each cell.

RAM analysis of DARTs

In Chapter 8, Section 8.1.3 the reflective array method (RAM) was used to analyze a reflective transducer with regular electrodes. A similar method can be used for the DART [3]. Considering one cell in a shorted transducer, the scattering is as given by eq. (8.4), with r_{s1} and r_{s2} replaced by R, t_s replaced by T, and p replaced by λ_0. The waves c_n and b_n on the right side of a cell are related to the waves c_{n-1} and b_{n-1} on the left by

$$c_n = (1/T)c_{n-1}\exp(-jk\lambda_0) + (R/T)b_{n-1},$$

$$b_n = -(R/T)c_{n-1} + (1/T)b_{n-1}\exp(-jk\lambda_0), \tag{9.15}$$

from eqs (8.5). Here c_n refers to waves propagating to the right, and b_n refers to waves propagating the left, as in Fig. 9.2. Starting from the left end, we take

$c_0 = 0$, so that there is no input, and cascade through the transducer to obtain the waves at the boundaries of each cell. This gives the waves c_N and b_N at the right, thus giving $P_{22} = c_N/b_N$ and $P_{12} = b_0/b_N$. The current entering the live electrode can be analyzed using the quasi-static theory. The electrode behaves approximately as an electrode in a regular array, as in Fig. 9.1b. From eq. (5.58), the current entering an electrode in a shorted regular array is

$$I_e = -j\omega W \phi_i \overline{\rho}_f(k),$$

where W is the beam width and $\overline{\rho}_f(k)$ is given by eq. (5.59). Here ϕ_i is the surface potential of the incident wave, with phase referred to the electrode center. For the RAM analysis, the wave amplitude c_n is defined such that $|c_n|^2/2$ equals the wave power, so we have $c_n = \phi[\omega W/(2\Gamma_s)]^{1/2}$, as in Chapter 8, Section 8.1.3. In the DART, the current must be calculated for the waves in both directions, allowing for phase changes due to the distances from the ports of the cell. From the total current I in the whole transducer, we have $P_{23} = I/b_N$. Cascading in the opposite direction gives P_{11}, P_{21} and P_{13}. Assuming no power loss, the acoustic conductance is given by $G_a = |P_{13}|^2 + |P_{23}|^2$ and the susceptance B_a is the Hilbert transform of G_a. The capacitance is obtained by electrostatic analysis, using the net charges Q_n of Section 5.4 in Chapter 5. We thus have the admittance $P_{33} = G_a + jB_a + j\omega C_t$. The above method can be applied to weighted DARTs, in which a_T and R can vary from one cell to another. The method takes account of the fact that a voltage applied to one cell produces electrostatic charges in neighboring cells.

Figure 9.3 shows the conductance G_a for a uniform DART on $37°Y–X$ quartz [3], with center frequency 71 MHz, length 80 periods and aperture 10 wavelengths. The film thickness h is 700 nm, giving $h/\lambda_0 = 1.58\%$. The theory uses the above RAM analysis and the coupling-of-modes (COM) analysis described below, giving similar results. The double-peaked curve is characteristic of many SPUDTs, and usually the two peaks have similar amplitudes. The slight difference experimentally is probably due to a non-ideal phase difference between reflection and transduction. The theory uses $R/j = -0.0115$, deduced from Ventura's finite-element method (FEM) results in Fig. 8.6. This is much larger than the value given by the first-order perturbation theory, seen on the same figure. The agreement with experiment supports the validity of the FEM results.

Figure 9.4 shows further theoretical results for the same DART, in dB. The directivity shown is $20 \log D$, and the acoustic reflection coefficient shown is $20 \log |S_{11}|$. The curves for conversion loss and reflection coefficient assume the presence of a parallel inductor, with the loading designed to give zero reflection at the center frequency as in eq. (9.4). The directivity curve is independent of

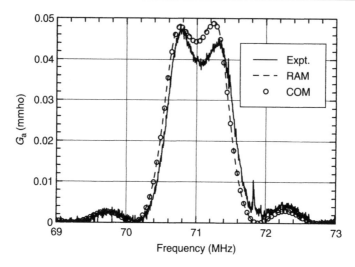

Figure 9.3 Conductance of a uniform DART on *ST–X* quartz. Experimental results courtesy of T. Thorvaldsson.

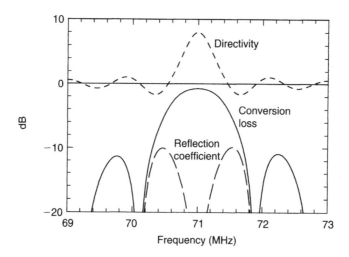

Figure 9.4 Theoretical curves for the DART of Fig. 9.3.

the loading, of course. It is noted that the directivity is strong only in a region near the center frequency, and even here it is only finite. It is therefore important not to consider the transducer as a two-port device; the radiation in the 'wrong' direction is of considerable significance.

COM analysis for uniform SPUDTs

The coupling-of-modes (COM) equations were derived in Chapter 8, Section 8.2. For transducer analysis, the P-matrix is given by eq. (8.40). These equations assume that the transducer is uniform, so they exclude weighting. However, a weighted transducer can be analyzed by cascading uniform sections, using eq. (D.16) in Appendix D. The uniform sections could simply be individual cells.

The COM parameters can be deduced from the simplified representation in Fig. 9.2. We consider the center frequency, where $\delta = 0$, and assume that $c_{12}\lambda_0$ is small. For one cell, eq. (8.40a) gives $P_{11} \approx -c_{12}^*\lambda_0$. Equating this to the RAM result in eq. (9.12) we find

$$c_{12}^* = -(R/\lambda_0)\exp[-2jk(d + d_0)]. \tag{9.16}$$

Similarly, the COM gives $P_{13} \approx a_1^*\lambda_0$, and equating this with the RAM expression gives

$$\alpha_1^* = -j(a_T/\lambda_0)\exp(-jkd_0). \tag{9.17}$$

For DARTs, a_T is given by eq. (9.11). For operation at the center frequency, and assuming no power loss, we set $\delta = 0$. Thus $k = k_c$, with $k_c = 2\pi/\lambda_0$. Equations (8.40a) and (8.40b) give

$$P_{11} = -(c_{12}^*/|c_{12}|)\tanh(|c_{12}|L);$$
$$P_{12} = \exp(-jk_cL)/\cosh(|c_{12}|L), \tag{9.18}$$

where L is the transducer length. These equations give $(1 + |P_{11}|)/|P_{12}| = \exp(|c_{12}|L)$. The COM expressions for P_{13} and P_{23} are given by eqs (8.40d) and (8.40e). Define θ and ϕ as the phases of α_1 and c_{12}, so that $\alpha_1 = |\alpha_1|\exp(j\theta)$ and $c_{12} = |c_{12}|\exp(j\phi)$. This gives [5]

$$\frac{P_{13}}{P_{23}} = -\left[\frac{\coth(|c_{12}|L/2) + e^{j(2\theta-\phi)}}{\coth(|c_{12}|L/2) - e^{-j(2\theta-\phi)}}\right]\exp[j(k_cL - 2\theta)]. \tag{9.19}$$

Maximum directivity occurs when $2\theta - \phi = 2n\pi$, so that $\exp[j(2\theta - \phi)] = 1$. We then find

$$D \equiv |P_{13}/P_{23}| = \exp(|c_{12}|L). \tag{9.20}$$

We thus have $D = (1 + |P_{11}|)/|P_{12}|$, in agreement with eq. (9.8). If D_1 is the directivity for one cell, the directivity for a transducer with N identical cells is $D_N = D_1^N$, in agreement with eq. (9.14). It is also found, from the COM equations, that P_{13}^2/P_{11} is negative real when $2\theta - \phi = 2n\pi$.

The COM method can be applied to SPUDTs in general. For uniform DART, it gives results similar to those of the RAM method.

9.3 BANDPASS FILTERING USING DARTS

Weighting

In Section 9.2 we have considered uniform DARTs, which have no weighting. For filtering applications these transducers are normally weighted to give more flexibility, in particular to reduce the sidelobes and to sharpen the skirts of the frequency response. In principle apodization could be used, but withdrawal weighting is usually preferred as in many transversal filters (Chapter 6). The reason is that device bandwidths are generally quite small, up to a few percent, and apodized DARTs would be more susceptible to diffraction effects.

Withdrawal-weighted DARTs are realized by modifying the reflectivity or transduction, or both, of individual cells [1, 6]. Figure 9.5 shows some modifications. As before, the broken lines indicate the cell boundaries. Type (a) is the normal DART, and type (b) has no reflection, these being as in Fig. 9.2 above. In types

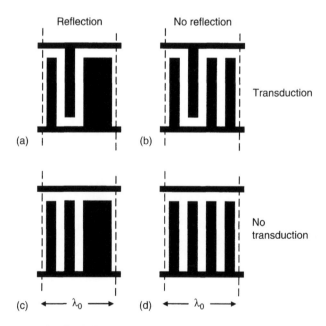

Figure 9.5 Modified DART cells. Types (a) and (c) include reflection, types (a) and (b) include transduction.

(c) and (d), the transduction has been eliminated by connecting the second electrode to the lower bus instead of the upper one. Thus, each cell can be with or without transduction and reflection. The DART can be analyzed as above, taking $a_T = 0$ or $R = 0$ as appropriate.

A further modification is to allow either of the transduction or reflection terms to change sign. The transduction sign can be changed by moving the live electrode to the right or left by an amount $\lambda_0/2$. For example, Fig. 9.5b could be changed so that the fourth electrode is live instead of the second. Strictly, this scheme is correct only for center-frequency operation, but in practice it is adequate because the bandwidth is quite small. Similarly, the reflection sign can be changed by shifting the wide electrode by $\lambda_0/4$. Any conflicts caused by this process can be resolved by further shifting a live electrode by λ_0, or a reflector electrode by $\lambda_0/2$. The introduction of reflector electrodes with different signs was a major development in DART design. Two reflectors with different signs form a weak resonator, with resonance at the center frequency. This 'resonant SPUDT', or 'RSPUDT', was found to be capable of much improved performance [7]. In particular, it enabled devices for a given electrical specification to be physically shorter, an important factor for mobile telephone handsets. A DART weighted in this way can be analyzed by a simple modification of the RAM, cascading the electrodes individually [8].

Design methods

For the non-reflective transversal filters of Chapter 6, the withdrawal weighting design starts from a continuous function representing the ideal time-domain response. A similar concept is applied to the DART, though here there are two functions – transduction and reflection. Each of these can be represented by a continuous analog function of distance, say $F_T(t)$ and $F_R(t)$, where $t = x/v$ is the time corresponding to a location x within the DART. For a uniform DART these functions are constants. In a weighted DART, each cell has a discrete transduction value which may be represented as 1, 0 or -1. These can be derived from the continuous function $F_T(t)$ by a withdrawal-weighting method such as that described in Chapter 6. Similarly, the cell reflection values, 1, 0 or -1, are obtained from $F_R(t)$.

The design problem is now one of the determining functions $F_T(t)$ and $F_R(t)$ so as to satisfy specifications on bandwidth, in-band ripple, skirt width, sidelobe rejection and so on. This is a difficult problem because the response depends on these functions in a complex non-linear fashion. Methods such as the Remez algorithm, commonly used for non-reflective transducers, are not applicable here. A relatively simple approach was suggested by Kodama *et al.* [2]. In eq. (9.3) we see that the transducer reflection coefficient S_{11} has two terms, one being P_{11} and the other being proportional to P_{13}^2. These terms need to cancel if S_{11} is to be zero. Assuming the reflectivity is weak, the acoustoelectric term P_{13} is approximately the Fourier transform of $F_T(t)$, and this can be designed to satisfy a specification, as for a non-reflective transducer. We now make the gross assumption that the term $P_{33} + Y_L$, which does not involve rapid phase changes, is approximately independent of frequency. Hence P_{11} needs to be proportional to P_{13}^2. In the time domain, the product of P_{13} with itself transforms to a convolution of $F_T(t)$ with itself, as shown by eq. (A.12) in Appendix A. Apart from constants, we have

$$F_R(t) \approx \int_{-\infty}^{\infty} F_T(2t - \tau)F_T(\tau)d\tau. \qquad (9.21)$$

A factor of 2 arises because a wave reflected at a position given by t is subject to a delay of $2t$. For example, if the transduction is unweighted, so that $F_T(t)$ is a rectangle, the corresponding $F_R(t)$ is a triangular function of the same length. Despite its simplicity, this approximate method is quite effective if the reflectivity is not too high. It was shown that the ripple due to the triple-transit signal can be substantially reduced in filters with 10 dB insertion loss [2].

Generally, a much more sophisticated approach is needed. In addition to the complexity of the DART behavior, there are other factors such as the electrical loading, the transducer aperture, and frequency variations of the reflection and transduction coefficients. For a device it is necessary to design two DARTs, which will not normally be the same. It is also necessary to take account of the multiple transits between these two transducers. Thus a 'global' design method is needed, taking account of all these variables. Highly sophisticated non-linear optimization methods are used for this purpose [1]. These depart radically from the original objective of minimizing the acoustic reflections. In a global design the reflections are not necessarily minimized – they are brought within limits such that the specification is satisfied. In fact, some designs actually exploit the triple-transit signal in order to sharpen the skirts of the response [9]. For the design process, an error function is defined, such that the response will be in specification if this function is small enough. The device design is to be such that this function is minimized. However, the optimization process cannot be done by continually reducing the function, because in general the function has many minima; a local minimum is not generally the optimal design. The algorithm used has to allow the error to increase sometimes. Optimization processes that do this include the genetic algorithm [10] and simulated annealing [11].

DART filters have been used extensively for intermediate-frequency (IF) applications in mobile telephony, typically with around 1 MHz bandwidth. Center

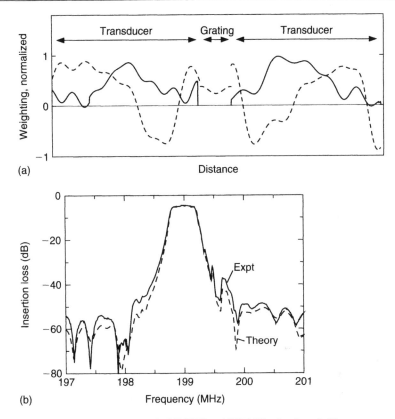

Figure 9.6 Design and performance of a SPUDT filter. (a) Weighting functions. Full lines: transduction, broken lines: reflection. (b) Frequency response. Full line: experimental, broken line: theoretical (courtesy of M. Solal, TriQuint Semiconductor).

frequencies are typically 85–230 MHz. Figure 9.6a shows the weighting for a filter consisting of two SPUDTs with a central grating [12]. Each SPUDT has weighted transduction and reflection functions, and the central grating also has weighted reflection. Thus, five weighting functions are used. In contrast, a non-reflective transversal filter (Chapter 6) has only one or two weighting functions. The experimental- and theoretical-frequency responses are shown in Fig. 9.6b. This device was designed to use a 'piston' waveguide mode, such that the mode amplitude has little variation across the transducer aperture. The sidelobes in the stop band above the passband are somewhat elevated because of unbound waveguide modes, which have been included in the analysis. These waveguide topics are mentioned later, in Sections 10.5 and 10.6 of Chapter 10.

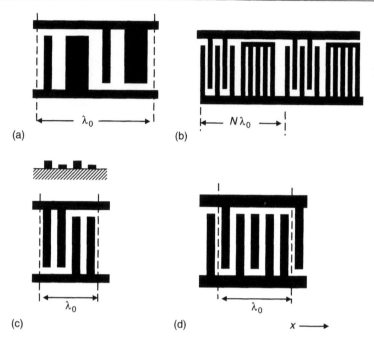

Figure 9.7 Other types of SPUDT. (a) Hanma–Hunsinger SPUDT. (b) Lewis group-type SPUDT. (c) Hartmann SPUDT. (d) Yamanouchi FEUDT. In each case, one cell of a periodic structure is indicated by the broken lines.

9.4 OTHER SPUDT STRUCTURES AND ANALYSIS FOR PARAMETERS

Figure 9.7 shows some other SPUDT types. The earliest SPUDT to be proposed was that of Hanma and Hunsinger [13] in Fig. 9.7a. Here the electrode widths are $\lambda_0/16$ and $3\lambda_0/16$, and all gaps are $\lambda_0/8$. The narrow electrodes limit the frequencies obtainable, but the transduction strength is stronger than that of the DART. Lewis [14] introduced the group type of Fig. 9.7b, where conventional transducers alternate with reflecting gratings. If the transducers are non-reflective, this gives SPUDT behavior if the spacing between a transducer center and the center of the first grating strip conforms to eq. (9.10). Unlike the DART, this is effective on either strongly or weakly piezoelectric substrates. If the transducers are the single-electrode type, with internal reflections, this SPUDT can still be made to work by adjusting the spacings. In this case, there is an additional advantage that the electrode widths are relatively large, at $\lambda_0/4$. However, the use of reflectivity and transduction in different regions is rather inflexible. Some 71 MHz filters on 112° rotated lithium tantalate gave insertion losses of 4 dB, bandwidth 300 kHz and 45 dB stop-band rejection [15].

Hartmann [5] introduced the SPUDT of Fig. 9.7c, which is simply a double-electrode transducer with alternating film thicknesses for the electrodes. It is evident that the transduction center is midway between the live electrodes, while the reflection center is at the center of a thick electrode. The separation of these centers is $\lambda_0/8$, an ideal value. Another type, due to Hartmann, is the EWC (Electrode Width Control) SPUDT [16]. This is almost the same as the DART, with the width of the wide electrode decreased from $3\lambda_0/8$ to $\lambda_0/4$.

Yamanouchi has introduced a wide variety of 'Floating Electrode Unidirectional Transducers' (FEUDTs), in which one or more electrodes is not connected to either bus bar [17, 18]. Many of these have regular electrodes (with constant width and pitch), as in the five-electrode FEUDT of Fig. 9.7d. Here the reflection center is at the center of the floating electrode, because this is a point of symmetry when the bus bars are shorted. For an applied voltage, the electrostatic solution is obtained by the usual method for regular electrodes (Chapter 5). The voltage on the floating electrode is determined by the condition that it should have zero net charge. A detailed analysis of a six-electrode FEUDT, based on Auld's perturbation theory, gave results agreeing well with accurate finite-element calculations [18]. These FEUDTs can also be analyzed using a quasi-static approach, giving results in good agreement [19]. The reflections arise because the incident wave induces voltages on the floating electrodes, and the voltages cause surface-wave generation. This gives a reflection coefficient proportional to $\Delta v/v$, so FEUDTs are generally effective only for strongly piezoelectric substrates such as lithium niobate. The five-electrode type performs well at the second harmonic, where the electrode width of $\lambda/5$ is not much less than the $\lambda/4$ width used in the DART at its fundamental frequency.

General analysis for uniform SPUDTs

The analysis methods considered above are adequate in the case of the DART on quartz, provided the accuracy needed is not too high. Generally, better methods are needed to analyze other SPUDT types, or SPUDTs on different substrate materials. For example, lithium niobate differs from quartz in that the reflections are strongly affected by electrical effects. For small film thickness the electrical effect dominates and a DART structure gives little reflectivity [20]. This is expected because the electric fields are very similar to a double-electrode transducer, as considered above. The reflectivity increases as the width of the wide electrode is reduced, and it is maximized at a width of $0.15\lambda_0$.

Assuming that transduction and reflection can be treated separately, the reflectivity can be ignored when analyzing the transduction. This implies that transduction can be deduced from the electrostatic charge density, as done in the quasi-static theory of Chapter 5. When unit voltage is applied to the SPUDT, the electrostatic charge density in one cell is $\rho_{e1}(x)$, say, with Fourier transform $\bar{\rho}_{e1}(\beta)$. According to eqs (5.33) and (5.34), the waves generated by this charge, in the $\pm x$ directions, have surface potentials $\phi_{s\pm}(x) = j\Gamma_s \bar{\rho}_{e1}(\mp k)\exp(\mp jkx)$. Here $\Gamma_s = (\Delta v/v)/\varepsilon_\infty$. Now, since $\rho_{e1}(x)$ is real, its Fourier transform has the property $\bar{\rho}_{e1}(-\beta) = \bar{\rho}_{e1}^*(\beta)$, so the functions $\bar{\rho}_{e1}(\mp k)$ have the same magnitude. It follows that we can choose a position for the origin, $x = 0$, such that the potentials $\phi_{s\pm}(x)$ are the same at $x = 0$. This position is a transduction center. With this new origin, the potentials become $\phi_{s\pm}(x) = j\Gamma_s |\bar{\rho}_{e1}(k)| \exp(\mp jkx)$. The parameter a_T of Section 9.2 is defined in terms of surface-wave power, and this gives

$$a_T = |\bar{\rho}_{e1}(k)| \sqrt{\omega W \Gamma_s/2}. \tag{9.22}$$

It is not necessary for the SPUDT to have geometrical symmetry about the point $x = 0$. The SPUDT capacitance can be obtained by integrating $\rho_{e1}(x)$ over the electrodes connected to one bus bar. An alternative approach is to define $\rho_{e1}(x)$ as the charge density obtained when a voltage is applied to one cell. This is a little different, because it will include charges in neighboring cells. However, the results are similar when a uniform SPUDT is analyzed, especially if its reflectivity is small.

For a periodic structure, the electrostatic solution is given exactly by Abbott and Hartmann [21] and Biryukov and Polevoi [22]. If N is the number of electrodes per period, the solution is

$$F(x) = \frac{\displaystyle\sum_{m=0}^{M} [a_m \cos(k\pi x/\lambda_0) + b_m \sin(k\pi x/\lambda_0)]}{j^s \displaystyle\prod_{n=1}^{N} \sqrt{|\sin[\pi(x - l_n)/\lambda_0]\sin[\pi(x - r_n)/\lambda_0]|}}. \tag{9.23}$$

Here the integer s is unity on the first electrode, and it increases by unity at each electrode edge. If N is even, $k = 2m$ and $M = (N - 2)/2$. If N is odd, $k = 2m + 1$ and $M = (N - 3)/2$. The parameters l_n and r_n are the locations of the left and right edges of electrode n. When x is in an unmetallized region, $F(x)$ is the parallel electric field $E_x(x)$. When x is in a metallized region, $F(x)$ is imaginary and $F(x)/j$ is the normal electric field. Equation (9.23) refers to electrodes in a vacuum. For a dielectric substrate, $F(x)/j$ is multiplied by $(\varepsilon_0 + \varepsilon_p)$ to obtain the charge density on the electrodes. The electrode potentials are obtained by integrating the parallel field in the gaps, and the constants a_m and b_m need to be determined such that the required electrode voltages are produced. The number of unknown coefficients is $N - 1$. This equals the number of voltage differences, which determine the fields. Thus, the voltages give $N - 1$ equations in the $N - 1$ unknowns.

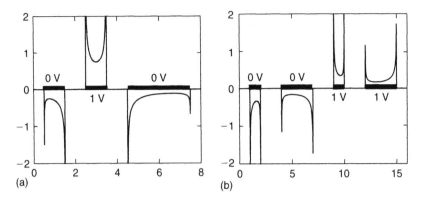

Figure 9.8 Electrostatic charge density. (a) DART. (b) Hanma–Hunsinger SPUDT.

Reflectivity is analyzed assuming the transducer to be shorted, so that transduction is absent. An approximate method for the reflectivity is to assume that reflection occurs at the edge of each electrode. This is consistent with Datta's reflection formula, as shown in Chapter 8, Section 8.1.4. Caution is needed because this method applies only if the reflectivity is dominated by mechanical effects. In addition, it is not accurate for larger film thicknesses, as shown by Fig. 8.6.

Figure 9.8 shows electrostatic charge densities for the DART and Hanma–Hunsinger transducers. For the DART, the charge density is almost symmetric about the center of the live electrode, showing that this point is approximately the transduction center as stated earlier. For the Hanma–Hunsinger SPUDT the field between the two live electrodes is relatively small, so the transduction center is approximately at the center of the area covered by these electrodes, that is, at a distance $5\lambda_0/16$ from the left edge of the narrow electrode. The fields are similar to those of a single-electrode transducer with metallization ratio 5/6, giving transduction strength larger than that of the DART. The reflection center is at the center of a wide electrode, which is a point of symmetry when the transducer is shorted. On this simple basis the distance between the two centers is $3\lambda_0/32$, not very different from the ideal value of $\lambda_0/8$.

A more sophisticated approach is that of Malocha and Abbott [23], in which the transduction is derived from electrostatic analysis as above. The reflection mechanism is a sum of electrical and mechanical effects. The electrical term is calculated from the electrostatic charge density induced in one cell when a surface wave is incident [22]. The waves generated by this charge are given by eqs (5.33) and (5.34) of Chapter 5, thus giving the electrical reflection coefficient. This is a quasi-static calculation, so it ignores the perturbation of the charges by the surface waves present. However, this is usually acceptable when considering one cell at a time. The total reflectivity can be calculated accurately using the finite-element method (FEM), as discussed in Chapter 8, Section 8.3.1. The mechanical term can then be obtained by subtracting the electrical term. This method gave an excellent agreement with experimental results for a variety of SPUDT geometries using quartz, lithium tantalate and langasite substrates.

9.5   OTHER SPUDT FILTERS

There are several other types of filter making use of SPUDTs to realize low insertion losses. It is worth noting that a simple SPUDT can be made by adding a reflector on one side of a conventional bidirectional transducer. In effect, this was the earliest type of SPUDT, using a transducer as a reflector [24]. A similar arrangement, using a grating reflector, is a common present-day surface-wave component. Two such SPUDTs form a type of resonator, as will be seen in Chapter 11.

Multi-track folded-path filters

For application in mobile phone handsets, SPUDT filters have come under much pressure to reduce the physical size needed. A variety of methods have been devised to address this question, and multi-track devices proved to be one approach. These devices are called 'folded-path' filters because the wave traverses the device more than once before producing an output signal. This implies that the signal has undergone more filtering than it would have in a single transit, and this improves the performance obtainable for a given length. Figure 9.9 shows two methods. The three-track device of Fig. 9.9a uses the fact that shorted SPUDTs have a large reflection coefficient, as noted in Section 9.1. Each track consists of two SPUDTs, and the three SPUDTs at each end are connected in parallel [1, 25]. The SPUDTs at each end are identical but the path lengths are incremented by $\lambda_0/3$, where λ_0 is the center-frequency wavelength. It is assumed that coupling between the tracks (due to diffraction, for example) is negligible. At the center frequency, waves generated at the left arrive at the right SPUDTs with phase increments of $120°$. Adding three equal-amplitude signals with relative phases $0°$, $120°$, $240°$ produces a total of zero. Hence no voltage is induced on the right SPUDTs, so these SPUDTs behave as if they were shorted, reflecting the waves strongly. The reflected waves arrive at the left SPUDTs with relative phases $0°$, $240°$, $480°$. Again, these add to give zero, so they are strongly reflected. Finally, the waves arrive at the right with relative phases $0°$, $360°$, $720°$, and now they are added in phase. With suitable electrical matching, a strong output signal is produced, with little reflection of the waves.

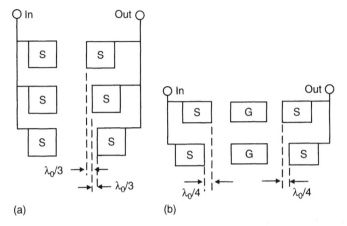

Figure 9.9 Folded-path SPUDT filters. (a) Three-track device. (b) Two-track device with central gratings. (S: SPUDT, G: grating).

In effect, the first-transit signal has been canceled and the main output of the device is due to the triple-transit signal. The output has been filtered twice by the conversion process in the SPUDTs, and also twice by the reflection process. The amount of filtering is therefore much more than in a one-track device. For example, the bandwidth or skirt width might be much smaller than that of a single-track device.

A two-track device [1, 26] is shown in Fig. 9.9b. Here each track consists of two SPUDTs plus a central grating, and the SPUDTs in one track are displaced by $\lambda_0/4$ relative to the other track. The operation of this device is rather complex, but again it is clear that the first-transit signal is canceled. It can be shown that the magnitude of the device response is essentially $C_s^2 \, R_s \, T_g \, R_g$, where C_s and R_s are the conversion and reflection coefficients of one SPUDT, and T_g and R_g are the transmission and reflection coefficients of one grating. Hence similar advantages are obtained, and the geometry is somewhat simpler because only two tracks are needed. For minimum loss, the grating needs to have a reflection coefficient of 3 dB ($R_g = 1/\sqrt{2}$) at the center frequency. The principle can be generalized to an arbitrary number N of tracks, and it can be shown that gratings are not necessary if N is odd, as in Fig. 9.9a. For both devices in Fig. 9.9, the SPUDTs can have weighted transduction and reflection functions and they can be designed by numerical optimization, as discussed earlier.

Other folded-path filters are shown in Fig. 9.10. Figure 9.10a shows a two-track type in which there are two weighted SPUDTs at each end, connected in parallel [9]. The two tracks have different designs. A special compact arrangement uses electrodes extending over the two tracks, so that there is no need for a central

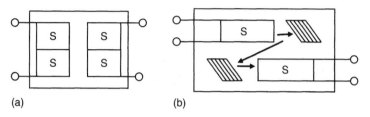

Figure 9.10 Further folded-path SPUDT filters. (a) Two-track SPUDT filter. (b) Z-path filter. S: SPUDT.

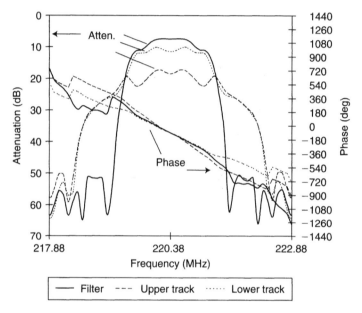

Figure 9.11 Response of two-track SPUDT filter. From Chamaly *et al.* [9], copyright 2000 IEEE, reproduced with permission.

bus bar. The responses of the two tracks can be made to partially cancel in the skirts of the response, thus reducing the skirt width. Simultaneously, they also cancel the triple-transit signal within the passband, reducing the ripple. This technique enables further reduction of device size. In Fig. 9.11 , the response of such a filter is shown by the bold lines. Other lines show the responses for the two tracks individually. The latter were obtained by applying attenuating material to the unwanted track. It can be seen that the responses of the two tracks combine to reduce substantially the signals in the skirt regions. Moreover, they also cancel passband ripples due to multiple-transit signals.

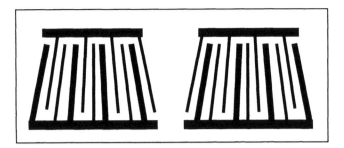

Figure 9.12 Filter using fanned SPUDTs.

Another folded-path technique, shown in Fig. 9.10b, is the Z-path device using inclined reflective gratings [27]. Here a wave launched by one SPUDT is reflected by an inclined grating into the other track, where another grating reflects it toward the output SPUDT. The waves thus traverse the device twice. All of the devices in Figs. 9.9 and 9.10 have been extensively developed for IF filtering in systems such as mobile phone handsets for CDMA, typically with 1.2 MHz bandwidth and center frequencies 85–220 MHz. Typically, insertion loss is 10 dB and the stop-band suppression is 40 dB. Quartz substrates are used because the narrow skirt width demands good temperature stability.

Fanned filters

Another modification of the SPUDT is to vary the transducer pitch in the transverse direction, as shown in Fig. 9.12. The pitch is smallest at one side of the device, and largest at the other side. This principle can also be applied to bidirectional transducers. Waves are generated and received with most efficiency at transverse locations where the pitch corresponds to the wavelength. This method is effective for generating a response flat within a specified passband, with sharp skirts and good stop-band rejection. When SPUDTs are used, the fanned arrangement increases the bandwidth obtainable. Bausk *et al.* [28] show examples using bidirectional transducers. Martin and Steiner [29] demonstrated SPUDT filters, giving 7 MHz bandwidth centered at 70 MHz with 10 dB insertion loss, 2 MHz skirt width and 30 dB stop-band rejection, using a $Y–Z$ lithium niobate substrate.

Table 9.1 summarizes the performance of the above filter types, and also other types described below.

Table 9.1 Performance of some low-loss filters.

Type	Substrate	Center frequency (MHz)	Bandwidth (MHz)	Skirt width (MHz)	Insertion loss (dB)	Rejection (dB)	Reference
Group SPUDT	$X-112°Y$ LT	71	0.3	0.3	4	45	[15]
Two-track SPUDT	$ST-X$ quartz	220.38	1.5	0.35	10	40	[9]
Z-path	$ST-X$ quartz	210.38	1.6	1.0	8	45	[27]
Fanned SPUDT	$Y-Z$ LN	70	7	2	10	30	[29]
Three-transducer device	$36°Y-X$ LT	900	30	20	8	40	[15]
Ring	$128°Y-X$ LN	150	3	3	1	30	[30]
Reflector filter	$128°Y-X$ LN	200	20	10	4	40	[31]
IIDT	$36°Y-X$ LT	880	20	20	1.6	30	[32]

LT: lithium tantalate and LN: lithium niobate.

9.6 OTHER LOW-LOSS TECHNIQUES

In addition to the use of SPUDTs, described above, a wide range of other methods have been investigated for low-loss devices [33]. Methods using resonators are considered later, in Chapter 11. In this section we describe a variety of other methods for obtaining low loss. These are less common in practical devices, so they are described here in less detail.

Three-transducer device

Several methods use a symmetrical transducer with identical waves incident from both sides. The term 'bilateral' is used here for these devices. With this arrangement, the transducer conversion loss is ideally zero if it is electrically matched. Figure 9.13a shows a three-transducer filter [34] using a central symmetrical output transducer. The two input transducers are symmetrical about the device center and they are connected electrically, so that identical waves are incident on both sides of the output transducer. Using the P-matrix, eq. (9.1), the symmetrical central transducer has $P_{23} = P_{13}$, and it can be shown that $(P_{11} + P_{12}) = -P_{13}/P_{13}^*$. This follows from eq. (9.2). For a load with admittance Y_L we have $I/V = -Y_L$. Setting $A_{i2} = A_{i1}$ and using $G_a = 2|P_{13}|^2$, the reflection coefficient is found to be

$$\frac{A_{t1}}{A_{i1}} = \frac{A_{t2}}{A_{i2}} = \frac{P_{13}}{P_{13}^*}\left[\frac{2G_a}{P_{33} + Y_L} - 1\right] \tag{9.24}$$

When the transducer is electrically matched we have $Y_L = P_{33}^*$, giving $P_{33} + Y_L = 2G_a$ and hence $A_{t1} = 0$. This result does not depend on the output

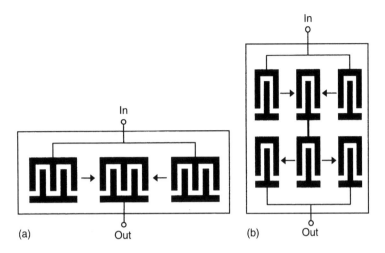

Figure 9.13 Low-loss devices using bilateral transducers. (a) Three-transducer filter. (b) Extra track added using self-resonant transducers.

transducer being uniform or non-reflective. The three-transducer filter of Fig. 9.13a ideally gives 3 dB insertion loss and no triple-transit signal, when electrically matched. Over a finite bandwidth, the triple-transit suppression and insertion loss can be improved by using SPUDTs for the outer transducers.

Figure 9.13b shows an adaptation where two tracks are coupled by means of identical central transducers. These are simply uniform transducers connected directly to each other. They will be electrically matched if they are *self-resonant*, that is, if the imaginary part of the admittance Y_t is small so that $Y_t \approx G_a$. This is considered in Chapter 6, Section 6.1. The track-changing transducers give very low loss, and they introduce extra frequency filtering. Hodé *et al.* [15] show some 900 MHz filters using these methods. In this case, each SPUDT consisted of a bidirectional transducer within a U-shaped multistrip coupler. The devices gave 8 dB insertion loss and 30 MHz bandwidth. The substrate was $36° Y–X$ lithium tantalate, which supports a leaky surface wave (described in Chapter 11).

Ring filter

Figure 9.14a shows a type of *ring filter*. Here the input and output transducers are coupled by means of components called 'reflecting trackchangers'. Each trackchanger transfers the surface wave from one track to another, and also changes its direction. There are several ways of doing this, one being a type of multistrip coupler (m.s.c.) with strip spacing $\lambda_0/3$, as shown. In the

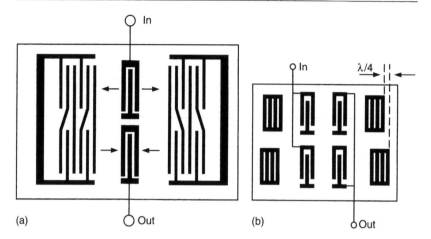

Figure 9.14 Further bilateral devices. (a) Ring filter using reflecting multistrip trackchangers. (b) Two-track reflector filter.

upper track, the phases of the wave at the m.s.c. strips are $0°$, $120°$, $240°$, ... (at the center frequency), giving corresponding voltages on the strips. The connections are such that the phases of the strip voltages in the lower track are $0°$, $240°$, $120°$, ..., equivalent to $0°$, $-120°$, $-240°$, ... These couple efficiently to a surface wave propagating in the opposite direction. The output transducer receives identical waves from both sides, so it does not reflect if it is electrically matched. Similarly, any waves reflected by the output transducer are absorbed without reflection by the input transducer, if it is matched. Filters of this type on $128° Y$–X lithium niobate had 150 MHz center frequency and 1 dB insertion loss [30].

The two-track reflector filter, Fig. 9.14b, has input and output transducers in each track, and one of the four transducers has its connections reversed. Thus, the signal due to direct surface-wave transit from input to output is canceled. Reflecting gratings are added, with positions differing by $\lambda/4$ at each end, such that the waves reflected by the gratings arrive at the output transducers in antiphase. The output transducers have identical waves incident from both directions, so they behave bilaterally. The device output is produced after conversion in both transducers and reflection at a grating, so this device is another folded-path device. This method can give wide bandwidths, as shown by a device with 10% bandwidth and 4 dB insertion loss [31].

Interdigitated interdigital transducer

The principle of the bilateral transducer is also present in the *interdigitated interdigital transducer*, or IIDT [35]. As shown in Fig. 9.15a, this consists

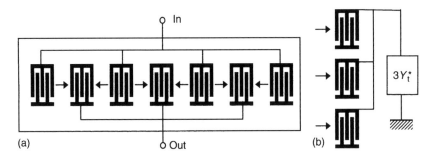

Figure 9.15 (a) The interdigitated interdigital transducer (IIDT). (b) Three connected transducers, illustrating the principle of the IIDT.

of a sequence of conventional transducers alternately connected to the input and output bus bars. For simplicity, assume that there are seven transducers, as shown, all identical and with admittance Y_t. Consider initially three identical transducers in different tracks, as in Fig. 9.15b. Any one transducer, with an individual load, will be electrically matched if the load admittance is Y_t^*. For three transducers, connected in parallel and with identical surface waves incident, the matching condition requires a load admittance of $3Y_t^*$. This is easily seen since the three transducers act as one transducer with aperture three times as large, with an admittance $3Y_t$. If surface waves are incident from both directions, the acoustic reflections will cancel, as argued before. In the IIDT, the output transducers behave in the same manner. Each output transducer has identical surface waves incident from both sides, and all of the power of these waves is transferred to the load because the transducers are matched. Because the power is totally absorbed, each output transducer is acoustically isolated from the others, so it behaves as one of the transducers in Fig. 9.15b.

A similar argument applies for the input transducers on Fig. 9.15a, except for those at the ends where there is some loss because of waves radiated outwards. Ideally, the insertion loss, in dB, is $10\log\left[(N_t-1)/(N_t+1)\right]$, where N_t is the number of transducers, typically 5–9. The description above assumes ideal matching, and a practical load can supply this only at one frequency, of course. At other frequencies the waves are not fully absorbed and the response becomes very complex because there are many paths for the waves. Appendix D gives a method for analyzing the device. Some authors have demonstrated a two-track version, with tracks coupled by means of self-resonant transducers in the manner of Fig. 9.13b. Yatsuda *et al.* [32] describe a practical filter with center frequency 880 MHz, 1.6 dB insertion loss, 20 MHz bandwidth, 20 MHz skirt width and 30 dB stop-band rejection. The substrate was $36°Y$–X lithium tantalate.

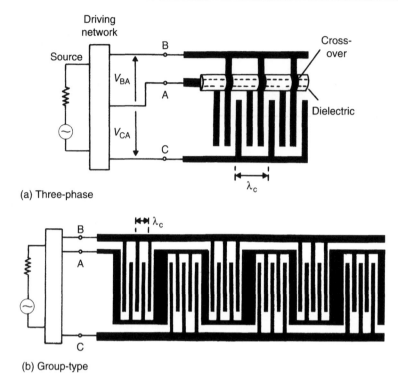

(a) Three-phase

(b) Group-type

Figure 9.16 Multiphase transducers. (a) Three-phase type. (b) Group type.

Multi-phase transducers

Finally, we briefly consider multi-phase transducers. Figure 9.16 shows two types. The three-phase type has regular electrodes connected sequentially to the three bus bars, and this necessitates a dielectric crossover. The bus bars are connected to the source via a driving network consisting of a few lumped components. This is designed to provide voltages differing in phase by $120°$, at the center frequency. At this frequency, waves in the 'forward' direction are reinforced while those in the reverse direction are canceled, so that the directivity is ideally total ($D = \infty$). The group-type transducer of Fig. 9.16b is a sequence of conventional transducers connected to three bus bars, with the spacing from one group to the next being $(N \pm 1/4)\lambda$. In this case one bus bar takes the form of a central meander line, regarded as the ground connection, and the other bus bars have voltages differing in phase by $90°$.

These multi-phase types require rather clumsy-driving circuits to provide the necessary phases. They have been superseded by the more recent SPUDTs, which eliminate the need for such circuits. However, the multi-phase types do have some advantages. They can provide total directivity ($D = \infty$) and can have quite wide bandwidths, whereas SPUDTs are often limited to bandwidths of a few percent and directivities of 10 dB or so. For the three-phase type, the dielectric crossover is awkward to fabricate, but this problem does not arise in the group type. Some description of the analysis and driving networks is given elsewhere [36].

REFERENCES

1. J.M. Hodé, J. Desbois, P. Dufilié, M. Solal and P. Ventura. 'SPUDT-based filters: design, principles and optimization', *IEEE Ultrason. Symp.*, **1**, 39–50 (1995).
2. T. Kodama, H. Kawabata, Y. Yasuhara and H. Sato. 'Design of low-loss SAW filters employing distributed acoustic reflection transducers', *IEEE Ultrason. Symp.*, **1**, 59–64 (1986).
3. D.P. Morgan. 'Reflective array modeling for reflective and directional SAW transducers', *IEEE Trans. Ultrason. Ferroelect. Freq. Contr.*, **45**, 152–157 (1998).
4. B.P. Abbott, C.S. Hartmann and D.C. Malocha. 'Transduction magnitude and phase for COM modeling of SAW devices', *IEEE Trans. Ultrason. Ferroelect. Freq. Contr.*, **39**, 54–60 (1992).
5. C.S. Hartmann, P.V. Wright, R.J. Kansy and E.M. Garber. 'An analysis of SAW interdigital transducers with internal reflections and the application to the design of single-phase unidirectional transducers', *IEEE Ultrason. Symp.*, **1**, 40–45 (1982).
6. C. Ruppel, R. Dill, J. Franz, S. Kurp and W. Ruile. 'Design of generalised SPUDT filters', *IEEE Ultrason. Symp.*, **1**, 165–168 (1996).
7. P. Ventura, M. Solal, P. Dufilié, J.M. Hodé and F. Roux. 'A new concept in SPUDT design: the RSPUDT', *IEEE Ultrason. Symp.*, **1**, 1–6 (1994).
8. D.P. Morgan. 'Reflective array method for analysis and design of weighted DART transducers and filters', *IEEE Ultrason. Symp.*, **1**, 133–137 (2001).
9. S. Chamaly, P. Dufilie, P. Blanc-Benon and M. Doisy. 'A new generalized two-track RSPUDT structure for SAW filter size reduction: design and performance', *IEEE Ultrason. Symp.*, **1**, 407–412 (2000).
10. D.E. Goldberg. *Genetic Algorithms in Search, Optimisation and Machine Learning*, Addison-Wesley, 1989.
11. W.H. Press, B.P. Flannery, S.A. Teukolsky and W.T. Vetterling. *Numerical Recipes – The Art of Scientific Computing (Fortran version)*, Cambridge University Press, 1989.
12. M. Solal, O. Holmgren and K. Kokkonen. 'Design modeling and visualization of low transverse modes R-SPUDT devices', *IEEE Ultrason. Symp.*, **1**, 82–87 (2006).
13. K. Hanma and B.J. Hunsinger. 'A new triple transit suppression technique', *IEEE Ultrason. Symp.*, 1976, pp. 328–331.
14. M.F. Lewis. 'Low loss SAW devices employing single stage fabrication', *IEEE Ultrason. Symp.*, 1983, pp. 104–108.
15. J.-M. Hode, M. Doisy and P. Dufilie. 'Application of low loss SAW filters to RF and IF filtering in digital cellular radio systems', *IEEE Ultrason. Symp.*, **1**, 429–434 (1990).
16. C.S. Hartmann and B.P. Abbott. 'Overview of design challenges for single phase unidirectional SAW filters', *IEEE Ultrason. Symp.*, **1**, 79–89 (1989).
17. K. Yamanouchi, C.S. Lee, K. Yamamoto, T. Meguro and H. Odagawa. 'GHz-range low-loss wide band filters using new floating electrode type unidirectional transducers', *IEEE Ultrason. Symp.*, **1**, 139–142 (1992).

18. M. Takeuchi and K. Yamanouchi. 'Coupled-mode analysis of SAW floating electrode type unidirectional transducers', *IEEE Trans. Ultrason. Ferroelect. Freq. Contr.*, **40**, 648–658 (1993).

19. D.P. Morgan. 'Investigation of novel floating-electrode unidirectional SAW transducers (FEUDTs)', *IEEE Ultrason. Symp.*, **1**, 15–19 (2000).

20. J. Koskela, J. Fagerholm, D.P. Morgan and M.M. Salomaa. 'Self-consistent analysis of arbitrary 1-D SAW transducers', *IEEE Ultrason. Symp.*, **1**, 135–138 (1996).

21. B.P. Abbott and C.S. Hartmann. 'An efficient evaluation of the electrostatic fields in IDTs with periodic electrode sequences', *IEEE Ultrason. Symp.*, **1**, 157–160 (1993).

22. S.V. Biryukov and V.G. Polevoi. 'The electrostatic problem for SAW interdigital transducers in an external electric field – Part II: Periodic structures', *IEEE Trans. Ultrason. Ferroelect. Freq. Contr.*, **43**, 1160–1170 (1996).

23. S. Malocha and B.P. Abbott. 'Calculation of COM parameters for an arbitrary IDT cell', *IEEE Ultrason. Symp.*, **1**, 267–270 (2002).

24. W.R. Smith, H.M. Gerard, J.H. Collins, T.M. Reeder and H.J. Shaw. 'Design of surface wave delay lines with interdigital transducers', *IEEE Trans.*, **MTT-17**, 865–873 (1969).

25. D.P. Morgan and T. Thorvaldsson. 'A new low-loss SAW filter technique', *IEEE Ultrason. Symp.*, **1**, 23–26 (1994).

26. M. Solal and J.-M. Hodé. 'A new compact SAW low loss filter for mobile radio', *IEEE Ultrason. Symp.*, **1**, 105–109 (1993).

27. S. Friesleben, A. Bergmann, U. Bauernschmidt, C. Ruppel and J. Franz. 'A highly miniaturized recursive Z-path SAW filter', *IEEE Ultrason. Symp.*, **1**, 347–350 (1999).

28. E. Bausk, R. Taziev and A. Lee. 'Synthesis of slanted and quasi-slanted SAW transducers', *IEEE Trans. Ultrason. Ferroelec. Freq. Contr.*, **51**, 1002–1009 (2004).

29. G. Martin and B. Steiner. 'SAW filters including one-focus slanted interdigital transducers', *IEEE Trans. Ultrason. Ferroelec. Freq. Contr.*, **50**, 94–98 (2003).

30. S.A. Dobershtein and V.A. Malyukhov. 'SAW ring filters with an insertion loss of 1 dB', *IEEE Trans. Ultrason. Ferroelec. Freq. Contr.*, **44**, 590–596 (1997).

31. G. Muller, J. Machui, L. Reindl, R. Weigel and P. Russer. 'Design of a low loss SAW reflector filter with extremely wide bandwidth for mobile communication systems', *IEEE Trans. Microwave Theory and Tech.*, **41**, 2147–2155 (1993).

32. H. Yatsuda, T. Inaoka, Y. Takeuchi and T. Horishima. 'IIDT-type low loss SAW filters with improved stopband rejection in the range of 1 to 2 GHz', *IEEE Ultrason. Symp.*, **1**, 67–70 (1992).

33. C. Ruppel, R. Dill, A. Fischerauer, G. Fischerauer, W. Gawlik, J. Machui, F. Muller, L. Reindl, G. Scholl, I. Schropp and K. Wagner. 'SAW devices for consumer communication applications', *IEEE Trans. Ultrason. Ferroelec. Freq. Contr.*, **40**, 438–452 (1993).

34. M.F. Lewis. 'Triple-transit suppression in surface-acoustic-wave devices', *Electron. Lett.*, **8**, 553–554 (1972).

35. M.F. Lewis. 'SAW filters employing interdigitated interdigital transducers, IIDT', *IEEE Ultrason. Symp.*, **1**, 12–17 (1982).

36. D.P. Morgan. *Surface-Wave Devices for Signal Processing,* Elsevier, 1991, Chapter 7.

10

WAVEGUIDES AND TRANSVERSELY
COUPLED RESONATOR FILTERS

The term 'waveguide' refers to a structure that guides the wave along a particular path, preventing energy from escaping at the sides so that ideally the wave propagates without attenuation. This may be used, for example, to counter attenuation due to diffraction, discussed in Chapter 4. For surface waves, many types of waveguide have been shown to be effective [1]. For present-day technology a particular type is relevant, consisting of a strip of material which reduces the wave velocity. The strip is laid in the intended propagation direction, and it may be either metal or dielectric. It may reduce the velocity by electrical or mechanical loading or a combination of both. This is often called a 'strip waveguide', or '$\Delta v/v$ waveguide'. More complex structures operate in a similar manner. In particular, an interdigital transducer can have waveguide characteristics because the wave velocity within it is lower than the velocity outside. However, this case is more complex because there are three velocities involved – in the electrodes, the bus bars and the free surface.

Characteristically, a waveguide will support multiple surface-wave modes with different phase velocities, and the modes are all dispersive. Typically, all modes except the lowest-velocity one can only exist above their respective 'cutoff' frequencies.

In many surface-wave devices, waveguide effects are too weak to be of any practical concern. However, they are significant in several areas:

(a) A simple metal strip can form a waveguide. This has been used extensively in convolvers using narrow surface-wave beams, as described in Chapter 7. Without the waveguiding, these devices would suffer severe attenuation due to diffraction.

(b) The waveguiding in conventional transducers can cause distortion in the device response. Thus applies particularly for resonators, which can exhibit a series of resonances corresponding to several waveguide modes.
(c) A particular type of bandpass filter, the transversely coupled resonator (TCR) filter, exploits the waveguide properties of transducers and gratings. This device has two resonances arising from two waveguide modes, and these form two poles in the frequency response.

In this chapter we first consider simple waveguides, consisting of a single or double strip, in Section 10.1. Section 10.2 is concerned with excitation of waveguide modes by sources such as transducer electrodes, and Section 10.3 gives a general waveguide analysis for an arbitrary number of waveguide regions. Following this we consider the TCR filter in Section 10.4. This filter makes use of waveguide concepts discussed in earlier sections, while resonators are considered in more detail later, in Chapter 11. The rest of this chapter deals with topics that are of more specialized interest. These are unbound modes, Section 10.5, and waveguides incorporating reflecting gratings, Section 10.6.

10.1 BASIC STRIP WAVEGUIDES

Figure 10.1 illustrates the type of waveguide considered here. It consists simply of a strip of width $2W$ deposited on the surface. The surface is normal to the z axis, and the strip is oriented in the x direction. Normally we are concerned with metallic strips, but dielectric strips are also possible. This structure was

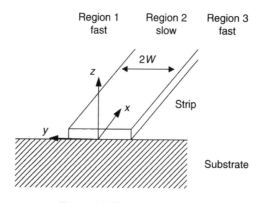

Figure 10.1 Single-strip waveguide.

investigated by, for example, Tiersten [2] and Schmidt and Coldren [3]. A rigorous analysis of the structure would be very complex. For example, the stresses in the strip will vary within it, and complications also arise from anisotropy. Consequently, a simplified model is normally used. In particular, the wave amplitude is represented by a scalar $\psi(x, y)$, as for diffraction analysis (Chapter 4). The amplitude $\psi(x, y)$ and its differential $d\psi/dy$ can be taken to be continuous [2].

We first consider plane waves on a uniform surface, in the absence of the waveguide. The amplitude is assumed to satisfy the wave equation $\partial^2\psi/\partial x^2 + \partial^2\psi/\partial y^2 = (1/v^2)\partial^2\psi/\partial t^2$, where v is the velocity. For a plane wave with arbitrary propagation direction, the amplitude has the form $\psi(x, y) = \exp(-jk_x x) \exp(-jk_y y)$, where k_x and k_y are the components of the wave vector $\mathbf{k}$, so that $k_x^2 + k_y^2 = k^2 \equiv |\mathbf{k}|^2$. A factor $\exp(j\omega t)$ is implicit, as usual. The wavenumber k will vary with propagation direction because of anisotropy. Assuming parabolic anisotropy, we write $k = k_0/(1 + \alpha_a\theta^2/2)$, where θ is the propagation direction and $\tan\theta = k_y/k_x$. The velocity is proportional to $(1 + \alpha_a\theta^2/2)$. The parameter α_a specifies the anisotropy and it is zero for an isotropic material. For ST–X quartz we have $\alpha_a = 0.38$. The wavenumber k_0 refers to propagation with $\theta = 0$, so that $\mathbf{k}$ is in the x direction. We then find

$$(1 + \alpha_a)k_y^2 + k_x^2 = k_0^2, \tag{10.1}$$

where a term k_y^4/k_x^2 has been ignored because practical cases have $k_y^2 \ll k_x^2$. Equation (10.1) assumes that there is no beam steering. It corresponds to eq. (4.10), used for diffraction analysis, with $b = 1 + \alpha_a$.

A waveguide has regions with different plane-wave velocities, and solutions will be found by assuming that the waves in each region are sums of plane waves with the above form. The anisotropy α_a is assumed to be the same everywhere. For a waveguide mode, k_x must be the same everywhere, and it will be denoted by β. However, k_y and k_0 will generally be different in different regions, so they are given an additional subscript n, the region number. For region n, eq. (10.1) becomes

$$\beta^2 = k_{0n}^2 - (1 + \alpha_a)k_{yn}^2 \tag{10.2}$$

Here k_{0n} is simply ω/v_{0n}, where v_{0n} is the velocity of plane waves propagating along x in region n. If β and k_{0n} are specified, eq. (10.2) gives two solutions for k_{yn}. These may be imaginary, and in this case it is convenient to define $\kappa_{yn} = k_{yn}/j$ so that κ_{yn} is real. Thus, eq. (10.2) can also be expressed as:

$$\beta^2 = k_{0n}^2 + (1 + \alpha_a)\kappa_{yn}^2. \tag{10.3}$$

When k_{yn} is imaginary the wave is no longer plane, but the solution still satisfies the wave equation. For a waveguide mode, the wave amplitude in the outer regions ($|y| > W$) needs to decay with y so that energy is confined within the guide. This requires k_y to be imaginary, and hence β is greater than k_0 in this region. To obtain this condition, the plane-wave velocity in the strip region needs to be lower than that of the outer regions.

For the waveguide of Fig. 10.1, the central 'slow' region $|y| < W$ is region 2. The outer regions 1 and 3 are 'fast' and have $v_{03} = v_{01}$, so that $k_{03} = k_{01} < k_{02}$. In region 2, the k_{y2} values are given by eq. (10.2), which has two solutions. Taking $k_{y2} > 0$, the amplitude has the form $\exp(\pm j\,k_{y2}\,y)$. To find a mode symmetric in y, these are added to give $\psi = A\cos(k_{y2}\,y)$. All amplitudes also vary as $\exp(-j\beta x)$, and this term is omitted. Outside the strip, the 'fast' regions 1 and 3 have $|y| > W$. Here the amplitude needs to decay with y, and it is taken to be $\psi = B\exp[-\kappa_{y1}(|y| - W)]$, where κ_{y1} is positive real and equal to κ_{y3}. Here A and B are constants. The boundary conditions are that ψ and $d\psi/dy$ are continuous at the boundaries $y = \pm W$. Setting ψ continuous gives $B/A = \cos(k_{y2}W)$, and the additional continuity of $d\psi/dy$ gives the dispersion relation

$$\tan(k_{y2}W) = \kappa_{y1}/k_{y2}, \quad \text{for symmetric modes.} \qquad (10.4)$$

For antisymmetric modes, the amplitude in region 2 is taken to be $\psi = A\sin(k_{y2}\,y)$, and for regions 1 and 3 we set $\psi = B\,\text{sgn}(y)\exp[-\kappa_{y1}(|y| - W)]$. Using the boundary conditions gives $B/A = \sin(k_{y2}W)$, and the dispersion relation is

$$\cot(k_{y2}W) = \kappa_{y1}/k_{y2}, \quad \text{for antisymmetric modes.} \qquad (10.5)$$

These equations give separate families of modes. A more general form could have $\psi = \cos(k_{y2}y + \phi)$ in region 2, where ϕ is some constant. However, this is a sum of sine and cosine terms, giving two modes with different velocities. It is therefore sufficient to consider symmetric and antisymmetric solutions. These solutions are sometimes referred to as 'bound' modes, to distinguish them from some unbound modes described later (Section 10.5).

Experimental results confirm these equations [3]. Figure 10.2 shows velocities for the first three modes of a single-strip guide. Here $v_{02}/v_{01} = 0.995$, which is a typical value for ST–X quartz. The phase velocity v is expressed in the normalized form $v_{\text{norm}} = (v - v_{02})/(v_{01} - v_{02})$, so that $v_{\text{norm}} = 1$ when $v = v_{01}$. The lowest-order mode (the one with smallest phase velocity) is symmetric, and subsequent modes alternate between antisymmetric and symmetric. At zero frequency, the first mode has phase velocity v_{01}, the plane-wave velocity of the 'fast' region. The other modes have this velocity at their cutoff frequencies. This mode velocity corresponds to $\kappa_{y1} = 0$, so that $\tan(k_{y2}W) = 0$ or $\cot(k_{y2}W) = 0$. Figure 10.3 shows profiles of the first three modes for a normalized frequency $W/\lambda_0 = 8$, where $\lambda_0 = v_{02}/f$. These have been scaled so that their extrema are at ± 1.

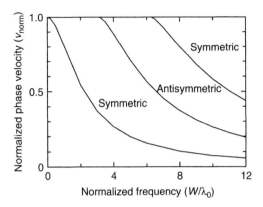

Figure 10.2 Phase velocities and profiles for first three modes in single-strip waveguide. $\lambda_0 = v_{02}/f$ is the wavelength of plane waves in the 'slow' region.

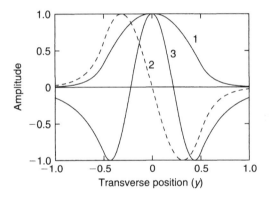

Figure 10.3 Profiles for the first three modes of the waveguide in Fig. 10.2, for $W/\lambda_0 = 8$. The modes are numbered in order of increasing phase velocity, and mode 2 is antisymmetric. The abscissa is normalized such that the strip edges are at $y = \pm 1/2$.

Double-strip waveguide

In some instances there is interest in two parallel strip waveguides, illustrated in Fig. 10.4. The strips are identical, with width W, and the spacing g is small enough to give some acoustic coupling between them. Regions 2 and 4 are the 'slow' regions, where the strips are located, with $g/2 < |y| < g/2 + W$. The transverse wavenumbers $k_{y2} = k_{y4}$ are real. Regions 1, 3 and 5 are 'fast' regions where $\kappa_{y1} = \kappa_{y3} = \kappa_{y5}$ are real. This system can be analyzed by the method used above for single waveguides, though it is more complex of course. The boundary conditions need to be applied at the four boundaries $y = \pm g/2, \pm(g/2 + W)$.

This leads to the dispersion relation

$$k_{y2}W = \tan^{-1}(\kappa_{y1}/\kappa_{y2}) + \tan^{-1}[(\kappa_{y1}/\kappa_{y2})\tanh(\kappa_{y1}g/2)], \qquad (10.6)$$

for symmetric modes. For antisymmetric modes, the result is the same except that the tanh is replaced by coth. Section 10.3 below gives an alternative derivation.

If the gap g is relatively large, the effect of the coupling is such that each mode of a single strip is split into two modes, one symmetric and the other antisymmetric, with similar velocities. Within each strip, the mode profile $\psi(y)$ is similar to that of a single strip. This can be seen from the profiles shown in the lower part of Fig. 10.4, which assumes an ST–X quartz substrate and $g/W = 3$. If a wave is launched in the left strip only, this can be represented as a sum of two modes, symmetric and antisymmetric, which reinforce in the left strip and cancel in the right strip. The modes travel with wavenumbers β_s and β_a, say. After traveling a distance $x_c = \pi/|\beta_s - \beta_a|$ they reinforce in the right strip and cancel in the left strip, so the energy has been transferred from one strip to the other. The same phenomenon occurs for Rayleigh waves in a parallel-sided plate, Section 2.2.5 in Chapter 2. The case of Fig. 10.4 gives $x_c = 62\,500$ wavelengths, so the coupling is too weak to be significant practically. However, stronger coupling occurs in the TCR filter (Section 10.4), which is related to the double-strip waveguide. In this case g is much smaller and the coupling is strong enough to be useful.

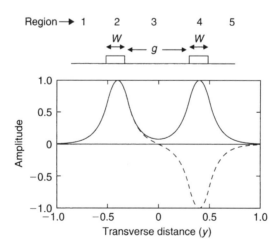

Figure 10.4 Double-strip waveguide. The upper part shows the geometry. The lower part shows the profiles of the lowest-order symmetric and antisymmetric modes, taking $g/W = 3$ and $W/\lambda_0 = 5$. Other data are as for Fig. 10.3. The abscissa is normalized so that the outer edges of the strips are at $y = \pm 1/2$.

In lithium niobate, larger velocity differences are obtained and the coupling between the parallel strips is stronger. In particular, the dual-channel waveguide convolver of Section 7.6.3 in Chapter 7 needs to be designed such that the coupling is minimized.

10.2 WAVEGUIDE MODES IN INTERDIGITAL DEVICES

An interdigital transducer can exhibit waveguide modes because the surface-wave velocity within it is less than that in the free surface outside. Here we consider the coupling to these modes. The profile for mode m is denoted by $\psi_m(y)$. The amplitude will be $\psi_m(y) \exp(-j\beta_m x)$, where β_m is the wavenumber of mode m. Section 10.3 below gives a general method for calculating $\psi_m(y)$. It also shows that $\psi_m(y)$ has phase independent of y, and here we take $\psi_m(y)$ to be real without losing generality. The modes are known to be orthogonal [4], so that

$$\int_{-\infty}^{\infty} \psi_n(y)\psi_m(y)\mathrm{d}y = 0, \quad \text{for } n \neq m. \tag{10.7}$$

Figure 10.5 shows a line source and a line receiver embedded in a waveguide structure. The source represents one element of a launching transducer, which may be taken as the active electrode overlap between two electrodes. Similarly, the receiver represents an active overlap in a receiving transducer. Electrode reflections are ignored here. The waveguide structure is taken to be uniform, so the pattern of the electrodes must be continuous throughout (except for the electrode connections to the bus bars). The source generates a wave with amplitude variation $S(y)$. It is assumed that this wave can be represented as a sum of the waveguide modes, so that

$$S(y) = \sum_m a_m \psi_m(y), \tag{10.8}$$

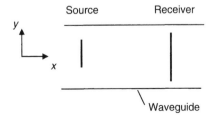

Figure 10.5 Source and receiver in a waveguide structure.

where the a_m are constants, independent of y. Multiplying both sides by $\psi_n(y)$, integrating and using orthogonality, this gives

$$\int_{-\infty}^{\infty} S(y)\psi_n(y)dy = a_n \int_{-\infty}^{\infty} [\psi_n(y)]^2 dy$$

The source amplitude $S(y)$ is taken to be unity over some range of y and zero elsewhere. The left side of this equation can be written as $\int_s \psi_n(y)dy$, where the integral is over the y range occupied by the source. We then have

$$a_n = \frac{\int_s \psi_n(y)dy}{\int_{-\infty}^{\infty} [\psi_n(y)]^2 dy}. \qquad (10.9)$$

The current produced by the receiving transducer, assumed to be shorted, can be written as

$$I = \gamma \sum_m a_m \exp(-j\theta_m) \int_r \psi_m(y)dy \qquad (10.10)$$

Here the integral is over the y range of the receiving element. The parameter γ is some constant independent of y, and θ_m is a phase corresponding to the transit time for mode m.

If there were no waveguiding, the output current would be $I_0 = \gamma W_s \exp(-j\theta_0)$, where W_s is the source aperture and θ_0 is the phase. It is assumed that the receiver fully overlaps the source, which implies that its aperture W_r exceeds W_s. Using eq. (10.10), the ratio of currents can be expressed as

$$I/I_0 = (W_r/W_s)^{1/2} \sum_m \exp[j(\theta_0 - \theta_m)]K_{tr}^{(m)} K_{ts}^{(m)} \qquad (10.11)$$

Here $K_{ts}^{(m)}$ and $K_{tr}^{(m)}$ are transduction parameters for mode m, defined by

$$K_{ts}^{(m)} = \frac{\int_s \psi_m dy}{\sqrt{W_s \int_{-\infty}^{\infty} \psi_m^2 dy}}; \quad K_{tr}^{(m)} = \frac{\int_r \psi_m dy}{\sqrt{W_r \int_{-\infty}^{\infty} \psi_m^2 dy}} \qquad (10.12)$$

Equation (10.11) includes a sum over all the mode numbers m, giving the response between one source and one receiver. For the complete device response, it also needs to be summed over the sources in the launching transducer and the receivers in the receiving transducer. The function $K_{ts}^{(m)}$ is unity in the case where there is no waveguiding. This can be seen by assuming that the source generates only one mode, with $\psi(y) = 1$ when y is within the source region, and $\psi(y) = 0$ elsewhere. Similarly, $K_{tr}^{(m)}$ is unity if there is no waveguiding. In this case, eq. (10.11) becomes $I/I_0 = \sqrt{W_r/W_s}$.

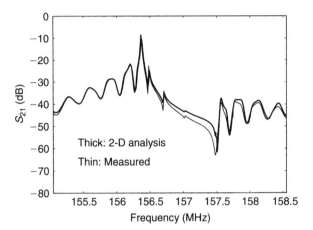

Figure 10.6 Response of a two-port resonator on *ST−X* quartz, with aperture $40\lambda_0$ and film thickness $h/\lambda_0 = 2\%$, showing ripples due to waveguide modes. Thin lines: experimental; thick lines: theoretical. From Hirota and Nakamura [5], copyright 2001 IEEE, reproduced with permission.

Often, we are concerned with a symmetrical waveguide and a symmetrical source distribution, usually a source in an unapodized transducer. This usually couples most strongly to the fundamental waveguide mode. Coupling to asymmetric modes is precluded by the symmetry, and coupling to higher symmetric modes is relatively small because the mode profiles have an oscillatory character.

The total device response is a sum of contributions corresponding to the various modes. These contributions will have different delays because of the different mode velocities. In practical devices these contributions usually overlap in frequency, so that they are not seen separately. However, for a high-Q resonator the individual modes give rise to sharp peaks which can sometimes be distinguished. The main response, due to the lowest-order mode, gives a major peak, and small subsidiary peaks due to higher-order modes are observed at higher frequencies. Figure 10.6 shows some experimental and theoretical data for a resonator response [5]. This case includes the added complication of electrode reflections, which will be considered later, Section 10.6. In resonators, the coupling to higher modes can be reduced by transducer apodization, such that the electrode overlap is largest at the transducer center and small at the edges. The transducer then generates a wave with its maximum amplitude at the center ($y = 0$), and this reduces the coupling to higher waveguide modes.

Some reservations need to be noted. Firstly, the waveguide analysis generally yields only a small number of modes, typically 10. To represent a general function as a sum of mode profiles an infinite sum is necessary, in a manner similar to Fourier synthesis. Since eq. (10.8) can only be summed over the finite number of modes, it gives $S(y)$ only approximately. However, this is often adequate if the transducers are relatively long, for example 100 wavelengths or more. Secondly, in reflective transducers and gratings the electrode reflectivity needs to be allowed for. Thirdly, the above approach assumes that the waveguide properties are uniform everywhere. This implies that components such as transducers and gratings have similar geometries, for example similar bus bar widths. The first two points will be considered further in Sections 10.5 and 10.6.

10.3 ANALYSIS FOR GENERAL WAVEGUIDES

In general, waveguides can be much more complex than the cases considered above, with more complex velocity distributions. For example, an interdigital transducer has different velocities in the electrode region, the bus regions and the free surface on either side. The TCR filter is even more complex. To analyse such waveguides, a general approach called the stack matrix can be used [6]. The following analysis [7] is based on this method. The waveguide is represented as in Fig. 10.7, which refers to the TCR filter. The waveguide is taken to have N regions, with $N = 7$ for Fig. 10.7.

Assuming that a value for β has been chosen, the transverse wavenumber k_{yn} in region n is determined by eq. (10.2). The amplitude in region n is written as

$$\psi_n(y) = A_n[\exp(jk_{yn}\, y) + B_n \exp(-jk_{yn}\, y)], \tag{10.13}$$

where a common factor $\exp(-j\beta x)$ is omitted. The constants A_n and B_n may be complex. As before we may also write $k_{yn} = j\kappa_{yn}$, and eq. (10.13) becomes

$$\psi_n(y) = A_n[\exp(-\kappa_{yn}\, y) + B_n \exp(\kappa_{yn}\, y)] \tag{10.14}$$

This form is convenient if k_{yn} is imaginary. We define the function

$$F_n(y) = k_{yn}\frac{1 + \alpha_n \tan(k_{yn}\, y)}{\alpha_n - \tan(k_{yn}\, y)} = k_{yn} \tan(\varepsilon_n + k_{yn}\, y), \tag{10.15}$$

where

$$\alpha_n = j(1 + B_n)/(1 - B_n), \tag{10.16}$$

and $\varepsilon_n = \cot^{-1}\alpha_n$. It is shown below that $F_n(y)$ is always real, irrespective of whether k_{yn} is real. The parameter α_n is real when k_{yn} is real, and imaginary when k_{yn} is imaginary.

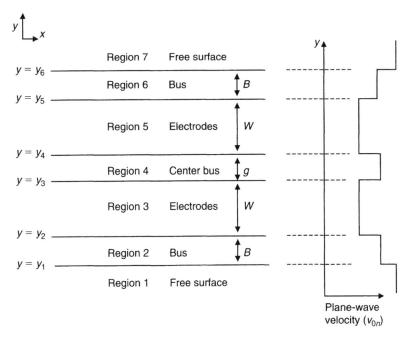

Figure 10.7 Geometry for a waveguide representing the TCR filter. The right part of the figure represents the distribution of plane-wave velocities v_{0n}.

It is shown below that the amplitude $\psi(y)$ can be taken to be real for all y (for a particular x). The boundary conditions are that ψ and $d\psi/dy$ are continuous, with these functions given by eq. (10.13) and its derivative. These boundary conditions are found to be satisfied by making $F(y)$ continuous. Thus, for a boundary at $y = y_n$, between region n (where $\psi = \psi_n$) and region $n+1$ (where $\psi = \psi_{n+1}$), we have

$$F_{n+1}(y = y_n) = F_n(y = y_n). \tag{10.17}$$

To find the dispersion relation we assume a value for β and start by considering region 1, the lower outside region. Here the amplitude is proportional to $\exp(\kappa_{y1}\, y)$, and from eq. (10.14) we have $B_1 = \infty$, so that $\alpha_1 = -j$. In region 2 we have $F_2(y)$ given by eq. (10.15), and using the continuity condition of eq. (10.17) at $y = y_1$ gives the parameter α_2. This process is repeated throughout the structure, finally giving the value of α_N which applies in the upper outer region $y > y_{N-1}$. Here the amplitude needs to be proportional to $\exp(-\kappa_{yN}\, y)$, and this requires $B_N = 0$ and hence $\alpha_N = j$. The calculation gives this result only if the correct value of β is chosen, so trial values of β are used until α_N is correct. In general there will be several correct β values, since there are several modes. Once β has been found, the α_n values give B_n by using eq. (10.16), and continuity of ψ can be used to find the A_n. Equations (10.13) or (10.14) then give the mode profile $\psi(y)$.

To show that ψ has the same phase at all y (for a particular x), suppose that $\psi_n(y)$ is real at some arbitrary point in region n. If k_{yn} is real the two terms in eq. (10.13) are conjugates, giving $B_n = A_n^*/A_n$. If k_{yn} is imaginary, eq. (10.14) shows that A_n and B_n are real. In either case, $\psi_n(y)$ is real for all y in region n. Since ψ is continuous across each boundary, it follows that it must be real for all y. Using eq. (10.16), we also find that α_n is real if k_{yn} is real, and α_n is imaginary if k_{yn} is imaginary. It also follows that $F_n(y)$ is real everywhere.

Derivation for double-strip waveguide

For illustration, the method is applied to the symmetrical double waveguide of Fig. 10.4. For a symmetric mode, the amplitude in the central region 3 has the form $\cosh(\kappa_{y3} y)$ and eq. (10.14) shows that $B_3 = 1$, giving $\alpha_3 = j\infty$. Noting that $\kappa_{y3} = \kappa_{y1}$, this leads to $F_3(y) = -\kappa_{y1} \tanh(\kappa_{y1} y)$. For region 2 we find $F_2(y) = k_{y2} \tan(\varepsilon_2 + k_{y2} y)$. In region 1, $\psi(y)$ has the form $\exp(\kappa_{y1} y)$, giving $\alpha_1 = -j$ and $F_1(y) = -\kappa_{y1}$. Applying the boundary conditions $F_1(-W - g/2) = F_2(-W - g/2)$ and $F_2(-g/2) = F_3(-g/2)$ gives two equations involving ε_2, and eliminating this gives the dispersion relation of eq. (10.6). For an antisymmetric mode the method is very similar, taking the amplitude in region 3 to have the form $\sinh(\kappa_{y3} y)$.

10.4 TRANSVERSELY-COUPLED RESONATOR (TCR) FILTER

The TCR filter combines the concepts of waveguiding and resonances to realize a bandpass filter with exceptionally small fractional bandwidth, typically 0.1%. This is much smaller than the bandwidth of other surface-wave filters (Table 1.2). Moreover, the insertion loss can be small, good stop-band rejection can be obtained and the device is physically very compact. These features are well suited to the requirements for intermediate frequency (IF) filtering in mobile telephone handsets for the GSM system. The basic scheme is due to Tiersten and Smythe [8], and practical devices were developed by Tanaka et al. [9] and others [7, 10, 11].

As shown in Fig. 10.8, the device is basically two resonators located physically close together, each resonator consisting of a single-electrode transducer between two gratings. A wave generated in one resonator takes the form of one

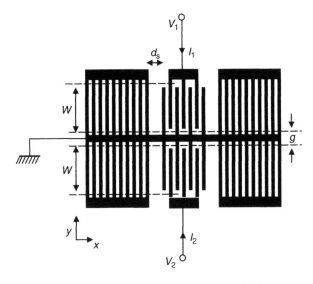

Figure 10.8 Transversely-coupled resonator (TCR) filter.

or more waveguide modes extending laterally beyond the transducer area, rather like the mode profiles in Fig. 10.3. Consequently, some energy of this wave is within the region of the second resonator, and coupling takes place. To obtain adequate coupling, the transducer aperture W is made quite small, typically 3–6 wavelengths, and the gap g is also small, typically one wavelength. Usually the substrate is $ST–X$ quartz and the metallization is aluminum, with normalized thickness typically $h/\lambda \approx 2\%$. The transducers and gratings are quite long, usually with 200 electrodes or more.

The operation is complicated by the fact that each strip reflects the waves, so the device has reflection and waveguiding effects simultaneously. For the present, we assume that these effects can be considered separately, though this point will be discussed further in Section 10.6. Consider first the waveguide modes, assuming temporarily that reflections can be ignored. The waveguide structure is very complicated, consisting of seven regions which are indicated on Fig. 10.7. Regions 3 and 5 are the electrode regions, of width W and with the same plane-wave velocity so that $v_{03} = v_{05}$. Regions 2, 4 and 6 are bus regions, with width $g/2$ and B and with the same plane-wave velocity. Regions 1 and 7 are free-surface regions extending to $y = \pm\infty$. The plane-wave velocities can be evaluated from the formula in Chapter 8, Section 8.1, eq. (8.18). As indicated in Fig. 10.7, the electrode regions have the lowest velocities, even though they have less metallized area than the bus bars. The reason is that the

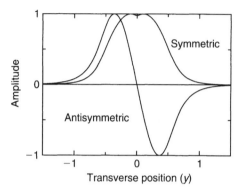

Figure 10.9 Profiles of the first two modes of a TCR filter waveguide on ST–X quartz, with $W = 6\lambda_0$, $g = \lambda_0$ and $B = 4\lambda_0$, where $\lambda_0 = v_{03}/f$. The abscissa is normalized such that the outer edges of the electrodes are at $y = \pm 1/2$.

'stored energy' term, involving the coefficient K_V, is significant here. For ST–X quartz substrates, typical velocity ratios are $v_{03}/v_{02} = 0.998$ and $v_{02}/v_{01} = 0.997$. The waveguide modes can be deduced using the analysis of Section 10.3 above. Actually, it is often the case that the amplitude decays quite rapidly in the outer bus bars, regions 2 and 6, and the waveguide can be approximated by taking these regions to have infinite width, so that $B = \infty$. In this case, the waveguide simplifies to the two-strip form of Fig. 10.4, and the mode velocities are given by eq. (10.6). However, this approximation is not always adequate for practical devices.

The mode velocities have forms similar to those of a single-strip waveguide, shown in Fig. 10.2. Typically, the third and higher modes are either absent or weakly coupled, so that they have little effect. The present discussion therefore considers only the first mode (symmetric) and the second mode (antisymmetric). Typical profiles of these modes are in Fig. 10.9.

To analyse the filter, it is convenient to consider symmetric and antisymmetric voltages at the ports, as shown in Fig. 10.10. For the symmetric case, voltages V are applied to each port and the currents entering each port are denoted by I_s. For the antisymmetric case, voltages $\pm V$ are applied, and the currents are $\pm I_a$. In each case there is only one waveguide mode present, so the analysis is simplified. The device Y-matrix is defined as usual by $I_i = \Sigma_j Y_{ij} V_j$, with I_i and V_j as on Fig. 10.8, and this gives

$$Y_{11} = Y_{22} = (I_s + I_a)/(2V); \quad Y_{12} = Y_{21} = (I_s - I_a)/(2V) \qquad (10.18)$$

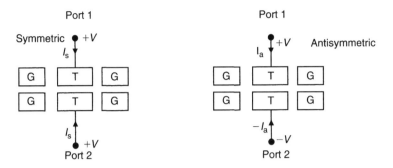

Figure 10.10 Analysis of the TCR filter for symmetric and antisymmetric modes separately. T = transducer, G = grating.

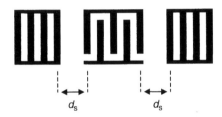

Figure 10.11 One-track resonator, equivalent to the TCR filter for one waveguide mode.

For the two cases of Fig. 10.10, the device behaves like a single-track resonator, indicated in Fig. 10.11. Resonators of this type have already been mentioned in Chapter 1, Section 1.6. In the present case, there is a complication that the central transducer is the single-electrode type and so its response is affected by electrode reflections. This complication is considered later, in Chapter 11, Section 11.1.2, where it is shown that the transducer reflectivity can be compensated by adjusting the transducer-grating distance d_s. In addition, it is usual to set the grating Bragg frequency at the frequency where the transducer conductance is maximized, somewhat less than the transducer Bragg frequency. These factors need to be allowed for in order to minimize stop band signals in the TCR filter response.

The device analysis, for one mode, is almost the same as that for a device with no waveguiding. The P-matrices of the transducer and gratings can be found using coupling-of-modes (COM) theory (Chapter 8), and these are cascaded to obtain the device response as in Appendix D. For no waveguiding, the parameters needed for the COM theory are considered in Chapter 8. Waveguiding introduces two new factors. Firstly, the transduction efficiency is modified, as

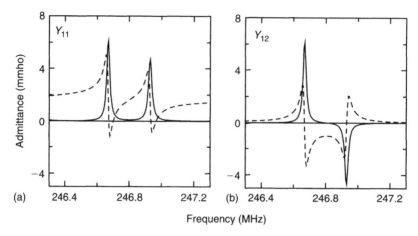

Figure 10.12 Theoretical admittance for a typical TCR filter. (a) Y_{11} and (b) Y_{12}. Solid lines show the real parts, broken lines show the imaginary parts.

considered in Section 10.2 above. To allow for this, the transduction parameter α_1 in the COM equations is multiplied by the transduction function $K_{ts}^{(m)}$ of eq. (10.12). Here the integral can be taken over a source in one track, and we have $W_s = W$. This factor refers to mode m. For the TCR filter it is the same for all sources because the transducer is unapodized. The second factor is that the mode profile modifies the reflection coefficient. For mode m the effective reflection coefficient of an electrode is $r_e' = K_r^{(m)} r_e$, where r_e is the coefficient for no waveguiding. The function $K_r^{(m)}$ is evaluated later, in Section 10.6. For COM theory, the reflection parameter c_{12} is multiplied by $K_r^{(m)}$. The factors $K_{ts}^{(m)}$ and $K_r^{(m)}$ are usually quite close to unity, but they do cause small distortions in the frequency response. Using COM parameters determined in this way, the theory agrees well with the responses of experimental devices [7]. Alternatively, the required COM parameters can be derived from experimental measurements [10].

The two single-mode devices represented by Fig. 10.11 have very similar responses. However, they give slightly different resonant frequencies because of the different mode velocities. The fractional difference of mode velocities is typically half of the fractional difference between v_{02} and v_{03}, the plane-wave velocities in the bus bars (v_{02}) and electrodes (v_{03}). Typically we have $v_{03}/v_{02} = 0.997$, so the fractional difference of mode velocities is typically 0.15%. The fractional bandwidth of the filter is approximately equal to this figure. The resonance of each device contributes a pole to the response of the filter, so that the filter is a two-pole device. Figure 10.12 shows the admittance parameters Y_{11} and Y_{12} of a typical filter, calculated from eq. (10.18).

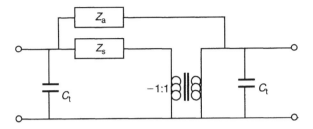

Figure 10.13 Equivalent circuit of a TCR filter.

An equivalent circuit is shown in Fig. 10.13, assuming that the device is well designed so that its response is dominated by two clear resonances. The boxes marked Z_s and Z_a are L-C-r series resonant circuits for the symmetric and antisymmetric modes, respectively. The capacitors C_t are the transducer capacitances. The derivation for one resonance is analogous to that for a longitudinal resonator, described later in Section 11.1.2 of Chapter 11, For the two-pole TCR filter, the two series resonant circuits are simply combined. The circuit applies only in the frequency region where the gratings reflect strongly. At other frequencies the weak reflectivity of the gratings reduces the response, and consequently the stop-band rejection is better than that of an ideal two-pole filter.

A practical response is shown in Fig. 10.14. For this filter two identical devices, such as that of Fig. 10.8, have been fabricated on the same substrate and cascaded electrically in order to improve the stop-band rejection. Because the individual devices have very low loss, the overall insertion loss remains small. A device cascaded in this way is often called a four-pole TCR filter. The insertion loss for Fig. 10.14 is 4 dB.

10.5 UNBOUND WAVEGUIDE MODES

In the previous sections, we have considered waveguide modes in which the amplitude decays in the outer regions and the wavenumber β is real. However, these assumptions are somewhat restrictive. Other mode types can exist, relevant to practical devices [4, 6, 12]. For simplicity, this topic will be considered in the context of the one-strip waveguide of Fig. 10.1. The previous analysis, in Section 10.1, assumed that the amplitude outside the strip decayed exponentially with distance, so that the energy is confined to the strip region. A solution such as this is called a 'bound' mode. For this discussion we introduce a new variable k_T defined by

$$k_T^2 = (k_{02}^2 - k_{01}^2)/(1 + \alpha_a), \tag{10.19}$$

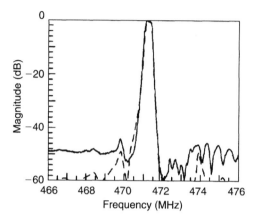

Figure 10.14 Practical TCR filter response, from Ruppel et al. [18]. The insertion loss is 4 dB. Copyright 1993 IEEE, reproduced with permission.

taking $k_T > 0$. This quantity is a constant (for a particular frequency). As before, $k_{0n} = \omega/v_{0n}$ where v_{0n} are the plane-wave velocities for x propagation. The strip region (region 2) has the lower velocity, so that $k_{02} > k_{01}$, and hence k_T is real. The symmetric-mode dispersion relation of eq. (10.4) can be rearranged as

$$\cos(k_{y2}W) = k_{y2}/k_T \tag{10.20}$$

where k_{y2}, defined by eq. (10.2), is taken to be real. For antisymmetric modes, given by eq. (10.5), the cosine in eq. (10.20) is replaced by a sine. The solutions are bound modes, which are seen to exist only for $k_{y2} \le k_T$. The cutoff points occur when $k_{y2} = k_T$.

More generally, the amplitude in the outer regions could have a sinusoidal variation. For this case, we consider

$$\psi(y) = B\cos[k_{y1}(|y| - W) + \phi]$$

in region 1 ($y < W$) and region 3 ($y > W$). Here k_{y1} is assumed to be real. From eq. (10.19) we find $k_{y1}^2 = k_{y2}^2 - k_T^2$, which shows that $k_{y2} \ge k_T$. For region 2 we take $\psi(y) = A\cos(k_{y2}\,y)$ as before, assuming the solution to be symmetric. To find the dispersion relation, we use the boundary conditions that ψ and $d\psi/dy$ are continuous. This gives

$$\tan^2(k_{y2}\,W) = [1 - k_T^2/k_{y2}^2]\tan^2\phi \tag{10.21}$$

When $k_{y2} = k_T$ we have $\tan(k_{y2}W) = 0$. This condition also applies for the cutoff frequencies of the bound modes. For $k_{y2} > k_T$ the term in square brackets is between 0 and 1. A solution can be found for any value of k_{y2} by choosing a suitable value of ϕ. Hence the modes in this region form a continuum. In contrast, the bound modes occur only at discrete values of k_{y2}.

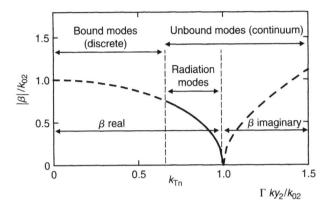

Figure 10.15 Range of existence for bound and unbound modes, for single-strip waveguide.

The value of β is given by eq. (10.2). This shows that β^2 is positive if $0 < k_{y2} < k_{02}/\sqrt{1+\alpha_a}$. We then have β real, giving a non-lossy mode called a *radiation mode*. Because k_{y1} is real, the mode velocity ω/β is greater than the 'fast' plane-wave velocity v_{01}. If $k_{y2} > k_{02}/\sqrt{(1+\alpha_a)}$, we have β^2 negative, so β is imaginary. In this case the amplitude decays exponentially with x. Identical remarks apply for the antisymmetric modes, which have dispersion as in eq. (10.21) but with the tangents replaced by cotangents. It might be thought that the radiation modes cannot occur in a surface-wave device because they imply that energy is incident on the guide from the sides. However, a practical situation generally involves a sum of modes, and these can combine in such a way that the total energy flow is not toward the guide [12]. Some examples of unbound modes are described by Jungwirth *et al.* [6].

Figure 10.15 shows the situation schematically. The figure is essentially a plot of β as a function of k_{y2}, with β normalized as $y = |\beta|/k_{02}$ and k_{y2} normalized as $x = \Gamma k_{y2}/k_{02}$. Here Γ is defined as $\Gamma = \sqrt{(1+\alpha_a)}$. The relationship is simply $y^2 = 1 - x^2$, which is another form of eq. (10.2) with $n = 2$. The radiation modes occur in the region $k_{Tn} < x < 1$, where k_{Tn} is a normalized version of k_T defined by $k_{Tn} = \Gamma k_T/k_{02} = \sqrt{1 - k_{01}^2/k_{02}^2}$.

In Section 10.2 above it was stated that a general source distribution can be represented in terms of mode profiles, as in eq. (10.8). This equation is approximate because it only includes the bound modes, which are finite in number. The equation becomes more accurate if the unbound modes are included. Because the unbound modes form a continuum, the contribution from them is an infinite sum, expressed as an integral. The concept of mode orthogonality still applies

[4, 6, 12], and it may be used to determine the distribution of mode amplitudes. However, a simpler analysis using only the bound modes is often adequate, particularly if the components are relatively long, for example 100 wavelengths or more. This is usually the case for the TCR filter, as in Section 10.4.

If the surface is uniform, so that $k_{01} = k_{02}$, we have $k_T = 0$. In this case there are no bound modes. The unbound modes have amplitudes which are sums of terms such as $\exp[-j(k_y y + \beta x)]$, where β is either real or imaginary. These can be identified as the plane waves that form the basis of the diffraction theory in Chapter 4, Section 4.1.

10.6 WAVEGUIDES INCLUDING ELECTRODE REFLECTIVITY

In a single-electrode transducer the electrodes can reflect the wave strongly, as considered in Chapter 8. This causes substantial complications if the transducer is also behaving as a waveguide. This situation is of practical significance in one-track resonators, in the TCR filter and in single-phase unidirectional transducers (SPUDTs) (Chapter 9). There have been several approaches to the analysis of this complex topic. Haus [13, 14] used a COM-based theory to show that reflective electrodes can themselves guide the wave, without any background velocity shift. A second approach is to analyze the reflection and transmission of waveguide modes by each electrode, using cascading to analyze the whole structure [12, 15]. A third method is to divide the whole structure into two-dimensional rectangular finite elements, of length Δx and width Δy. The wave amplitudes in these elements can be related by a matrix which takes account of transduction, reflection and propagation effects, leading to equations which can be solved for the waveguide behavior [4, 16].

Mode scattering by a reflecting electrode

Consider a reflecting electrode within a waveguide structure. In the absence of waveguiding, the electrode reflection coefficient for waves with normal incidence is r_e. This can be evaluated as in Chapter 8, Section 8.1.4. The waveguide is assumed to support several bound modes, and one of these, mode m, is incident on the electrode. The mode amplitude, at the electrode location, is $\psi_m(y)$. At each y on the electrode, the reflected wave amplitude $\psi_r(y)$ is proportional to r_e and to the incident wave amplitude. Thus,

$$\psi_r(y) = \begin{cases} r_e \psi_m(y) & \text{when } y \text{ is on the electrode} \\ 0 & \text{for other } y. \end{cases}$$

We now expand $\psi_r(y)$ in terms of the bound modes, assuming that this is an adequate approximation. We write $\psi_r(y) = \Sigma_n a_n \psi_n(y)$, where $\psi_n(y)$ are the bound mode profiles. Multiply both sides by $\psi_k(y)$, integrate over the range $y = \pm\infty$ and use orthogonality (eq. 10.7). This gives

$$\frac{a_k}{r_e} = \frac{\int_e \psi_m(y)\psi_k(y)dy}{\int_{-\infty}^{\infty} \psi_k^2(y)dy} \qquad (10.22)$$

where the integral in the numerator is over the y range occupied by the electrode. The a_k give the expansion coefficients for the wave generated in the reverse direction, when the incident wave is mode m with profile $\psi_m(y)$.

In the case of the TCR filter, it was assumed in Section 10.4 that only two modes can propagate, one symmetric and one antisymmetric. Since the geometry is symmetric, the symmetric mode cannot be reflected into the antisymmetric one, and vice versa. Hence the reflected wave consists only of mode m, the incident mode. The reflection coefficient of this mode is a_m, which is written as $a_m = r_e K_r^{(m)}$. Here $K_r^{(m)}$ equals the right side of eq. (10.22), with $k = m$. As before, unbound modes have been neglected but this approximation is expected to be reasonable if the transducers and gratings are long. The reflection coefficients of the two modes are slightly different, and this can have the effect of increasing the filter bandwidth a little.

Transmission can be considered in a similar manner, representing the transmitted wave as a sum of waveguide modes. However, for the two-track TCR filter of Fig. 10.8, adequate results can be obtained without considering this. Analysis is usually done using the COM method, and this implicitly makes the transmission and reflection coefficients consistent, as discussed in Chapter 8, Section 8.2.

For the analysis of the TCR, Section 10.4, the COM method was applied for the individual modes separately. In general there could be several modes present simultaneously. If so, a more sophisticated approach is called for. One approach to this problem is to first analyze the waveguide modes as if there were no reflections, and then develop a scattering matrix for each reflecting electrode. For the modes, the velocity v_{03} in the electrode region can be taken to refer to the lower band edge frequency f_1, so that $v_{03} = 2\,pf_1$ [12]. The matrix needs to account for excitation of all relevant modes when a voltage is applied. It also expresses reflection and transmission phenomena in terms of scattering of each incident mode into all other waveguide modes. Given such a matrix, it is possible to cascade so as to obtain the response of the device as a whole [12, 15]. This is a generalization of the approach described in Section 10.4 above. Solal [12] has applied this method to a novel type of TCR filter, in which there are three tracks instead of two. This gives a frequency response with three poles, improving the sharpness of the skirts. However, for this structure it was found that there were unexpected signals in the stop band of the response, above the passband. The theory agreed well with experiment when the multi-mode scattering matrix was used, with the unbound modes included. This showed that the unexpected signals were due to the unbound modes.

Two-dimensional coupled-mode analysis

An alternative approach, due to Haus [13, 14], is based on a two-dimensional version of the COM equations. The geometry is taken to be similar to the single-strip waveguide of Fig. 10.1, where the outer regions $|y| > W$ are free-surface regions. In the central region where $|y| < W$ there is an infinite periodic set of reflecting strips parallel to the y axis. In addition, the effective velocity v_{e2} in this region can differ from the plane-wave velocity v_{01} in the free-surface region. Here v_{e2} is the plane-wave velocity which would be obtained in region 2 in the absence of the reflectivity.

This system can be described by coupled-mode equations related to those of Section 8.2 in Chapter 8, but developed to allow for variation the y direction as well as the x direction. Using the usual boundary conditions leads to a complicated dispersion relation with multiple solutions. This shows that guiding can occur even when $v_{e2} = v_{01}$, so that the guiding is soleley due to the reflectivity of the strips. However, this is difficult to confirm experimentally because of the difficulty of making the two velocities equal. Typically, the dispersion is rather like that of an unguided reflective structure, shown in Fig. 8.10, in the low-frequency region where $\delta < 0$. Thus, the wavenumber s is real at low frequencies. However, s is imaginary for all higher frequencies ($\delta > 0$), so that the wave is attenuated. In practical situations we usually have $v_{e2} < v_{01}$, so that the velocity difference is an additional factor causing waveguiding. For a relatively large velocity difference, this factor becomes dominant and the solutions are similar to those described in Section 10.1 above.

Hirota and Nakamura [5] extended Haus's work to include anisotropy and transduction, needed for analysis of transducers. The theory gave very good agreement with experimental results for two-port resonators on quartz, as illustrated by Fig. 10.6 above.

Solutions similar to those given by the two-dimensional COM theory are also given by the mode scattering theory [15] and by the theory using two-dimensional finite elements [16].

Avoidance of waveguiding effects

As mentioned in Section 10.2, waveguiding problems in resonators are sometimes minimized by apodizing the transducers. In some devices, such as SPUDT filters, it may be possible to minimize the effects by modifying the target frequency response so as to compensate for the expected distortion due to waveguiding. A third approach, shown to be effective for unapodized transducers, is to design the waveguide to produce a 'piston' mode [17]. This mode has a constant amplitude over the region occupied by the electrodes, with decay outside this region. To obtain this mode, the velocity in the bus region needs to be reduced to a specific value, and this is obtained by breaking the bus into discrete strips so that the stored energy term reduces the velocity. Figure 9.6 refers to a SPUDT filter using this mode.

REFERENCES

1. A.A. Oliner. 'Waveguides for surface waves', in A.A. Oliner (ed.), *Acoustic Surface Waves*, Springer, 1978, pp. 187–223.
2. H.F. Tiersten. 'Elastic surface waves guided by thin films', *J. Appl. Phys.*, **40**, 770–789 (1969).
3. R.V. Schmidt and L.A. Coldren. 'Thin film acoustic surface waveguides on anisotropic media', *IEEE Trans. Son. Ultrason.*, **SU-22**, 115–122 (1975).
4. S. Rooth and A. Rønnekleiv. 'SAW propagation and reflection in transducers behaving as waveguides in the sense of supporting bound and leaky modes', *IEEE Ultrason. Symp.*, **1**, 201–206 (1996).
5. K. Hirota and K. Nakamura. 'Analysis of SAW grating waveguides using 2-D coupling-of-modes equations', *IEEE Ultrason. Symp.*, **1**, 115–120 (2001).
6. M. Jungwirth, N. Pöcksteiner, G. Kovacs and R. Weigel. 'Analysis of multi-channel planar waveguides', *IEEE Trans. Ultrason. Ferroelect. Freq. Contr.*, **49**, 519–527 (2002).
7. D.P. Morgan, S. Richards and A. Staples. 'Development of analysis for SAW transversely-coupled waveguide resonator filters', *IEEE Ultrason. Symp.*, **1**, 177–181 (1996).
8. H.F. Tiersten and R.C. Smythe. 'Guided acoustic-surface-wave filters', *Appl. Phys. Lett.*, **28**, 111–113 (1976).
9. M. Tanaka, T. Morita, K. Ono and Y. Nakazawa. 'Narrow bandpass filters using double-mode SAW resonators on quartz', *38th Ann. Freq. Contr. Symp.*, 1984, pp. 286–293.
10. D.P. Chen, M.A. Schwab, C. Lambert, C.S. Hartmann and J. Heighway. 'Precise design technique of SAW transversely coupled resonator filters on quartz', *IEEE Ultrason. Symp.*, **1**, 67–70 (1994).
11. S.V. Biryukov, G. Martin, V.G. Polevoi and M. Weinacht. 'Consistent generalization of COM equations to three-dimensional structures and the theory of the SAW transversely coupled waveguide resonator filter', *IEEE Trans. Ultrason. Ferroelect. Freq. Contr.*, **42**, 612–618 (1995).
12. M. Solal. 'A P-matrix based model for the analysis of SAW transversely coupled resonator filters, including guided modes and a continuum of radiated waves', *IEEE Trans. Ultrason. Ferroelect. Freq. Contr.*, **50**, 1729–1741 (2003).
13. H.A. Haus. 'Modes in SAW grating resonators', *Appl. Phys. Lett.*, **48**, 4955–4961 (1977).
14. H.A. Haus and K.L. Wang. 'Modes of grating waveguides', *Appl. Phys. Lett.*, **49**, 1061–1069 (1978).

15. M. Solal, V. Laude and S. Ballandras. 'A P-matrix based model for SAW grating waveguides taking into account mode conversion at the reflections', *IEEE Trans. Ultrason. Ferroelect. Freq. Contr.*, **51**, 1690–1696 (2004).

16. M. Mayer, G. Kovacs, A. Bergmann and K. Wagner. 'A powerful novel method for the simulation of waveguiding in SAW devices', *IEEE Ultrason. Symp.*, **1**, 720–723 (2003).

17. M. Mayer, A. Bergmann, G. Kovacs and K. Wagner. 'Low loss recursive filters for basestation applications without spurious modes', *IEEE Ultrason. Symp.*, **1**, 1061–1064 (2005).

18. C.C.W. Ruppel, R. Dill, A. Fischerauer, W. Gawlik, J. Machui, F. Muller, L. Reindl, W. Ruile, G. Scholl, I. Schropp and K.C. Wagner. 'SAW devices for consumer communications applications', *IEEE Trans. Ultrason. Ferroelect. Freq. Contr.*, **40**, 438–452 (1993).

11

RESONATORS AND RESONATOR FILTERS

Surface-wave resonators have been investigated since the early 1970s, with particular regard to the applications as stabilizing elements for oscillators. This application is similar to the familiar use of bulk acoustic-wave resonators for the same purpose. Surface-wave devices are applicable at much higher frequencies than bulk-wave devices, though the resulting oscillator stability is not so high. Another role for surface-wave resonators emerged in the 1980s, when it was realized that they could provide the low insertion losses being demanded for bandpass filtering in mobile telephones. There are several implementations of this – the impedance element filter (IEF), longitudinally coupled resonator (LCR) filter and transversely coupled resonator (TCR) filter. TCR filter is described in Chapter 10.

For bandpass filter applications there has been a demand for substrate materials giving both strong piezoelectric coupling and moderate temperature stability. The conventional Rayleigh-wave materials do not satisfy this requirement, and other possibilities were explored. For many RF filters, the chosen solution to this problem is the use of rotated Y-cut lithium tantalate substrates, with rotation $36°$ to $42°$. In this case, the wave is a *leaky* wave, not a true surface wave in the usual sense. Because of its significance for resonator filters, the topic of leaky waves is discussed in this chapter. Where not stated, it is assumed that the only acoustic wave present is a piezoelectric Rayleigh wave.

It is assumed in this chapter that the natural single-phase unidirectional transducer (N-SPUDT) effect is absent, so that a geometrically symmetric transducer will generate identical surface waves in both directions. For most common surface-wave materials, the symmetry is high enough to ensure this. The effect is discussed in Chapter 8, Section 8.2.3, and resonators using N-SPUDT substrates are considered elsewhere [1].

11.1 RESONATOR TYPES

11.1.1 Gratings and cavities

As for many microwave or optical resonators, a prime requirement for a surface-wave resonator is an efficient reflector of the waves. Two such reflectors can be used to form a resonant cavity. In optics or microwaves, a localized mirror can be used, but for surface waves there is no practical well-localized reflector. Although any surface perturbation, such as a deposited strip, will reflect surface waves, it will in general produce bulk waves as well, leading to unacceptable losses. The key element for most resonators is a reflecting *grating*, as proposed by Ash [2]. In a grating each strip reflects weakly, but if there are many strips the total reflection coefficient can be close to unity when the pitch p is equal to half the wavelength. If the grating is many wavelengths long, we expect that strong coupling between the contra-directed surface waves can only occur over a small range of frequencies. Similarly, the surface- to bulk-wave coupling also occurs for a small range of frequencies, and this will be centered at a different frequency because the velocities are different. If the grating is long enough, there will be little overlap between these ranges. Hence we can expect the surface-wave reflection process to be relatively free of losses due to bulk-wave generation.

Most frequently, the strips in the grating are shorted metal electrodes which are conveniently fabricated in the same process as the transducer(s). However, grooves are often used for high-Q resonators, and open-circuit metal electrodes are also possible.

Figure 11.1 shows a basic resonant cavity, consisting of two reflecting gratings which are taken to be identical. The behavior of a grating is deduced from the coupling-of-modes (COM) equations in Chapter 8, and expressed using the

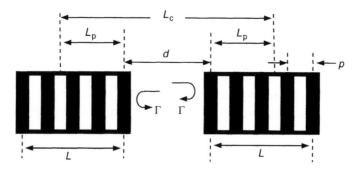

Figure 11.1 Surface-wave cavity resonator consisting of two reflecting gratings. The strips are usually shorted or open-circuit metal electrodes, or grooves.

P-matrix of Section 5.3 in Chapter 5. Here we ignore imperfections such as propagation loss, diffraction and waveguiding. In Chapter 8 the strip reflection coefficients were denoted by r_{s1} and r_{s2}, referring to waves incident on the acoustic ports 1 and 2, respectively. In this chapter the strip geometry is taken to be symmetrical and the N-SPUDT effect is assumed to be absent. Hence these coefficients are the same. We define r_s as the strip reflection coefficient for waves incident from either direction, so that $r_{s1} = r_{s2} = r_s$. This reflection coefficient is referred to the center of the strip. As explained in Chapter 8, the strip transmission coefficient is assumed to be positive real, and from power conservation we find that r_s is imaginary. Its value is considered in Chapter 8, Section 8.1.4.

From the symmetry, the two grating reflection coefficients P_{11} and P_{22} are equal, and here we write $P_{11} = P_{22} = \Gamma$. From the COM theory, these are given by eqs (8.40a) and (8.40c), where c_{12} is the COM reflectivity parameter. The ports are at the centers of the end strips, as shown by broken lines on Fig. 11.1. The Bragg frequency, at which the pitch p equals half the wavelength, is $f_0 = 2v/p$. Here v is the velocity the wave would have if the strips did not reflect. This is close to the free-surface velocity v_f, and its value is considered in Chapter 8, Section 8.1.4. The detuning parameter is $\delta = k - k_0$, where k_0 is the wavenumber at the Bragg frequency, and δ is real because we assume no losses. In the present case, $c_{12} = r_s/p$, as shown by eq. (8.41). As shown in Chapter 8, Section 8.1.4, r_s typically has magnitude 1–3%, and since it is imaginary its phase is $\pm\pi/2$.

The magnitude and delay of Γ are illustrated in Fig. 11.2. At frequency f_0 we have $\delta = 0$, and the magnitude of the reflection coefficient is

$$|\Gamma| = \tanh(|c_{12}|L) \approx \tanh(N|r_s|) \quad \text{at } f = f_0, \tag{11.1}$$

where $L = (N - 1)p$ is the length and N is the number of strips. Hence, $|\Gamma|$ is close to unity at $f = f_0$ if $N|r_s| \gg 1$. When this is so, $|\Gamma|$ is close to unity for a frequency range $\Delta f = 2f_0|r_s|/\pi$. For an infinite grating ($N = \infty$), $|\Gamma|$ is unity over this range.

For a resonator, the required spacing of the gratings depends on the phase of Γ, denoted by ϕ_g. From eq. (8.40a), ϕ_g is given by

$$\tan(\phi_g - \phi_r) = -(\delta/\sigma)\tanh(\sigma L), \tag{11.2}$$

where ϕ_r is the phase of r_s. Since r_s is imaginary, we have $\phi_r = \pm\pi/2$. Also, $\sigma \equiv s/j = \sqrt{(|c_{12}|^2 - \delta^2)}$. In the stop band, σ is real and it is equal to the attenuation coefficient of the grating mode. At the Bragg frequency f_0 we have $\sigma = \pm|c_{12}|$ and $\delta = 0$, so that $\phi_g = \phi_r$. Thus, the grating reflection coefficient

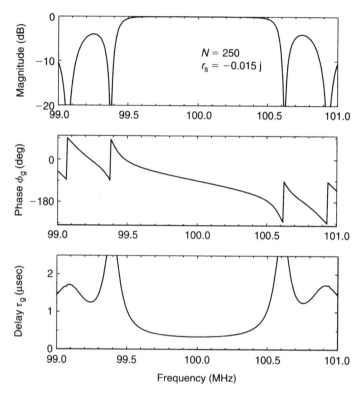

Figure 11.2 Grating reflection coefficient: magnitude, phase and delay. Bragg frequency is 100 MHz.

has the same phase as the reflection coefficient of one strip. This is not surprising because all the possible path lengths for the reflected waves, starting from the port and including those involving multiple reflections, are multiples of the wavelength.

The reflection occurs effectively at some distance into the grating. This can be found from the delay τ_g of the reflected wave, which is $\tau_g = -d\phi_g/d\omega = -(1/v)\,d\phi_g/dk = -(1/v)d\phi_g/d\delta$. This assumes v to be constant, a good approximation in practice. The strip reflection phase ϕ_r is also a constant. At frequency f_0, where $\phi_g = \phi_r$, eq. (11.2) gives

$$\tau_g = \tanh(|c_{12}|L)/[v|c_{12}|] \quad \text{for } f = f_0. \tag{11.3}$$

The wave is effectively reflected at a penetration depth $L_p = \tau_g v/2$. The grating can be thought of as having a mirror at this depth. For resonator applications the gratings usually have L large enough to give strong reflectivity, and the tanh can be set to unity. Hence, at f_0 we have $\tau_g = 1/[v|c_{12}|]$ and

$$L_p = p/(2|r_s|) \quad \text{at } f = f_0. \tag{11.4}$$

Results similar to the above can also be deduced from the reflective array method (RAM) theory of Section 8.1 in Chapter 8.

For a resonator, the phase ϕ_g of Γ determines the spacing required. Consider a 'round trip' in which a wave starts between the gratings and is reflected twice before arriving back at its starting point. The phase change for this journey is $2\phi_g - 2\omega d/v$, where d is the space between the gratings. As shown in Fig. 11.1, d is measured between the ports, at the centers of the end strips. Resonance occurs when the phase change is a multiple of 2π. Noting that $\omega/v = 2\pi/\lambda$, where λ is the wavelength, resonance occurs when

$$d = (n + \phi_g/\pi)\lambda/2. \tag{11.5}$$

At the Bragg frequency we have $\phi_g = \phi_r = \pm \pi/2$, and the condition for resonance becomes

$$d = (n \pm 1/2)\lambda/2 \quad \text{for } f = f_0, \tag{11.6}$$

where the sign is the same as the sign of ϕ_r. It is assumed here that v is the same everywhere, so that λ is also the same everywhere. In practice v has slightly different values in different regions, but the spacings are easily adapted to account for this. These differences can have practical significance because the performance of a high-Q resonator depends critically on the velocities. The modes given by eq. (11.6) are symmetric for n even and antisymmetric for n odd.

To find the spacing of the resonant modes we consider the cavity length L_c. As noted above, the waves are effectively reflected at a penetration depth $L_p = \tau_g v/2$. An effective cavity length L_c can be defined by adding the penetration depths to the grating spacing d, so that

$$L_c(\omega) = d + 2L_p(\omega) = d + v\tau_g(\omega). \tag{11.7}$$

The frequency dependence of τ_g implies that L_c is frequency dependent. For simplicity, the mode spacing is estimated by taking L_c to be constant. Taking the mode frequencies to be such that $L_c = n\,\lambda/2$, the spacing $(\Delta f)_m$ is given by $(\Delta f)_m/f_0 = \lambda/(2L_c)$. Using the Bragg-frequency value of L_p, eq. (11.4), we find $(\Delta f)_m/f_0 \approx 1/(d/p + 1/|r_s|)$. If d is small, this is approximately $|r_s|$. As seen earlier, the bandwidth of the gratings (over which the reflection coefficient is strong) is given by $\Delta f/f_0 = 2|r_s|/\pi$. As a consequence, surface-wave resonators can usually be designed to show only one resonance, even though the cavity length L_c may be many wavelengths. This is usually the preferred situation for an oscillator application. As d is increased, the cavity will generally have a larger number of resonances.

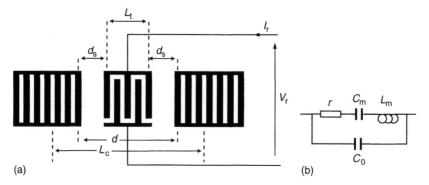

Figure 11.3 (a) One-port resonator. (b) Butterworth-van Dyke equivalent circuit.

11.1.2 Single-port resonator

A single-port resonator is formed by adding a central transducer to the cavity discussed above, giving the device shown in Fig. 11.3. For simplicity the transducer is assumed to have a symmetrical geometry, and to be centrally located in the cavity. The transducer P-matrix is denoted by P_{ij}, as defined in Appendix D, so that

$$[A_{t1}, A_{t2}, I]^T = [P_{ij}][A_{i1}, A_{i2}, V]^T \qquad (11.8)$$

where A_{in} and A_{tn} are amplitudes of waves incident on and leaving the transducer, respectively, and $n = 1, 2$ indicates the acoustic port. The superscript T indicates a transpose. The transducer symmetry implies that $P_{22} = P_{11}$ and $P_{23} = P_{13}$. Taking the resonator voltage and current to be V_r and I_r, the admittance $Y_r = I_r/V_r$ can be found by using the cascading techniques of Appendix D. We first allow for the spacings by defining

$$\Gamma' = \Gamma \exp(-2jkd_s)$$

where d_s is the transducer-grating distance on each side, measured between the ports. This has the effect of translating the grating port to a point coincident with the adjacent transducer port. Using the cascading formulae of Appendix D gives [3]

$$Y_r = P_{33} + \frac{4P_{13}^2 \Gamma'}{\Gamma'(P_{11} + P_{12}) - 1}. \qquad (11.9)$$

Resonances occur when the denominator approaches zero. As usual, P_{33} is the admittance of the transducer when isolated, so that the gratings are absent.

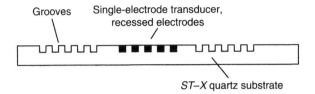

Figure 11.4 One-port resonator with groove gratings and recessed transducer electrodes.

Non-reflective transducer

Early resonator designs [4–8], dating from the 1970s, often used transducers in which internal reflections were not significant. These could be a non-reflective type (with $P_{11} = 0$) such as the double-electrode transducer, or the single-electrode type with a limited number of electrodes. Alternatively, a single-electrode transducer can be used with its electrodes recessed into the surface by depositing them in an array of grooves, as in Fig. 11.4. This minimizes the electrode reflectivity. For best performance, this is combined with the use of grooves in the gratings [6].

For a non-reflective transducer, P_{12} is associated with an undisturbed transit from port 2 to port 1, so it simply has the form $P_{12} = \exp(-jkL_t)$, where L_t is the transducer length. The denominator of eq. (11.9) becomes $\Gamma \exp(-jkd) - 1$, where $d = 2d_s + L_t$ is the distance between the grating ports. Resonances occur when the phase of $\Gamma \exp(-jkd)$ is a multiple of 2π, and this gives

$$d = (2n + \phi_g/\pi)\,\lambda/2. \qquad (11.10)$$

This corresponds to half of the cavity modes given by eq. (11.5), because a symmetrical resonator can only respond to symmetric modes. For a resonance at the grating Bragg frequency, ϕ_g/π can be replaced by $\pm 1/2$, as in eq. (11.6).

Assuming power conservation and symmetry, we apply eq. (D.6) with $P_{11} = 0$ and find that $2P_{13}^2 = -G_a \exp(-jkL_t)$, where L_t is the transducer length. Equation (11.9) then gives the resonator admittance as

$$Y_r = j\,\text{Im}\{P_{33}\} + G_a(1 + \gamma)/(1 - \gamma), \qquad (11.11)$$

where $\gamma = \Gamma \exp(-jkd) = |\Gamma| \exp(j\phi_g - jkd)$. It is now assumed that the frequency is within the stop band where the gratings reflect strongly, so that $|\Gamma| \approx 1$. We can then write $\gamma \approx \exp(j\theta_c)$, where $\theta_c = \phi_g - kd$ is the phase change for one transit of the cavity. Resonances occur for $\theta_c = 2n\pi$. The function $(1 + \gamma)/(1 - \gamma)$ becomes $j \cot(\theta_c/2)$, which is imaginary with a first-order pole at the resonance frequency ω_r. This makes it suitable for approximation as a series resonant L–C circuit, assuming that the device has only one resonance.

As shown in Fig. 11.3b, this branch has motional inductance and capacitance L_m and C_m, with $\omega_r = 1/\sqrt{(L_m C_m)}$, and a series resistance r. The remaining part of Y_r is j Im$\{P_{33}\}$, and this is represented by a capacitance C_0. Assuming that B_a is small, C_0 is approximately the transducer capacitance C_t. The whole circuit is known as a Butterworth-van Dyke circuit, with admittance

$$Y_r = j\omega C_0 + [j\omega L_m - j/(\omega C_m) + r]^{-1}.$$

For the circuit we find $dX/d\omega = 2L_m$ at the resonant frequency, where X is the reactance of the series branch. For the surface-wave device, $G_a X = -\tan(\theta_c/2)$. Ignoring the frequency variation of G_a, we find [4]

$$L_m = T_c(\omega_r)/[4G_a(\omega_r)], \qquad (11.12)$$

where $T_c(\omega) = L_c/v = -d\theta_c/d\omega$ is the cavity length in time units. This applies for any value of ω_r. The resistance in the circuit is found by evaluating the real part of the impedance of the resonant branch, at the resonance frequency, giving

$$r = (1 - |\Gamma|)/[2G_a(\omega_r)]. \qquad (11.13)$$

Multiple resonances can be represented by multiple resonant branches in the equivalent circuit, connected in parallel with each other. This is valid because each resonant circuit has small admittance at frequencies near the resonances of other circuits.

Figure 11.5 shows the admittance of a one-port resonator, using eq. (11.9) with COM analysis for the gratings and transducers. The small ripples about 0.7 MHz from the main peak are due to extra weakly coupled resonant modes. The peak frequency for Im$\{Y_r\}$ is the resonance frequency, ω_r, and Im$\{Y_r\}$ is zero at the antiresonance frequency ω_a. The difference between these frequencies is related to the capacitance ratio C_m/C_0. Assuming that r is small, and noting that $C_m \ll C_0$, analysis of the circuit gives

$$(\omega_a - \omega_r)/\omega_r \approx 0.5 C_m/C_0. \qquad (11.14)$$

In many applications it is desirable to maximize this ratio. Assuming a single-electrode non-reflective transducer, we have $G_a(\omega_c)/(\omega_c C_t) = 2.87 N_p \Delta v/v$, where ω_c is the center frequency, from the quasi-static analysis of Section 5.5.3 in Chapter 5. Assuming that the resonance occurs at the same frequency, the capacitance ratio is found to be

$$C_m/C_0 \approx 1.8(L_t/L_c)\Delta v/v. \qquad (11.15)$$

Usually, the ratio L_t/L_c is a little less than unity, so the maximum value for C_m/C_0 is around 1.5 $\Delta v/v$.

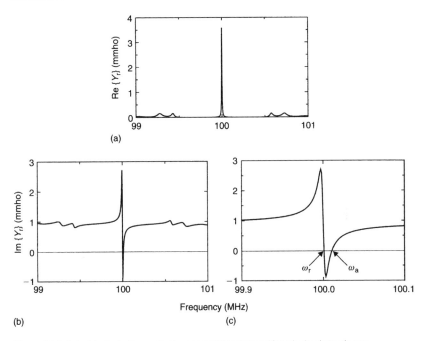

(a)

(b)

Frequency (MHz)

(c)

Figure 11.5 Calculated admittance Y_r of a one-port resonator with a single-electrode non-reflective transducer. Transducer has 61 electrodes, with center frequency 100 MHz. Grating design as in Fig. 11.2. Grating spacing d is 110.75 wavelengths. Part (c) shows a detail of the imaginary part of Y_r.

From the equivalent circuit of Fig. 11.3b, the quality factor of the resonator is $Q = \omega_r L_m / r$. If the only source of loss is the finite grating reflectivity, then r is given by eq. (11.13) and we find $Q = 0.5 \omega_r T_c(\omega_r) / (1 - |\Gamma|)$. Hence, the Q factor increases with the cavity length. In practice there are many other possible loss mechanisms, such as propagation loss, diffraction, bulk-wave generation, waveguiding and resistive losses in the electrodes. Each type of loss can be accounted for by a value of r. If the total loss is small the equivalent circuit remains valid, with r taken as the sum of the individual contributions.

Some loss is due to bulk-wave excitation at the ends of gratings and transducers, and this can be reduced by modifying the electrode pitches such that the grating-transducer gaps are eliminated [9].

Reflective transducer

In many one-port resonators the transducer is a single-electrode type, chosen because the relatively wide electrodes are easier to fabricate at high frequencies. This transducer generally has significant electrode reflections, so we cannot take $P_{11} = 0$. From eq. (11.9) it can be seen that the admittance is formally the same, except that $(P_{11} + P_{12})$ has replaced P_{12}. It is shown in Appendix D, eq. (D.9), that $(P_{11} + P_{12})$ has magnitude unity for a symmetrical transducer with no loss. Consequently, the reflections do not substantially affect the resonator operation, though the phase of the term $(P_{11} + P_{12})$ affects the resonance frequency. An alternative explanation is that a shorted transducer fully transmits the waves, so that $|A_{t1}| = |A_{t2}| = |A_{i1}| = |A_{i2}|$. This can be deduced from the P-matrix definition, eq. (11.8), setting $V = 0$, $A_{i2} = A_{i1}$ and $A_{t2} = A_{t1}$ because of the symmetry, and using symmetry and power conservation in the transducer. This gives $|P_{11} + P_{12}| = 1$.

Writing $P_{11} + P_{12} = \exp(j\phi_{12})$, the resonance condition becomes

$$d = [2n + (\phi_g + \phi_{12})/\pi]\lambda/2 + L_t. \tag{11.16}$$

For a non-reflective transducer we have $\phi_{12} = -2\pi L_t/\lambda$, and eq. (11.16) becomes the same as eq. (11.10). The distortion due to the electrode reflections has been studied by Uno and Jumonji [10], who confirmed that this modifies the grating spacing needed for resonance at a specified frequency. It is often desirable to make the resonance occur at the frequency where the transducer conductance G_a is maximized, since this minimizes L_m [eq. (11.12)] and so maximizes C_m/C_0. Hence the grating Bragg frequency is made less than that of the transducer [10]. These considerations are of practical importance in the design of transversely-coupled resonator (TCR) filters, described in Chapter 10.

A simplified expression for the admittance [11] can be derived by assuming that the frequency is in the range where the gratings reflect strongly, so that $|\Gamma| \approx 1$. The real part of the admittance is small except in the immediate vicinity of the resonance. The imaginary part can be approximated as

$$\text{Im}\{Y_r\} \approx \text{Im}\{P_{33}\} + G_a(\omega)\cot(\theta_c/2). \tag{11.17}$$

Here θ_c is the phase of $\Gamma'(P_{11} + P_{12})$, so that resonances occur for $\theta_c = 2n\pi$. This corresponds to eq. (11.11) if $P_{11} = 0$. Equation (11.17) gives results agreeing well with accurate calculations [11]. From eq. (11.17), the motional inductance is again found to be $L_m = T_c(\omega_r)/[4G_a(\omega_r)]$, with T_c defined as $T_c = -d\theta_c/d\omega$. Thus, this formula applies irrespective of whether the transducer has internal reflections. The reflections do affect the response, of course – they affect the values of P_{13} and G_a in the above formulae.

11.1.3 Two-port resonator

A resonator with two transducers, shown in Fig. 11.6, is often used for oscillator applications. Here the two transducers are taken to be identical

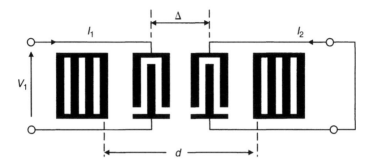

Figure 11.6 Two-port resonator.

and symmetrical, and the gratings are also identical. The behavior of this device is more complicated than that of the one-port resonator. It can be analyzed using the cascading methods in Appendix D, but a simplified approximate approach is considered here because it leads to equivalent circuits. It is assumed here that the transducers are the non-reflecting type, giving $P_{11} = P_{22} = 0$ and $P_{12} = \exp(-jkL_t)$, where L_t is the length. As before, we have $2P_{13}^2 = -G_a \exp(-jkL_t)$. We define d as the distance between grating ports and Δ as the distance between transducer centers. Since the transducers are non-reflective the resonance condition is given by eq. (11.5), as before.

Consider applying a voltage V_1 to the left transducer, with the right transducer shorted. In this situation, the right transducer does not affect the current I_1 taken by the left transducer. This current can be found by using the P-matrix, eq. (11.8), or using the methods of Appendix D. After some manipulation, I_1 is found to be given by

$$I_1/V_1 = Y_{11} = P_{33} + 2G_a\gamma[\gamma + \cos(k\Delta)]/(1 - \gamma^2), \tag{11.18}$$

where P_{33} is the admittance of each transducer and G_a is its acoustic conductance. The parameter γ is defined by $\gamma = \Gamma \exp(-jkd)$. We also have $Y_{22} = Y_{11}$ by symmetry. The current I_2 in the right transducer is given by

$$I_2/V_1 = Y_{12} = G_a[\exp(-jk\Delta/2) + \gamma\exp(jk\Delta/2)]^2/(1 - \gamma^2). \tag{11.19}$$

Resonances can occur for $\gamma \approx \pm 1$. We assume that only one resonance mode is present, and this will be symmetric if $\gamma \approx 1$ and antisymmetric if $\gamma \approx -1$. The parameter Δ affects the coupling between the transducers and the resonant mode.

Antisymmetric mode

For $\gamma \approx -1$ the coupling is strongest if the transducer spacing is set such that $\cos(k\Delta) = -1$, so that the transducer centers are spaced by $n + 1/2$ wavelengths.

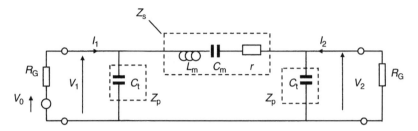

Figure 11.7 Equivalent circuit for two-port resonator, antisymmetric mode.

Assuming that this term does not vary very rapidly with frequency, the above equations give

$$Y_{11} = P_{33} - \frac{2\gamma G_a}{\gamma + 1}; \quad Y_{12} = G_a \frac{\gamma - 1}{\gamma + 1} \quad \text{for antisymmetric mode.} \quad (11.20)$$

As shown in Fig. 11.7 the equivalent circuit is taken to have a series branch of impedance Z_s with a parallel branch of impedance Z_p on each side. The circuit is equivalent in the sense that its Y-matrix is the same as that of the surface-wave device. The series branch has impedance $Z_s = -1/Y_{12}$, represented by a resonant circuit with motional inductance L_m. To find this, we can ignore losses by setting $\gamma = \exp(j\theta_c)$, where $\theta_c = \phi_g - kd$, which gives $d\theta_c/d\omega = -L_c/v$. Defining $X \equiv \text{Im}\{Z_s\}$ we find $X = G_a^{-1}\cot(\theta/2)$, and with $L_m = (1/2)[dX/d\omega]_{\omega_r}$ this gives

$$L_m = T_c/[4G_a(\omega_r)] \quad (11.21)$$

as in the case of the single-port resonator, eq. (11.12). The resistance r is obtained from $\text{Re}\{Z_s\}$ at resonance, giving $r = 0.5(1 - |\Gamma|)/G_a(\omega_r)$ as before. For the parallel branch we find $Z_p^{-1} = Y_{11} + Y_{12}$, and using eqs (11.20) this gives $Z_p^{-1} = P_{33} - G_a$. Assuming that B_a is small, Z_p can be represented by C_t, which is the capacitance of each transducer.

Symmetric mode

For this mode a similar circuit can be used but an ideal $-1:1$ transformer is added to account for the change of symmetry, as shown in Fig. 11.8. We have $\gamma \approx 1$ for resonance, so it is assumed that the device is designed with $\cos(k\Delta) = 1$, and that this factor varies little over the frequencies of interest. Equations (11.18) and (11.19) give

$$Y_{11} = P_{33} + \frac{2G_a\gamma}{1 - \gamma}; \quad Y_{12} = G_a \frac{1 + \gamma}{1 - \gamma} \quad \text{for symmetric mode.} \quad (11.22)$$

The circuit of Fig. 11.8 gives $Z_p^{-1} = Y_{11} - Y_{12}$, leading to $Z_p^{-1} = P_{33} - G_a$, so the parallel branches are capacitors as before. For the series branch

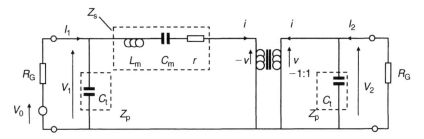

Figure 11.8 Equivalent circuit for two-port resonator, symmetric mode.

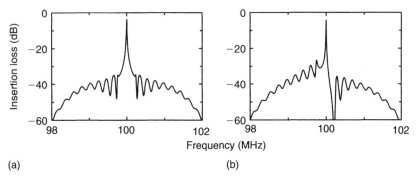

Figure 11.9 Calculated insertion loss of two-port resonators. (a) with non-reflective transducers. (b) synchronous type with reflective transducers.

we find $Z_s = 1/Y_{12}$. A calculation similar to the antisymmetric case gives $L_m = T_c/[4G_a(\omega_r)]$ as before, with $r = 0.5(1 - |\Gamma|)/G_a(\omega_r)$.

If the two ports are connected together, this device becomes a one-port resonator responding to symmetric modes, and the admittance is found to be in agreement with eq. (11.11).

Fig. 11.9a shows the calculated insertion loss of a two-port resonator on *ST–X* quartz, with non-reflective single-electrode transducers. In addition to the sharp spike due to the resonance, there is a slowly varying background signal which arises from direct transit between the two transducers. The equivalent circuit gives the resonant peak but not the background signal.

Synchronous resonators

An alternative two-port design is shown in Fig. 11.10a, using shorted metal strip gratings and single-electrode transducers. All strips have the same width and spacing, and the device resonates at the common Bragg frequency. If a

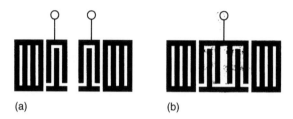

Figure 11.10 Resonators using electrodes of width $\lambda_0/4$. (a) Two-port synchronous resonator. (b) One-port hiccup resonator.

transducer is shorted, its electrodes behave the same as the strips in a grating. Each transducer is located so as to continue the electrode pattern of the adjacent grating, effectively extending its length. The resonance condition is given by eq. (11.10) as before, but taking d as the distance between the transducer ports. This design is convenient for fabricating high-frequency devices because recessing is not needed and the strips are all identical, with width $\lambda/4$. Q-values exceeding 2000 are obtainable at 2 GHz [12]. Figure 11.9b illustrates the response obtained. Compared with Fig. 11.9a there is some distortion, but this is acceptable in oscillator applications.

If the two transducers of this device are connected together electrically, it becomes a synchronous one-port resonator as shown in Fig. 11.9b. Here there are no transducer-grating gaps. The central gap has width $(2n \pm 1/2)\lambda/2$, as in eq. (11.10), causing the resonance to occur at the Bragg frequency. This is known as a 'hiccup' resonator. It has been investigated in several forms, including two-pole devices realized by including two gaps in the design [13].

11.1.4 Single-electrode transducer as resonator

Another type of surface-wave resonator is simply a long single-electrode transducer. This is illustrated in Fig. 11.11, where the solid lines show the admittance for a transducer on Y–Z lithium niobate, calculated from COM theory. Electrode reflections cause the peak of the conductance to become sharper and to shift downwards in frequency, as shown before in Fig. 8.11. If the transducer is long enough the conductance peak becomes very sharp, and the admittance is similar to that of a resonator.

Physically, this is related to the behavior of an infinite-shorted grating, considered in Chapter 8. At the edges of the first stop band, eq. (8.7) shows that the grating mode has wavenumber $\gamma = \pi/p$. This shows that all the fields have periodicity $2p$. This also applies to the charge density, and hence the currents entering the electrodes are alternating in sign. The same applies for a

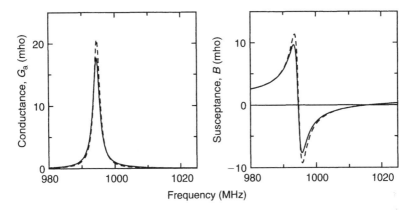

Figure 11.11 Conductance and susceptance of a 400-period single-electrode transducer on Y–Z lithium niobate, with Bragg frequency 1000 MHz. Electrode reflection coefficient is $r_s = -0.017\,j$. Solid lines: from COM theory. Broken lines: from Plessky formula, eq. (11.23).

single-electrode transducer with the bus bars shorted. Since the electrodes are connected alternately, the corresponding currents in the short circuit have the same phase. Thus, for an infinite-length transducer, a finite-amplitude wave is associated with an infinite current. For a finite but long transducer, this behavior is essentially a resonance. This usually happens only at the lower band edge – at the upper band edge the current is eliminated by symmetry as noted in Chapter 8, Section 8.3.2.

Plessky [14] showed that the admittance can be approximated well by simplifying the COM expression for P_{33}, eq. (8.40f). Assuming that the transducer length L is large enough and that the wave has finite attenuation, terms that increase with L will dominate. The terms including P_{31} and P_{32} are therefore excluded. Noting that α_1 and c_{12} are imaginary, we find $K_1 = K_2 = j\alpha_1/(\delta - j|c_{12}|)$. Assuming that r_s is negative imaginary, the admittance Y_t is found to be

$$Y_t \approx -j\frac{2N_p}{\pi} \frac{4|\alpha_1 p|^2}{(\omega - \omega_0)/\omega_0 + (2|r_s| - j\alpha)/(2\pi)} + j\omega C_t \qquad (11.23)$$

where $\omega_0 = \pi v/p$ is the Bragg frequency, α_1 is the COM transduction parameter (eq. 8.42), N_p is the number of periods and α is the propagation loss in Nepers per wavelength. If the attenuation is small, Y_t may be approximated as

$$Y_t \approx -8j(N_p/\pi)|\alpha_1 p|^2 \omega_r/(\omega - \omega_r) + j\omega C_t, \qquad (11.24)$$

where $\omega_r = \omega_0(1 - |r_s|/\pi)$ is the resonance frequency, equal to the lower band edge frequency. This leads to the interesting conclusion that the response is not affected by the electrode reflectivity r_s, except for the shift in resonance frequency and the fact that $|r_s|$ must be large enough for the formula to be valid. Clearly, Y_t is infinite at the resonance frequency ω_r. Y_t is zero at the antiresonance frequency ω_a. Using eq. (8.42) for α_1 it is found that

$$(\omega_a - \omega_r)/\omega_r \approx 0.9\Delta v/v, \tag{11.25}$$

for a metallization ratio $a/p = 0.5$. This is similar to the result for a conventional one-port resonator, eq. (11.15). The admittance has a capacitive term and a term giving a first-order pole, and it can therefore be represented by the Butterworth-van Dyke circuit of Fig. 11.3. The capacitance ratio is given by eq. (11.14).

In Fig. 11.11, the admittance calculated from eq. (11.23) can be seen to agree well with the results of the COM. It is noted that the resonance frequency is below the Bragg frequency of 1000 MHz. The distance between the resonance and antiresonance frequencies is large because the coupling is strong, in agreement with eq. (11.25).

This type of resonator has been widely used in radio frequency impedance element filters, considered in Section 11.3 below. It can also be modified by adding an extra component with a spacing agreeing with eq. (11.6). This gives two poles, one due to the space and the other associated with the long transducer. In this way, some two-pole bandpass filters have been realized [15].

11.2   SURFACE-WAVE OSCILLATORS

As in the case of bulk acoustic wave resonators, the excellent stability and reproducibility of surface-wave resonators on quartz makes them attractive as controlling elements for oscillators [6, 16, 7]. Surface-wave resonators have the advantage of high-frequency operation, up to 1 GHz, while bulk-wave devices are limited to about 30 MHz. On the other hand, bulk-wave oscillators can give higher stability.

A common configuration consists of a two-port surface-wave resonator with an amplifier providing feedback from the output to the input, as in Fig. 11.12. This arrangement will oscillate provided the loop gain exceeds unity and the phase change around the loop is a multiple of 2π. The resonator response has narrow bandwidth and rapid phase variation, so this condition is only satisfied at one frequency. The phase of the loop can be adjusted by means of a phase shifter, which enables the frequency to be adjusted. Alternatively, a two-transducer delay line can be used in place of the resonator [16]. If a delay line is used the bandwidth must be small enough so that the loop gain can exceed unity at only one of the frequencies where the loop phase is $2n\pi$. This condition is obtained

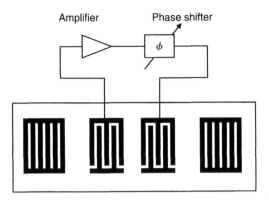

Figure 11.12 Surface-wave oscillator using a two-port resonator with amplified feedback. The phase shifter can be adjusted to change the oscillation frequency.

if one or both of the transducers has a length comparable to the space between the transducer centers.

A prime consideration is the short-term stability. This concerns rapid random fluctuations of frequency, which cause the oscillator output waveform to have a spectrum of finite width. This is a very complex topic [6], so only a brief discussion is given here. The stability depends on the loss and delay of the surface-wave device. An approximate analysis, assuming that the noise is thermal and originates outside the device, shows that the 3 dB width of the spectrum is [16]

$$\Delta\omega \approx 4k\theta(\mathrm{NF})G^2/[\tau^2 P_0], \tag{11.26}$$

where k is Boltzmann's constant, θ is absolute temperature and τ is the device delay. G, P_0 and NF are the amplitude gain, output noise power and noise figure for the amplifier, when saturated in the oscillation condition. This result applies for oscillators using either resonators or delay lines. For best stability we need to minimize $\Delta\omega$. This implies minimizing the insertion loss, which minimizes G, and maximizing the delay τ.

For a resonator we can take τ in eq. (11.26) to be the delay at the resonance frequency, since the spectrum is very narrow. This is related to the Q-factor of the device. For a shorted resonator, the circuit of Fig. 11.7 gives the Q-factor as $Q_u = \omega_r L_m/r$. This is called the *unloaded Q-factor*. The resistance r represents the total power loss, which may arise from several phenomena. It is given by eq. (11.13) if the loss is mainly due to the finite reflectivity of the gratings. More generally, we can consider a device connected to a source and load, both with resistive impedance R_G, as in Fig. 11.7. It is not usually necessary to consider inductive tuning because a resonator can have low loss without tuning. For simplicity, we ignore the transducer capacitances C_t, on the assumption that they have reactance much larger than R_G. The circuit has resistance $r + 2R_G$, so the Q-factor is

$$Q_l = \omega_r L_m / (r + 2R_G) = Q_u r / (r + 2R_G). \tag{11.27}$$

This is the *loaded Q-factor*. Analysis of Fig. 11.7 gives $V_2/V_0 = R_G/(Z_s + 2R_G)$, where Z_s is the impedance of the series branch. If ϕ is the phase of V_2/V_0, the delay is $\tau(\omega) = -d\phi/d\omega$. After some manipulation, we find

$$\tau(\omega_r) = 2Q_l/\omega_r. \tag{11.28}$$

This shows that to increase τ we need to increase Q_l. However, eq. (11.27) shows that this involves reducing R_G and hence increasing the insertion loss. We then need to increase G in eq. (11.26), which tends to decrease the stability. The optimum situation is an intermediate case, typically with an insertion loss of about 6 dB [17]. This corresponds to $R_G = r/2$ and $Q_l = Q_u/2$. Figure 11.13 shows the insertion loss of a practical 425 MHz two-port resonator. The form of the phase curve shows that τ is maximized at a resonance. In an oscillator, the resonances at ± 0.6 MHz do not give oscillation because of the phase differences.

For the best quality resonators using grooves, the unloaded Q can be up to $10^7/f$, where f is the frequency in MHz [6]. Thus, $Q_u = 10^5$ at 100 MHz, or 10^4 at

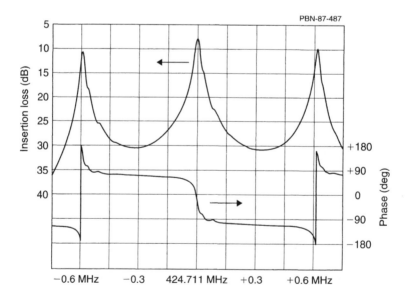

Figure 11.13 Insertion loss of a 425 MHz two-port resonator. From Parker and Montress [6], copyright 1988 IEEE, reproduced with permission.

1000 MHz. For the synchronous resonator, Fig. 11.10a, with all components made from aluminum film, Q_u values exceeding 2000 can be obtained at 2 GHz [12]. Compared with a delay line, the resonator gives a larger delay for the same device length. On the other hand, it cannot tolerate as much input power, so P_0 has to be less.

The short-term stability has been the subject of extensive studies [6, 18] going beyond the approximate result of eq. (11.26). In general there can be several different phenomena present, giving rise to characteristic regions in the spectrum. The spectrum can have regions where the power density is proportional to $(\Delta f)^{-2}, (\Delta f)^{-1}$ or $(\Delta f)^0$, where (Δf) is the frequency deviation from the peak. The $(\Delta f)^{-1}$ term is called 'flicker' noise, known to be affected by the active area of the device. The $(\Delta f)^0$ term is simply thermal noise independent of the resonance. The stability is often expressed as the single-sideband power density $L[(\Delta f)]$, measured in dBc/Hz, where dBc refers to power in dB relative to the carrier. For example, a 500 MHz oscillator gave [6] $L[(\Delta f)] = -45$ dBc/Hz at $\Delta f = 1$ Hz, -105 dBc/Hz at $\Delta f = 100$ Hz and -180 dBc/Hz at $\Delta f = 10^5$ Hz. Resonators using surface transverse waves (STWs) on quartz can give Q_u values of 8000 at 1 GHz. The STW is a type of leaky wave which can tolerate higher power levels than surface waves on quartz, giving the prospect of better short-term stability [18].

Apart from short-term stability, discussed above, the medium- and long-term stability are also important. Medium-term stability refers to temperature effects, usually minimized by using $ST-X$ quartz substrates, as discussed in Chapter 4. Long-term stability refers to ageing, that is, frequency changes over periods of months or years. The mechanisms involved in these processes are not understood, but experimental evidence shows that packaging and mounting methods have an important influence. Parker [6] has shown that a package made of fused quartz can reduce ageing, giving stabilities of 1 ppm per year or better.

The above discussion concerns two-port devices, though one-port resonators can also be used in oscillator applications [6, 19].

11.3 IMPEDANCE ELEMENT FILTERS

The impedance element filter (IEF) is a common type of bandpass filter for RF filtering in mobile phone handsets, with center frequencies typically around 1 and 2 GHz [20, 21]. These applications demand very low insertion losses, and therefore cannot be satisfied by the transversal filters described in Chapter 6. Other requirements are that the filters need to be physically very small, within a few mm, and to have ability to handle moderately large power levels, up to

2 W. The SPUDTs described in Chapter 10 are not attractive here because they use narrow electrodes, inconvenient to fabricate at high frequencies.

The use of resonators in bandpass filters is familiar in several technologies, notably L–C circuits, waveguide resonators and dielectric resonators. Piezoelectric acoustic resonators, using bulk or surface waves, can also be used, but the presence of the parallel capacitor in the equivalent circuit limits the bandwidth obtainable. For this reason, the surface-wave devices usually rely on different design concepts. The IEF has individual resonators connected electrically, without any acoustic interaction.

In addition, these devices normally use a leaky surface wave, rather than the usual Rayleigh-type surface wave. The leaky wave is chosen because it gives high-piezoelectric coupling combined with moderate temperature stability and a high velocity. The commonest material choice is $42° Y–X$ lithium tantalate. The leaky wave has a character rather different from that of the Rayleigh wave, but it can be generated by interdigital transducers, and reflected by gratings, in a similar manner. These leaky-wave devices can be analyzed approximately by using methods normally used for surface-wave devices, with modified parameters. The COM theory has been commonly used. In this section, the operation of the filters will be described as if the waves were Rayleigh waves, and the details of leaky waves will be considered later.

For the IEF, the basic resonant element is simply a long uniform single-electrode transducer, as considered in Section 11.1.4 above. It is usual to add gratings at either end so as to reduce loss due to wave radiation here, so that the resonator appears as shown in Fig. 11.14. The gratings are comprised of shorted metal strips which continue the periodic pattern of the transducer, so that all strips have the same width and spacing; the Bragg frequency is the same everywhere. Such a device is called a 'synchronous' resonator. This scheme is used to minimize loss, particularly for leaky-wave devices. Because the electrode pattern is continuous, the wave behaves much like a wave in an infinite uniform grating, which tends to trap the wave at the surface. Discontinuities, which would tend to cause loss because of bulk-wave generation, are avoided. Typically, there are several hundred electrodes in the transducer and a few tens of electrodes in each grating. The admittance is approximately given by the equations in Section 11.1.4, as shown by the example of Fig. 11.11.

For a bandpass filter, several such resonators are constructed on one substrate and connected electrically in a ladder arrangement, alternating between series and parallel connection as shown Fig. 11.15 [21]. The series and parallel

Figure 11.14 Synchronous resonator as impedance element.

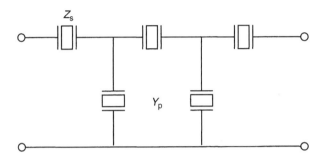

Figure 11.15 Impedance element filter principle. Each element is a resonator such as that of Fig. 11.14.

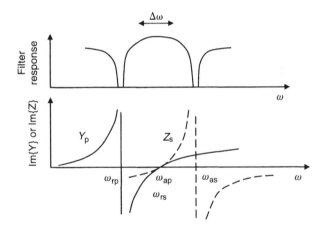

Figure 11.16 Top: schematic of IEF response. Bottom: Resonator characteristics.

resonators are given different Bragg frequencies. In the required passband, the series resonators give low impedance and the parallel resonators give high impedance, so that the input signal is passed with little loss. Figure 11.16 shows schematically a filter response, with the admittance Y_p of a parallel resonator and the impedance Z_s of a series resonator. The figure assumes these functions to be imaginary. The real parts, which are small because the Q values are high, cause some loss but do not affect the principles. The admittances are maximized at the resonant frequencies ω_{rp} and ω_{rs}, and minimized at the anti resonant frequencies ω_{ap} and ω_{as}. The frequency separation, $\omega_a - \omega_r$, is similar for the two resonators because it is determined by the coupling, as shown by eq. (11.25). In the passband, Z_s and Y_p are both small so there is little loss. At ω_{rp} the filter response has a null due to the pole in Y_p, and at ω_{as} there is a null due to the pole in Z_s. Outside this frequency range there is little acoustic activity, and the response is determined mainly by the static capacitances of the resonators. The stop-band attenuation may be increased by increasing the number of resonators.

The resonators are designed with $\omega_{ap} \approx \omega_{rs}$, and this gives a filter bandwidth of typically $\Delta\omega \approx (\omega_{as} - \omega_{rp})/2$, leading to $\Delta\omega/\omega_{rs} \approx \Delta v/v$ using eq. (11.25). This relation is not very rigid, but it can be concluded that typically the bandwidth will be limited to a few percent. For this reason, leaky waves are attractive since they give strong coupling with moderate temperature coefficients.

The filter response can be understood further by considering an infinite ladder filter in which the series and parallel components have impedances Z_s and Z_p. It can be shown that signals can propagate on this structure without attenuation if the impedance ratio is in the range $-4 < Z_s/Z_p < 0$. A negative ratio can only be obtained if Z_s and Z_p are both imaginary, a condition which is approximately true for high-Q resonators. The ratio is made negative by using $\omega_{ap} \approx \omega_{rs}$. Further, by considering the characteristic impedances of T and π sections, it can be shown that the required source and load impedances for low loss are approximately $[\omega_0^2 C_s C_p]^{-1/2}$, where ω_0 is the center frequency and C_s and C_p are the resonator capacitances [21]. This condition is not critical, so low loss can be obtained over an appreciable bandwidth.

Figure 11.17 shows an experimental result using the leaky wave on $42°Y\text{–}X$ lithium tantalate [22]. This refers to a duplexer for a mobile phone handset using the CDMA system. It has two filters in the same package, one for the receiver and one for the transmitter, both connected directly to the antenna. The insertion loss is about 2 dB. Each filter needs to have good rejection in the passband of the other. Good isolation is also important, minimizing the leakage of the transmitter signal through the filters to the receiver input. The whole duplexer is contained in a very small package of size 5×5 mm. For such filters, operating at high frequencies, it is very important to allow for stray components which affect the terminating impedances and give rise to unwanted 'feedthrough' signals due to electromagnetic coupling between the input and output. These effects can be modeled by sophisticated numerical methods [23], and considerable care is taken to allow for them when designing the device. A wide variety of similar devices have been developed with center frequencies around 2 GHz [24].

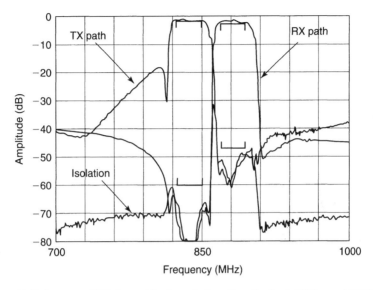

Figure 11.17 Response of IEFs in a mobile phone duplexer. From Peroix *et al.* [22], copyright 2001 IEEE, reproduced with permission.

Much attention has been given to the power handling capability, and for leaky waves this is limited by migration of the metal electrodes associated with the acoustic vibrations. To reduce this effect and thus increase the power handling, electrodes have been made of aluminum with other metals added, such as copper, and layered metals have also been used [25]. These methods have enabled 900 MHz devices to handle powers up to 2 W.

Temperature compensation techniques

For the PCS mobile phone system in the USA, the frequency bands specified for transmission and reception are unusually close together, with only a 20 MHz space between. The bands are of width 60 MHz and centered at 1880 and 1960 MHz. To meet this requirement, the bandpass filters need not only sharp skirts but also a small temperature coefficient, less than that of $36° Y–X$ lithium tantalate. This has been achieved by adding an SiO_2 layer [26], giving a temperature coefficient of delay (TCD) of about 16 ppm/°C. This technique has also been used for other devices, but for IEFs the thickness of the electrodes can cause problems and special fabrication techniques are needed.

Another method considered is a bonded substrate. In this technique, a lithium tantalate plate is bonded onto a glass substrate by direct bonding. An 836 MHz filter with a temperature coefficient of 6 ppm/°C was demonstrated [27]. Alternatively, lithium tantalate can be bonded onto silicon [28].

A SiO_2 film on lithium niobate can support a 'boundary wave' of the Maerfeld-Tournois type, with shear-horizontal (SH) motion, if the film thickness is a few wavelengths or more. This wave gives exponential decay into both the film and the substrate. The system can support two waves, the boundary wave and a Stoneley wave (with sagittal polarization), but the latter can be eliminated by choosing the orientation such that its coupling is zero. The boundary-wave coupling can be twice that of the leaky wave in $36°Y–X$ lithium tantalate. Low-loss IEF and LCR filters at 1 and 2 GHz have been demonstrated, with the advantage that the devices do not need encapsulation to protect them from contamination [29].

11.4 LEAKY WAVES

11.4.1 Leaky waves and surface-skimming bulk waves

In this book so far, the surface waves considered have nearly always been the piezoelectric Rayleigh type, which is non-leaky so that its wavenumber is real. It is characterized by the fact that its elastic displacement is in or near the sagittal plane and its velocity is less than that of the slowest bulk wave, which is normally a shear wave. For an isotropic bulk wave, with wavenumber k_b, a wave propagating at an angle θ to the surface has a wave vector with surface component $k_b \cos \theta$. This is always less that the surface-wave wavenumber k_s because $k_b < k_s$. A similar argument applies in the anisotropic case. Consequently, surface-wave solutions can exist without appreciable coupling to bulk waves.

In contrast to this argument, there are special cases in which a surface wave exists with velocity higher than that of the slowest bulk wave. This is possible because the surface-wave solution does not have elastic displacements in common with the slower bulk wave. A prime example is the Bleustein–Gulyaev wave, mentioned in Chapter 2, Section 2.3.3. The displacement is shear-horizontal (SH), that is, normal to the sagittal plane. The wave can exist in a piezoelectric material if the sagittal plane is normal to an even-order crystalline axis. The wave is bound to the surface only if the material is piezoelectric; otherwise the solution is a plane bulk wave.

A similar case is known as *surface-skimming bulk waves*, or SSBW. This is related to the observation that, in an isotropic material, a plane SH bulk wave can propagate parallel to the surface without violating the boundary conditions, that is, it has no associated stress on the surface. As mentioned in Chapter 2,

Section 2.2.3, the elastic displacement is shear horizontal. Related cases occur in anisotropic materials, for example in quartz [30]. In particular orientations, the boundary condition is almost satisfied and low-loss propagation between two transducers can be obtained. Such waves can have high velocities, making them attractive for high-frequency devices. An example is $36°Y–X+90°$ quartz, which gives a velocity of 5100 m/s and a temperature coefficient of delay (TCD) of zero. The delay is a quadratic function of temperature, as in eq. (4.33), and the constant c is about 60×10^{-9} (°C)$^{-2}$. Propagation is normal to the X axis. This case also gives zero piezoelectric coupling to a Rayleigh-type surface wave. Similar behavior is found in $41°Y–X$ and $64°Y–X$ lithium niobate and in $36°Y–X$ lithium tantalate. These orientations support leaky waves.

If we accept the analogy of an SH bulk wave, it might be expected that the amplitude of an SSBW will not be much affected by the presence of the surface. Thus, as for a cylindrical bulk wave in an infinite medium, we expect an amplitude variation as $x^{-1/2}$. This is indeed found experimentally when a transducer generates SSBWs in quartz [30]. In the above orientations of lithium niobate and tantalate, amplitude variations as $x^{-1/2}$ and $\exp(-\alpha x)$ are both seen [19]. These can be interpreted as SSBWs and leaky waves, respectively. Hashimoto [31] gives a theoretical interpretation, making the approximation that the material has 6 mm symmetry, showing that both types of x variation can occur. The theory shows that the SSBW amplitude falls as $x^{-1/2}$ at small x and as $x^{-3/2}$ at large x. In fact, these studies are not very relevant to practical devices, which generally make use of gratings so that the surface is not uniform.

In another development, Auld [32] showed that, in an isotropic material, a bulk SH wave can be guided along the surface by an array of grooves parallel to the wavefront. This concept can be applied to the SSBW in $36°Y–X+90°$ quartz, in which the wave can be trapped at the surface by the metal strips comprising reflecting gratings and single-electrode transducers. When used in this way, the wave is known as a *surface transverse wave* (STW) [18].

Another orientation of quartz gives an SSBW with exceptional temperature stability [30]. The orientation is $-50.5° -X + 90°$. The delay is approximately a cubic function of temperature, with a variation of 30 ppm from 0°C to 140°C. For comparison, $ST–X$ quartz gives a variation of 150 ppm over this range. Another quartz orientation, known as the LST cut, gives a leaky wave with exceptional temperature stability [33, 34], as shown in Table 4.2.

The numerical procedure for finding surface-wave solutions, discussed in Chapter 2, Section 2.3.2, looks for normal modes, that is solutions varying as $\exp(jkx)$. The wavenumber k is real for a conventional Rayleigh mode, but the method can be generalized so that k becomes complex, with an imaginary part giving the attenuation coefficient. An SSBW, with amplitude variation as $x^{-1/2}$, is not of this type. However, the SSBW orientations do give normal modes with low loss. Generally, the solution gives velocity higher than that of the slowest bulk wave, satisfying the boundary conditions but with some attenuation due to radiation of energy into the bulk. At particular orientations the attenuation may become insignificant because of decoupling between the wave and the lowest-velocity bulk wave, and then the wave may be suitable for practical devices. The term 'leaky surface acoustic wave' (LSAW), or 'pseudo-surface wave' (PSAW), is used for such cases when the attenuation is small or zero. In contrast, Rayleigh waves generally exist for all orientations. A wide variety of leaky-wave cases have been found, including the above orientations of quartz, lithium niobate and lithium tantalate [35]. Generally, the solutions can have the following characteristics:

- Low attenuation only at one specific orientation (at neighboring orientations, the attenuation increases).
- High velocity, higher than that of the slowest bulk wave, attractive for high-frequency applications.
- A grating is often necessary to trap the wave at the surface.
- Piezoelectric coupling higher than that of surface waves, dependent on the grating thickness.
- Temperature stability similar to that of surface waves, or better.

A particular example is $36° Y–X$ lithium tantalate. This has been extensively exploited in high-frequency low-loss bandpass filters, and hence the leaky-wave properties have been studied in detail. For this reason, the following section concentrates on this case, leaving other cases until later (Section 11.4.4).

11.4.2 Leaky waves in lithium tantalate

One prominent application of leaky waves has been in bandpass filtering for mobile telephones. This calls for radio frequency filters centered at about 950 MHz and above, with very low insertion loss and with narrow skirts in the frequency response. This application cannot be met by conventional Rayleigh waves because of the need for strong piezoelectric coupling and moderate temperature coefficients. Leaky waves in lithium tantalate have been used for this purpose.

Figure 11.18 shows theoretical velocities and attenuation for leaky waves in rotated Y-cut lithium tantalate, with propagation along the x axis. They are shown as functions of the rotation angle ψ, defined in Chapter 4, Fig. 4.7.

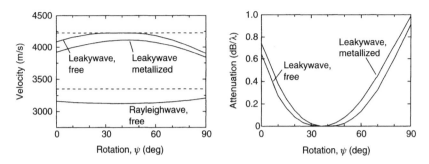

Figure 11.18 Velocities and attenuation for leaky surface waves in rotated Y-cut lithium tantalate. Broken lines indicate velocities for bulk shear waves.

The velocities of the wave on both free and metallized surfaces are greater than that of the slow shear wave. For the $36°Y-X$ orientation, where $\psi = 36°$, the attenuation becomes very small, theoretically less than 0.001 dB per wavelength, for both free and metallized surfaces [36, 37]. This attenuation would be almost undetectable in most practical devices, so the orientation is realistic for devices. At $36°$ the wave velocities are 4212 and 4111 m/s, giving a coupling parameter $\Delta v/v = 2.4\%$. The TCD is 32 ppm/°C. In contrast, Rayleigh waves in lithium tantalate generally have much smaller coupling while the temperature stability is little better. For example, the $X - 112°Y$ orientation gives $\Delta v/v = 0.35\%$ with TCD $= 18$ ppm/°C, as shown in Table 4.2. Rayleigh waves in lithium niobate have coupling comparable to that of the lithium tantalate leaky wave, but with much larger TCD. Hence the leaky wave offers strong coupling with moderate TCD, a combination not available using Rayleigh-wave orientations. The wave has elastic displacement almost normal to the sagittal plane, so it is essentially an SH wave. It is noted that there is also a Rayleigh-wave solution, with velocity around 3000 m/s as shown in Fig. 11.18, but this has very small piezoelectric coupling, with $\Delta v/v < 10^{-3}$. Consequently, the excitation of this wave by interdigital transducers is not significant in leaky-wave devices.

While the above considerations make the leaky-wave case attractive, it must be noted that the wave is physically quite different from a Rayleigh wave. To illustrate, Fig. 11.19 shows the effective permittivity $\varepsilon_s(\beta)$ for the $36°Y-X$ orientation, as a function of velocity $v = \omega/\beta$ with this taken as an independent variable. This is very different from the situation seen in a typical Rayleigh-wave substrate such as $Y-Z$ lithium niobate, shown in Fig. 3.2. For the tantalate case, $\text{Re}\{\varepsilon_s(\beta)\}$ has a pole at the metallized surface velocity $v_m = 4111$ m/s and a zero at the free-surface velocity $v_f = 4212$ m/s. The imaginary part is non-zero for all velocities greater than v_f. It also has a peak at v_m, and it is

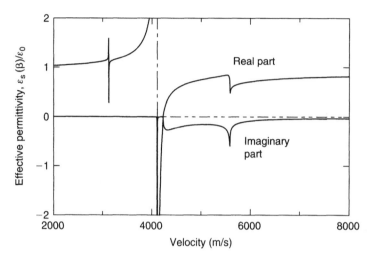

Figure 11.19 Effective permittivity for 36° Y–X lithium tantalate, calculated using program EPS [37].

zero for all velocities below this. There is also a pole at $v = 3128$ m/s corresponding to the non-leaky Rayleigh wave, and a longitudinal-wave pole at 5591 m/s.

For a Rayleigh wave the imaginary part would of course be zero in the region of both the pole and zero. Consequently, it was possible to approximate the function by Ingebrigtsen's approximation, eq. (3.28). For the leaky wave this is clearly not valid. The analysis for waves generated on a uniform free or metallized surface is very complex. However, this is not very relevant to practical devices, in which much of the surface is covered by the metal strips that form transducers and gratings.

The generation process for leaky waves is susceptible to generation of unwanted bulk shear waves, because the velocities are similar. For practical devices, it is important to take steps to avoid the energy loss due to bulk-wave generation. This can be done by using a reflective grating, as in the case of STWs mentioned above. The grating reduces the velocity of the leaky wave, but not that of the bulk wave. A grating with pitch approximately $\lambda/2$ traps the leaky wave at the surface. However, at higher frequencies the leaky wave can be coherently scattered into bulk waves, giving substantial loss, and this must be considered. This phenomenon also occurs for Rayleigh waves, but in this case the bulk-wave generation can be small, and it can occur outside the surface-wave passband so that it is of less consequence. In general, the 'grating' for a leaky-wave device can be either a shorted reflective grating or a single-electrode transducer, which behaves in a similar manner in this respect. For example, a resonator using the leaky wave can be realized using a uniform single-electrode transducer, as considered in Section 11.1.4 above.

Because of these considerations, the resonators in IEFs are usually the synchronous type, so that the electrode pattern in the transducer is continued into the grating, as in Fig. 11.14. From the point of view of the wave, the transducer and gratings form a continuous structure which guides the wave. There are no discontinuities to cause wave scattering into bulk waves, except at the ends of the structure.

Analogous leaky-wave solutions are found in rotated Y-cut lithium niobate, but in this case the rotation for minimum attenuation depends on the surface condition. For a free surface the required rotation is $41°$, while for a metallized surface it is $64°$ [35, 39]. The effective permittivity has a form similar to that of $36°Y-X$ tantalate [19]. The niobate cases give larger coupling than lithium tantalate, as shown by the $\Delta v/v$ value, but the large temperature coefficient has restricted their practical usage.

Table 11.1 summarizes the properties of leaky waves on uniform surfaces (free or metallized). The attenuation data should be treated with caution because in practice they are usually modified by the presence of a grating. The term LSAW refers to leaky surface waves, also called PSAW, as described above.

Table 11.1 Properties of leaky surface waves on uniform surfaces.

Material and Euler angles (°)	Wave type	Velocity (m/s)	$\Delta v/v$	Attenuation (dB/λ)	TCD (ppm/°C)	Reference
36°$Y-X$ LiTaO$_3$	LSAW	4226.7 f	2.8%	0 f	45 f	[40, 35]
(0, −54, 0)		4109.1 m		0 m	32 m	
41°$Y-X$ LiNbO$_3$	LSAW	4751.5 f	7.9%	0 f	78 f	[40, 35]
(0, −49, 0)		4378.1 m		0.01 m	80 m	
64°$Y-X$ LiNbO$_3$	LSAW	4691.9 f	5.1%	0.05 f	79 f	[40, 35]
(0, −26, 0)		4450.5 m		0.004 m	81 m	
Quartz, $ST-X$	LSAW	5078.9 f	0.015%	0.008 f		[39]
(0, −47.25, 0)		5078.1 m		0.008 m		
15°$Y-X$ LiNbO$_3$*	LSAW	4000 m	20%			[41]
(0, −75, 0)						
Li$_2$B$_4$O$_7$	LLSAW	6912.0 f	0.7%	0 f	−3 f	[40, 42]
(0, 47.3, 90)		6864.1 m		0.001 m	−23 m	
$Y-Z$ LiNbO$_3$	LLSAW	7285.2 f	1.5%	0.07 f		[40]
(0, 90, 90)		7177.7 m		1.5 m		

f = free surface; m = metallized surface; attenuation data are shown if greater than 0.001 dB/λ; TCD = temperature coefficient of delay.
*Estimated values from [41].

These are basically shear horizontal waves, with particle motion mainly in the direction normal to the sagittal plane. In addition, a variety of cases are known with basically *longitudinal* polarization, parallel to the wave vector. These generally have very high velocities. They are known as longitudinal leaky surface waves (LLSAW) or high-velocity pseudo-surface waves (HVP-SAW). DaCunha [40] gives details for these and many other cases, including polarization data.

11.4.3 Coupled-mode analysis of gratings and transducers

Because of the complexities of leaky-wave behavior, the analysis methods described in Chapters 5 and 8 for Rayleigh waves cannot be applied. Moreover, practical leaky-wave devices make substantial use of gratings in order to trap the waves at the surface, thus reducing the attenuation. Because of this, it is more suitable to consider the wave in the presence of the grating rather than on a uniform surface. Analysis assuming an infinite grating is sometimes adequate. From the grating analysis [3], or from experiments on long gratings, it is possible to estimate suitable parameters for the COM method of analysis. The COM analysis of Section 8.2 in Chapter 8 can then be used to analyze finite-length transducers and gratings. This approach assumes that end effects are negligible, though this is not always the case in practice.

Theoretical results for gratings have been shown by many authors [19, 43–45]. The methods used have been considered in Chapter 8, Section 8.3. Usually the finite element method (FEM) is used for the electrodes and the boundary element method (BEM) for the substrate. It is necessary to take account of the film thickness h and the metallization ratio a/p. Here we assume the common case $a/p = 0.5$. The COM parameters are taken to be independent of frequency. This implies that they can be deduced from the frequencies of the band edges, for shorted and open-circuit gratings, as explained in Chapter 8, Section 8.3.2. The electrode reflection coefficient r_s is imaginary because of symmetry and is given by $r_s = c_{12}p$, from eq. (8.41). Experimentally, r_s is known to be negative imaginary because it reduces the frequency of the conductance peak for a single-electrode transducer. Since the results are derived from grating analysis, r_s may be interpreted as a property of the grating mode, rather than an actual electrode reflection coefficient.

Figure 11.20 shows some results from the infinite-grating analysis [46, 47], as functions of the normalized film thickness $h/(2p)$. Here Dv/v is the fractional velocity change due to the electrodes, so that the velocity is $v = v_f(1 - Dv/v)$, where v_f is the free-surface velocity. Data for Dv/v and $|r_s|$ were confirmed very accurately by experiments of Rosler *et al.* [47].

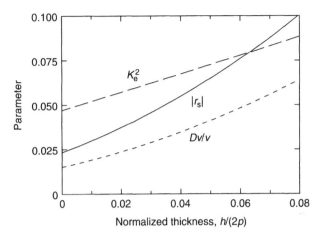

Figure 11.20 Theoretical parameters for leaky wave in a shorted aluminum grating on $36°Y–X$ lithium tantalate, with metallization ratio $a/p = 1/2$.

Numerical results often express the coupling in terms of the COM parameter α_1. To facilitate comparison with the Rayleigh-wave case, another form K_e^2 is used here. For Rayleigh waves, α_1 is given by eq. (8.42) with $\Gamma_s = (\Delta v/v)/\varepsilon_\infty = 1/2K^2/\varepsilon_\infty$. For the leaky wave, the numerical analysis gives a value for α_1, and the constant K_e^2 is defined such that α_1 is given by eq. (8.42) with $\Gamma_s = 1/2K_e^2/\varepsilon_\infty$. Thus K_e^2 becomes the usual coupling parameter K^2 in the case of a Rayleigh wave. In Fig. 11.15 we see that K_e^2 increases with film thickness $h/(2p)$, presumably because the wave becomes more tightly bound to the surface. This feature is quite unlike the behavior of Rayleigh waves, for which K^2 is independent of film thickness. It is notable that for $h = 0$ we have $K_e^2 \approx 2\Delta v/v$, as in the Rayleigh-wave case. The coupling can also be estimated from the difference of the resonance and antiresonance frequencies, as shown by eq. (11.25).

In addition to the above parameters, losses need to be considered. Leaky waves have been used mainly for IEFs and LCR filters. In both cases the propagation loss is significant, and it must be noted that the loss due to bulk-wave excitation is strongly frequency dependent. For IEFs, loss in the filter passband needs to be minimized in order to meet tight specifications. For the LCR filter, significant loss at frequencies above the filter passband affects the stop-band suppression.

Plessky *et al.* [48] have discussed limitations of the COM analysis, comparing with admittance measurements on synchronous one-port resonators using the leaky wave on $64°Y–X$ lithium niobate. The most important effect is loss due to generation of bulk shear waves, which have a velocity close to that of the leaky wave. At relatively low frequencies, the grating tends to trap the wave at the surface, reducing the loss. However, the grating can convert surface waves to bulk waves by synchronous scattering, and there is also bulk-wave generation by the transducer directly. Both mechanisms occur at frequencies a little higher than the resonance frequency. Studies of the longitudinal leaky wave on lithium tetraborate also show this [44]. For long transducers these losses begin quite sharply at an onset frequency determined by the velocities, so losses are small at lower frequencies. For short transducers the loss tends to be spread out in frequency. For this reason, long transducers can give lower losses in IEFs. A small amount of dispersion was also found to be relevant. Using empirical models, the bulk-wave loss was allowed for in the COM by using a frequency-dependent attenuation coefficient (giving an imaginary part to the detuning parameter δ), giving much better agreement with experiment.

The behavior of leaky waves in infinite gratings has been investigated very extensively, using sophisticated numerical analysis allowing for the finite thickness of the electrodes. The methods used are discussed in Chapter 8, Section 8.3. Suitable COM parameters can be deduced from the accurate-grating analysis and used to analyze finite-length devices. Figure 11.21 shows the experimental and theoretical admittance of a synchronous resonator on $36° Y$–X lithium tantalate. The grating stop band has frequency range 900 MHz (the resonance frequency) to 950 MHz. The antiresonance frequency is 930 MHz. Outside the stop band, where the waves can propagate with little attenuation, there are small ripples due to reflections at the ends of the device. At 950 MHz there is an unwanted blip associated with the upper edge of the stop band. In an IEF the response shows a corresponding perturbation. Usually the electrode reflection coefficient is made quite large so that the perturbation occurs outside the filter passband. Typically, this requires a large film thickness $h/(2p)$, in the region of 8%. This also has the merit that the coupling coefficient K_e^2 is increased (Fig. 11.20).

Plessky's approximate formula

An approximate formula for the dispersion in an infinite grating was deduced by Plessky [49, 19, 31]. The modeling is based on assuming that the displacement is purely transverse, and in the absence of the grating there would only be a bulk wave with velocity V_b. With p = electrode pitch, we define $k_c = \pi/p$ as the Bragg wavenumber and $\omega_b = V_b k_c$ as the Bragg frequency for the bulk wave. The wavenumber γ of the grating mode is normalized as $q = \gamma/k_c - 1$, and the

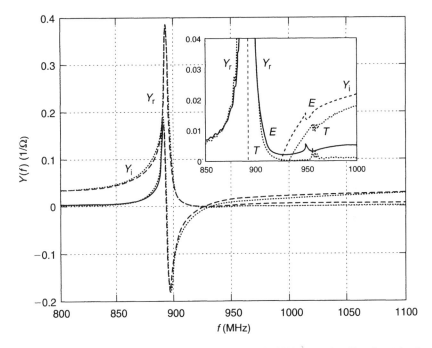

Figure 11.21 Admittance of a synchronous resonator on 36° *Y–X* lithium tantalate. Experimental and theoretical (COM) curves are shown. Normalised film thickness $h/(2p) = 8\%$. From Koskela *et al.* [45], copyright 1999 IEEE, reproduced with permission. Y_r and Y_i are real and imaginary parts of Y, respectively. E = Experimental, T = Theoretical.

frequency ω is normalized as $\Delta = \omega/\omega_b - 1$. Plessky's dispersion relation is

$$q^2 = \Delta^2 - \frac{1}{4}\left[\varepsilon^2 \pm 2\eta\sqrt{\Delta_b - \Delta}\right]^2 \tag{11.29}$$

where ε and η are two constants, to be determined by fitting to a more detailed analysis. These are related to the reflection and penetration of the waves, respectively. The sign is usually chosen to be $+$, which is assumed here. The frequency Δ_b is a frequency beyond which bulk-wave generation can occur, given by $\Delta_b = (2\varepsilon^2 - \eta^2)/4$. When Δ is less than Δ_b we have q^2 real, and the stop band occurs when q^2 is negative. The frequencies of the band edges are $\Delta_\pm$, obtained by setting $q^2 = 0$, giving

$$\Delta_\pm = -(\eta \mp \varepsilon)^2/2$$

The mean of these is the center of the stop band, given by $\Delta_c = -1/2(\eta^2 + \varepsilon^2)$.

In the COM analysis we have $\delta = k - k_c$, where k is the wavenumber of the undisturbed waves, with no reflection (for a leaky-wave grating mode, these

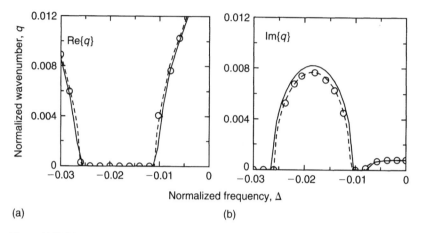

Figure 11.22 Dispersion relation for shorted aluminum grating on $36°Y$–X lithium tantalate, with $a/p = 1/2$ and with film thickness zero. (a): Re$\{q\}$, (b): Im$\{q\}$, where q is normalized wavenumber. Solid lines: from FEM analysis using FEMSDA [38], (courtesy of A. Shvetsov and S. Zhgoon). Broken lines: from Plessky formula, eq. (11.29). Points: from COM analysis fitted to Plessky formula.

waves can be notional). These waves are taken to have an effective velocity v_e. The center frequency ω_c is given by $\Delta_c = \omega_c/\omega_b - 1$, and $\omega_c = k_c v_e$, giving $v_e = V_b(1 + \Delta_c)$. Taking v_e to be constant, we have $k = (\omega/\omega_c) k_c = k_c (1 + \Delta)/(1 + \Delta_c)$, which gives

$$\delta = \delta_r \equiv (1 + \Delta)/(1 - \Delta_c) - 1.$$

This applies for $\Delta < \Delta_b$. When $\Delta > \Delta_b$ eq. (11.29) shows that q^2 has an imaginary part Im$\{q^2\} = \varepsilon^2 \eta [\Delta - \Delta_b]^{1/2}$. This gives $\delta^2 = \delta_r^2 + jk_c^2 \text{Im}\{q^2\}$ for $\Delta > \Delta_b$. For $\Delta < \Delta_b$ we take $\delta^2 = \delta_r^2$.

From the band edges we have $\Delta_+ - \Delta_- = 2\eta\varepsilon$, and from COM theory the band edges correspond to wavenumbers $k_\pm$, related by $k_+ - k_- = 2|c_{12}|$. Knowing that c_{12} is negative imaginary, this gives $c_{12} = -jk_c \eta\varepsilon/(1 + \Delta_c)$.

Figure 11.22 shows the dispersion relation for $36°Y$–X lithium tantalate, using the Plessky formula, eq. (11.29), and COM analysis. This refers to a shorted grating with film thickness $h = 0$ and metallization ratio $a/p = 1/2$, for which Plessky gives $\eta = 0.188$ and $\varepsilon = 0.041$. The attenuation due to bulk-wave generation at $\Delta > \Delta_b$ is clearly seen. The COM gives a good fit to Plessky's formula. The c_{12} value gives an electrode reflection coefficient of $|r_s| = 0.024$, which agrees well with the results in Fig. 11.20. For an open-circuit grating $\eta = 0.097$ and $\varepsilon = 0.06$, and from these the COM transduction parameter α_1 can be deduced, as shown in Chapter 8, Section 8.3.2. For non-zero film thickness,

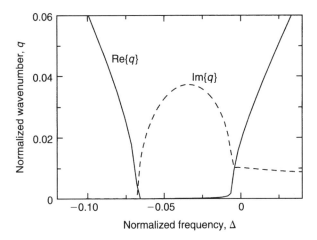

Figure 11.23 Dispersion curves for shorted aluminum grating on $42°Y-X$ lithium tantalate with $a/p = 1/2$ and $h/\lambda = 8\%$. FEM analysis using program FEMSDA [38] (courtesy of A. Shvetsov and S. Zhgoon).

fitting to FEM analysis [19] gives $\eta = 0.182 + 0.68\, h/(2p)$ and $\varepsilon = 0.039 + 1.24$ $h/(2p)$. A study of STWs by Abbott and Hashimoto [50] yielded a four-parameter dispersion relation which approximates to Plessky's [19].

Although many studies have concerned $36°Y-X$ lithium tantalate, it has been shown that this is not actually the optimum orientation for IEFs. This is because the attenuation is affected by the presence of the electrodes, which are relatively thick in this case. From theoretical and experimental studies [51], the orientation for minimum loss has a rotation angle of about $42°$, so this is normally used for practical filters. Apart from the attenuation, the properties are very similar to those of the $36°$ rotation, as shown by FEM analysis [51, 52]. Figure 11.23 shows dispersion curves calculated using FEM analysis, for a normalized film thickness of $h/\lambda = 8\%$, which is typical of RF filters using this material. Comparing with Fig. 11.22, the width of the stop band is much greater, and the peak value of $\mathrm{Im}\{q\}$ is also greater. These changes arise because the reflection coefficient $|r_s|$ increases with thickness, as shown in Fig. 11.20. Further investigation of IEF losses, considering the Q-values at the resonance and antiresonance frequencies, concluded [53] that lower losses can be obtained for a further rotation, to $48°$. This work also showed the variation of parameters with rotation angle.

11.4.4 Other leaky waves

Hashimoto [41] describes experiments using $15°Y-X$ lithium niobate, which gives a transverse leaky wave with very large piezoelectric coupling. The parameter $\Delta v/v$ shown in Table 11.1 has a large value of 20%, estimated from ω_r and ω_a using eq. (11.25). Using copper electrodes to obtain large reflectivity, a 1 GHz IEF with 15% fractional bandwidth and 0.5 dB loss was demonstrated, though it had ripples due to unwanted modes.

Kadota [54] showed that electrode reflection coefficients for SH leaky waves on quartz can be very large if heavy-metal electrodes are used. The orientation is $ST–X + 90°$, similar to the STW orientation. Using tantalum or tungsten metallization for the electrodes, efficient reflecting gratings can be made with only 10 electrodes, reducing the necessary length substantially. Resonator filters at 240 MHz were demonstrated. Alternatively, SH waves can be reflected efficiently at an abrupt substrate edge, made with high precision. This can be used to realize resonators without needing reflecting gratings. High-Q resonators were demonstrated at 2 GHz, using $36°Y–X$ lithium tantalate [55].

Other experiments used longitudinal leaky surface acoustic waves (LLSAW), with data given on Table 11.1. The high velocities make these waves attractive for high-frequency applications. In lithium tetraborate, the wave has a velocity of 6900 m/s and its behavior in reflective gratings has been studied theoretically in much detail [42]. A 1.2 GHz IIDT filter gave 3.2 dB insertion loss [56].

Another longitudinal leaky wave occurs in $Y–Z$ lithium niobate [57]. Again the high velocity, around 7000 m/s, is attractive for high-frequency devices. The complexities of the wave motion have been studied in detail [58, 59], and Fig. 8.16 shows an example of the electrode motion. Figure 11.24 shows the response of a 5.2 GHz IEF using this wave, giving 5 dB insertion loss and a bandwidth of 270 MHz [59]. Because of the high velocity, the frequency obtainable using commercial optical lithography exceeds the 3 GHz limit feasible using $36°Y–X$ lithium tantalate. This indicates the possibility of novel high-frequency applications, notably Bluetooth at 2.5 GHz and Hyperlan at 5.3 GHz.

11.5 LONGITUDINALLY-COUPLED RESONATOR (LCR) FILTERS

This device, also called a duble-mode SAW (DMS) filter offers an alternative way of realizing low loss devices at GHz frequencies, as shown by Morita [60]. Compared with the IEF it can provide better stop-band rejection, though the power handling and the shape factor are not as good. Figure 11.25 shows two versions of the device. On the left is a symmetrical resonator consisting of two identical transducers and two identical reflecting gratings. This is functionally similar to the two-port resonator of Fig. 11.6, considered earlier. In the LCR filter the transducers are the single-electrode type giving significant electrode

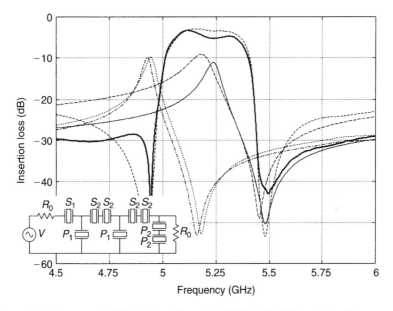

Figure 11.24 Response of 5 GHz bandpass filter using LLSAW on $Y-Z$ lithium niobate. From Makkonen *et al.* [59], copyright 2003 IEEE, reproduced with permission. Solid line: measured. Broken line: theoretical.

reflectivity, which complicates the mechanism. One way of regarding this is that the transducer and grating at each end must behave as a SPUDT, so that most of the surface-wave energy generated is directed toward the device center. For frequencies where the SPUDTs have strong reflectivity, the device Y-matrix can be approximated as [61]

$$Y_{11} = Y_{22} = j \operatorname{Im}\{Y_s\} + jG_a^s \cot \theta_c; \quad Y_{12} = Y_{21} = jG_a^s \operatorname{cosec} \theta_c \quad (11.30)$$

where Y_s is the admittance of each SPUDT, $G_a^s = \operatorname{Re}\{Y_s\}$ and θ_c is the phase corresponding to one transit of the cavity, so that resonances occur for $2\theta_c = 2n\pi$. If the space between the two SPUDTs is adjusted correctly, a two-pole response can be obtained, and this gives a flat passband with low loss if the appropriate loading is used.

The above device can respond to symmetric and antisymmetric modes. A modified version has three transducers with the outer transducers connected together, giving a symmetrical structure as shown in Fig. 11.25b. This can only respond to symmetric modes, and it gives a larger bandwidth because of the larger mode spacing. The three-transducer version is related to the one-port resonator of Section 11.1.2 above. The outer transducers and their adjacent gratings behave

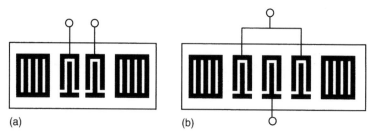

Figure 11.25 LCR filter. (a) Two-transducer type. (b) Three-transducer type.

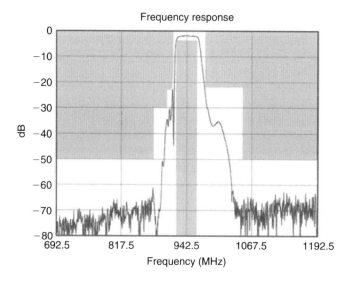

Figure 11.26 Response of a practical LCR filter on $42°Y–X$ lithium tantalate (courtesy of B.P. Abbott, Trinquint Semiconductor Inc.).

like SPUDTs, and the central transducer transmits the waves without attenuation (Section 11.1.2). When a high-coupling substrate such as lithium tantalate is used, the device does not require matching components. Such devices can be cascaded electrically in order to improve the stop-band rejection. In a further modification, the LCR may have five transducers connected alternately to the two ports.

A practical result is shown in Fig. 11.26. Typically, the response has a hump in the high-frequency skirt region. This is because the passband tends to be below the transducer Bragg frequency because of the shift due to electrode reflectivity; the transducer response remaining at higher frequencies explains the hump. The LCR therefore tends to give poor shape factors.

In comparison with the IEF, it is seen that the LCR does not use a synchronous structure. When used with a leaky-wave substrate, this implies the possibility of some loss due to bulk-wave generation at the ends of transducers or gratings. Although the loss is not generally very large, this topic has prompted studies in two ways. Firstly, there have been theoretical studies of finite-length structures using finite thickness electrodes and leaky-wave propagation [62, 63], extending the analysis of periodic structures considered in Chapter 8, Section 8.3.1. Secondly, losses have been reduced by filling in the gaps of the structure using 'distributed gaps', that is, interdigital transducer elements with different, or graded, pitch, such that the electrodes form an almost synchronous array. This technique was used, for example, in a five-transducer LCR filter at 1840 MHz with 1 dB insertion loss [64]. In a related technique used for Rayleigh-wave resonators, the electrode pitches are adjusted such that gaps are not necessary, and this reduces the loss due to bulk-wave excitation [9].

REFERENCES

1. D.P. Morgan, S. Zhgoon, A. Shvetsov, E. Semenova and V. Semenov. 'One-port SAW resonators using natural SPUDT substrates', *IEEE Ultrason. Symp.*, **1**, 446–449 (2005).
2. E.A. Ash. 'Surface wave grating reflectors and resonators', *IEEE MTT Symp.*, 1970, pp. 385–386.
3. V. Plessky and J. Koskela. 'Coupling-of-modes analysis of SAW devices', in C.C.W. Ruppel and T.A. Fjeldy (eds.), *Advances in Surface Acoustic Wave Technology, Systems and Applications*, Vol. 2, World Scientific, 2001, pp. 1–82.
4. Supriyo Datta. *Surface Acoustic Wave Devices*, Prentice-Hall, 1986.
5. E.J. Staples, J.S. Schoenwald, R.C. Rosenfeld and C.S. Hartmann. 'UHF surface acoustic wave resonators', *IEEE Ultrason. Symp.*, 1974, pp. 245–252.
6. T.E. Parker and G.K. Montress. 'Precision surface-acoustic-wave (SAW) oscillators', *IEEE Trans. Ultrason. Ferroelect. Freq. Contr.*, **35**, 342–354 (1988).
7. E.A. Ash. 'Fundamentals of signal processing devices', in A.A. Oliner (ed.), *Acoustic Surface Waves*, Springer, 1978, pp. 97–186.
8. D.T. Bell and R.C.N. Li. 'Surface acoustic wave resonators', *Proc. IEEE*, **64**, 711–721 (1976).
9. Y. Ebata. 'Suppression of bulk-scattering loss in SAW resonators with quasi-constant acoustic reflection periodicity', *IEEE Ultrason. Symp.*, **1**, 91–96 (1988).
10. T. Uno and H. Jumonji. 'Optimisation of quartz SAW resonator structure with groove gratings', *IEEE Trans. Ultrason. Ferroelect. Freq. Contr.*, **29**, 299–310 (1982).
11. D.P. Morgan. 'Simplified analysis of surface acoustic wave one-port resonators', *Electron. Lett.*, **39**, 1361–1362 (2003).
12. L.L. Pendergrass and L.G. Studebaker. 'SAW resonator design and fabrication for 2.0, 2.6 and 3.3 GHz', *IEEE Trans. Ultrason. Ferroelect. Freq. Contr.*, **35**, 372–379 (1988).
13. P.V. Wright. 'A review of SAW resonator technology', *IEEE Ultrason. Symp.*, **1**, 29–38 (1992).
14. V.P. Plessky. 'SAW impedance elements', *IEEE Trans. Ultrason. Ferroelect. Freq. Contr.*, **42**, 870–875 (1995).
15. J. Meltaus, V. Plessky and S. Hong. 'Double-resonance SAW filters', *IEEE Ultrason. Symp.*, **2**, 2301–2304 (2005).

16. M. Lewis. 'The surface acoustic wave oscillator – a natural and timely development of the quartz crystal oscillator', *Proc. 28th Ann. Symp. on Freq. Contr.*, 1974, pp. 304–314.

17. I.D. Avramov, private communication.

18. I.D. Avramov, F.L. Walls, T.E. Parker and G.K. Montress. 'Extremely low thermal noise floor high power oscillators using surface transverse wave devices', *IEEE Trans. Ultrason. Ferroelect. Freq. Contr.*, **43**, 20–29 (1996).

19. K.-Y. Hashimto. *Surface Acoustic Wave Devices in Telecommunications – Modelling and Simulation*, Springer, 2000.

20. M. Hikita, N. Shibagaki, K. Sakiyama and K. Hasegawa. 'Design methodology and experimental results for new ladder-type SAW resonator coupled filters', *IEEE Trans. Ultrason. Ferroelect. Freq. Contr.*, **42**, 495–508 (1995).

21. O. Ikata, T. Miyashita, T. Matsuda, T. Nishihara and Y. Sato. 'Development of low-loss bandpass filters using SAW resonators for portable telephones', *IEEE Ultrason. Symp.*, **1**, 111–115 (1992).

22. X. Peroix, M. Solal, J. Briot, S. Chamaly, M. Doisy and P. Girard. 'An accurate design and modeling tool for the design of RF SAW filters', *IEEE Ultrason. Symp.*, **1**, 75–80 (2001).

23. T. Makkonen, S. Kondratiev, V.P. Plessky, T. Thorvaldsson, J. Koskela, J.V. Knuuttila and M.M. Salomaa. 'Surface acoustic wave impedance element ISM duplexer: modeling and optical analysis', *IEEE Trans. Ultrason. Ferroelect. Freq. Contr.*, **48**, 652–665 (2001).

24. N. Shibagaki, N. Matsuura, K. Sakiyama and M. Hikita. 'An integrated SAW antenna duplexer for EGSM/DSC1800/PCS triple-band cellular systems', *IEEE Ultrason. Symp.*, **1**, 391–394 (2000).

25. Y. Satoh, T. Nishihara and O. Ikata. 'SAW duplexer metallisations for high power durability', *IEEE Ultrason. Symp.*, **1**, 17–26 (1998).

26. R. Takayama, H. Nakanishi, Y. Iwasaki, T. Inoue and T. Kawasaki. 'The approach to realise the characteristics of SAW resonators with temperature compensation and steepness for PCS duplexer', *IEEE Ultrason. Symp.*, **1**, 385–388 (2003).

27. H. Sato, K. Onishi, T. Shimamura and Y. Tomita. 'Temperature stable SAW devices using directly bonded LiTaO$_3$ glass substrates', *IEEE Ultrason. Symp.*, **1**, 335–338 (1998).

28. B.P. Abbott, J. Chocola, K. Lin, N. Naumenko and J. Caron. 'Characterization of bonded wafers for RF filters with reduced TCF', *IEEE Ultrason. Symp.*, **1**, 926–929 (2005).

29. H. Kando, D. Yamamoto, M. Mimura, T. Oda, A. Shimizu, K. Shimoda, E. Takata, T. Fuyutsume, R. Kubo and M. Kadota. 'RF filter using boundary acoustic wave', *IEEE Ultrason. Symp.*, **1**, 188–191 (2006).

30. M. Lewis. 'Surface-skimming bulk waves, SSBW', *IEEE Ultrason. Symp.*, 1977, pp. 744–752.

31. K.-Y. Hashimoto and M. Yamaguchi. 'Excitation and propagation of shear-horizontal-type surface and bulk acoustic waves', *IEEE Trans. Ultrason. Ferroelect. Freq. Contr.*, **48**, 1181–1188 (2001).

32. B.A. Auld, J.J. Gagnepain and M. Tan. 'Horizontal shear surface waves on corrugated surfaces', *Electron. Lett.*, **12**, 650–651 (1976).

33. Y. Shimizu, M.Tanaka and T. Watanabe. 'A new cut of quartz with extremely small temperature coefficient by leaky surface waves', *IEEE Ultrason. Symp.*, **1**, 233–236 (1985).

34. C.S. Lam and D.E. Holt. 'The temperature dependence of power leakage in LST-cut quartz surface acoustic wave filters', *IEEE Ultrason. Symp.*, **1**, 275–279 (1989).

35. K. Yamanouchi and M. Takeuchi. 'Applications for piezoelectric leaky surface waves', *IEEE Ultrason. Symp.*, **1**, 11–19 (1990).

36. M.F. Lewis. 'Acoustic wave devices employing surface skimming bulk waves'. U.S.A. patent 4,159,435 (1979).

37. K. Nakamura, M. Kazumi and H. Shimizu. 'SH-type and Rayleigh-type surface waves on rotated Y-cut LiTaO₃', *IEEE Ultrason. Symp.*, 1977, pp. 819–822.

38. K. Hashimoto and M. Yamaguchi. 'Free software products for simulation and design of surface acoustic wave and surface transverse wave devices', *IEEE Ann. Freq. Cont. Symp.*, 1996, pp. 300–307.

39. A. Takayanagi, K. Yamanouchi and K. Shibayama. 'Piezoelectric leaky surface waves in LiNbO₃', *Appl. Phys. Lett.*, **17**, 225–227 (1970).

40. M.P. da Cunha. 'Extended investigation on high velocity pseudo surface waves', *IEEE Trans. Ultrason. Ferroelect. Freq. Contr.*, **45**, 604–613 (1998).

41. K.-Y. Hashimoto, H. Asano, K. Matsuda, N. Yokoyama, T. Omori and M. Yamaguchi. 'Wideband Love wave filters operating in GHz range on Cu-grating/rotated *Y–X* LiNbO₃ substrate structure', *IEEE Ultrason. Symp.*, **2**, 1330–1334 (2004).

42. T. Sato and H. Abe. 'Propagation properties of longitudinal leaky surface waves on lithium tetraborate', *IEEE Trans. Ultrason. Ferroelect. Freq. Contr.*, **45**, 136–151 (1998).

43. R.C. Peach. 'A general Green function analysis for SAW devices', *IEEE Ultrason. Symp.*, **1**, 221–225 (1995).

44. T. Sato and H. Abe. 'Propagation of longitudinal leaky surface waves under periodic metal grating structure on lithium tetraborate', *IEEE Trans. Ultrason. Ferroelect. Freq. Contr.*, **45**, 394–408 (1998).

45. J. Koskela, V.P. Plessky and M.M. Salomaa. 'SAW/LSAW COM parameter extraction from computer experiments with harmonic admittance of a periodic array of electrodes', *IEEE Trans. Ultrason. Ferroelect. Freq. Contr.*, **46**, 806–816 (1999).

46. Y. Suzuki, M. Takeuchi, K. Nakamura and K. Hirota. 'Coupled-mode theory of SAW periodic structures', *Electron. Commun. Jap.*, Part 3, Vol. 76, 87–98 (1993).

47. U. Rosler, D. Cohrs, A. Dietz, G. Fischerauer, W. Ruile, P. Russer and R. Weigel. 'Determination of leaky SAW propagation, reflection and coupling on LiTaO₃', *IEEE Ultrason. Symp.*, **1**, 247–250 (1995).

48. V.P. Plessky, D.P. Chen and C.S. Hartmann. 'Patch improvements to COM model for leaky waves', *IEEE Ultrason. Symp.*, **1**, 297–300 (1994).

49. V.P. Plessky. 'A simple closed form dispersion equation for shear types of surface waves propagating in periodic structures', *IEEE Trans. Ultrason. Ferroelect. Freq. Contr.*, **40**, 421–423 (1993).

50. B.P. Abbott and K.-Y. Hashimoto. 'A coupling-of-modes formalism for surface transverse wave devices', *IEEE Ultrason. Symp.*, **1**, 239–245 (1995).

51. O. Kawachi, S. Mineyoshi, G. Endoh, M.Ueda, O. Ikata, K. Hashimoto and M. Yamaguchi. 'Optimal cut for leaky SAW on LiTaO₃ for high performance resonators and filters', *IEEE Trans. Ultrason. Ferroelect. Freq. Contr.*, **48**, 1442–1448 (2001).

52. S. Zhgoon, private communication.

53. N. Naumenko and B. Abbott. 'Optimum cut of LiTaO₃ for resonator filters with improved performance', *IEEE Ultrason. Symp.*, **1**, 385–390 (2003).

54. M. Kadota, T. Yoneda, K. Fujimoto, T. Nakao and E. Takata. 'Resonator filters using shear horizontal type leaky surface acoustic wave consisting of heavy-metal electrode and quartz substrate', *IEEE Trans. Ultrason. Ferroelect. Freq. Contr.*, **51**, 202–210 (2004).

55. M. Kadota, T. Kimura and D. Tamasaki. 'High frequency resonators with excellent temperature characteristics using edge reflection', *IEEE Ultrason. Symp.*, **3**, 1878–1882 (2006).

56. T. Sato and H. Abe. 'SAW device applications of longitudinal leaky surface waves on lithium tetraborate', *IEEE Trans. Ultrason. Ferroelect. Freq. Contr.*, **45**, 1506–1516 (1998).

57. V.I. Grigorievksi. 'Resonant properties of fast leaky surface acoustic waves on lithium niobate', *IEEE Ultrason. Symp.*, **1**, 197–200 (2001).

58. M. Solal, R. Lardat, T. Pastureaud, W. Steichen, V.P. Plessky, T. Makkonen and M.M. Salomaa. 'Existence of harmonic metal thickness mode propagation for longitudinal leaky waves', *IEEE Ultrason. Symp.*, **2**, 1207–1212 (2004).

59. T. Makkonen, V.P. Plessky, W. Steichen, S. Chamaly, C. Poirel, M. Solal and M.M. Salomaa. 'Properties of LLSAW on YZ-cut LiNbO$_3$; modelling and experiment', *IEEE Ultrason. Symp.*, **1**, 613–616 (2003).

60. T. Morita, Y. Watanabe, M. Tanaka and Y. Nakazawa. 'Wideband low loss double mode SAW filters', *IEEE Ultrason. Symp.*, **1**, 95–104 (1992).

61. D.P. Morgan. 'Idealised analysis of SAW longitudinally-coupled resonator (LCR) filters', *IEEE Trans. Ultrason. Ferroelect. Freq. Contr.*, **51**, 1165–1170 (2004).

62. P. Ventura, J.M. Hode and B. Lopes. 'Rigorous analysis of finite SAW devices with arbitrary electrode geometries', *IEEE Ultrason. Symp.*, **1**, 257–262 (1995).

63. R.C. Peach. 'Green function analysis for SAW devices with arbitrary electrode structure', *IEEE Ultrason. Symp.*, **1**, 99–103 (1997).

64. J. Meltaus, S. Harma, M.M. Salomaa and V.P. Plessky. 'Experimental results for a longitudinally-coupled 5-IDT resonator filter with distributed gaps', *IEEE Ultrason. Symp.*, **1**, 311–313 (2004).

Appendix A

FOURIER TRANSFORMS AND LINEAR FILTERS

The theory of surface-wave devices makes extensive use of Fourier transforms and the theory of linear filters. This appendix summarizes the results needed in this context. Many of the results are quoted without proof, and for more detail the reader is referred to the many texts available [1–4].

A.1   FOURIER TRANSFORMS

The Fourier transform of a function $g(t)$ is denoted $G(\omega)$, while the inverse transform of $G(\omega)$ is $g(t)$. The relation is expressed symbolically as

$$G(\omega) \leftrightarrow g(t). \tag{A.1}$$

The transforms are defined by the integrals

$$G(\omega) = \int_{-\infty}^{\infty} g(t) e^{-j\omega t} dt \quad \text{(forward transform)}, \tag{A.2}$$

$$g(t) = \frac{1}{2\pi} \int_{-\infty}^{\infty} G(\omega) e^{j\omega t} d\omega \quad \text{(inverse transform)}. \tag{A.3}$$

In this appendix, the variables are written as t and ω throughout, and these can have their usual meanings of time and frequency. Other variables can of course be used. In particular, we sometimes have x instead of t and 'wavenumber' β instead of ω. In the following, the use of t and ω specifies whether a forward or inverse transform is involved. In general, $g(t)$ and $G(\omega)$ may both be complex.

Several useful theorems follow directly from the above formulae. A basic property is that the transform is linear, so that the transform of the sum of two functions is the sum of their individual transforms. Useful symmetry theorems are:

(a) If $g(t)$ is even [so that $g(-t) = g(t)$], then $G(\omega)$ is even, and vice versa.
(b) If $g(t)$ is real and even, then $G(\omega)$ is real and even.
(c) If $g(t)$ and $G(\omega)$ are both real, then they are both even.

Transforms of conjugates are given by

$$g^*(t) \leftrightarrow G^*(-\omega); \quad G^*(\omega) \leftrightarrow g^*(-t). \tag{A.4}$$

Thus, if $g(t)$ is real we have $G(-\omega) = G^*(\omega)$, and if $G(\omega)$ is real we have $g(-t) = g^*(t)$. The scaling theorem states that

$$g(t/a) \leftrightarrow |a|G(a\omega), \tag{A.5}$$

where a is a real constant. With $a = -1$ we have $g(-t) \leftrightarrow G(-\omega)$. The shifting theorems are

$$g(t-a) \leftrightarrow e^{-ja\omega}G(\omega); \quad G(\omega - a) \leftrightarrow e^{jat}g(t), \tag{A.6}$$

and the modulation theorems are

$$g(t)\cos(\omega_0) \leftrightarrow \frac{1}{2}[G(\omega + \omega_0) + G(\omega - \omega_0)], \tag{A.7}$$

$$g(t)\sin(\omega_0 t) \leftrightarrow \frac{1}{2}j[G(\omega + \omega_0) - G(\omega - \omega_0)], \tag{A.8}$$

where ω_0 is a real constant. The differentiation theorems are

$$dg(t)/dt \leftrightarrow j\omega G(\omega); \quad dG(\omega)/d\omega \leftrightarrow -jtg(t). \tag{A.9}$$

Parseval's theorem states that

$$\int_{-\infty}^{\infty} g_1(t)g_2(t)dt = \frac{1}{2\pi}\int_{-\infty}^{\infty} G_1(-\omega)G_2(\omega)d\omega, \tag{A.10}$$

where $G_1(\omega)$ and $G_2(\omega)$ are the transforms of $g_1(t)$ and $g_2(t)$. In particular, when $g_1(t) = g_2^*(t)$ we have

$$\int_{-\infty}^{\infty} |g(t)|^2 dt = \frac{1}{2\pi}\int_{-\infty}^{\infty} |G(\omega)|^2 d\omega. \tag{A.11}$$

The convolution of two functions is denoted by an asterisk and is defined by

$$g_1(t) * g_2(t) = \int_{-\infty}^{\infty} g_1(\tau)g_2(t-\tau)d\tau = \int_{-\infty}^{\infty} g_1(t-\tau)g_2(\tau)d\tau$$

$$= g_2(t) * g_1(t). \tag{A.12}$$

Two convolution theorems are

$$g_1(t) * g_2(t) \leftrightarrow G_1(\omega)G_2(\omega), \tag{A.13}$$

$$G_1(\omega) * G_2(\omega) \leftrightarrow 2\pi g_1(t)g_2(t). \tag{A.14}$$

Delta function

The Dirac delta function $\delta(t)$ is defined such that

$$\int_{-\infty}^{\infty} \delta(t-a)g(t)dt = g(a), \tag{A.15}$$

where a is real and the function $g(t)$ is continuous at $t = a$. We can regard $\delta(t)$ as a function that is infinite at $t = 0$ and zero elsewhere. However, $\delta(t)$ is a type of 'generalized function', and not a function in the ordinary sense. Some resulting properties are, with a real,

$$\delta(t-a) * g(t) = g(t-a), \tag{A.16}$$

$$\delta(t-a)g(t) = \delta(t-a)g(a), \tag{A.17}$$

$$\delta(at) = \frac{1}{|a|}\delta(t). \tag{A.18}$$

If $f(t)$ is a continuous function and is zero at $t = a$, then

$$\delta(t-a)f(t) = 0. \tag{A.19}$$

Fourier transforms in the limit

Strictly speaking, two functions $g(t)$ and $G(\omega)$ can be a Fourier transform pair only if the integrals of eqs (A.2) and (A.3) converge. Generally, this requires $g(t)$ to approach zero for $t \to \pm\infty$ and $G(\omega)$ to approach zero for $\omega \to \pm\infty$. However, many functions of practical interest do not satisfy these conditions, for example the sinusoidal function $\cos(\omega_0 t)$. For such cases a special approach can often be used. To illustrate this, consider the output of a spectrum analyzer when the waveform $\cos(\omega_0 t)$ is applied. This gives two peaks at positions corresponding to the frequencies $\pm\omega_0$. The shapes of the peaks correspond to the response of the spectrum analyzer, because the latter has finite bandwidth. Provided the analyzer behaves linearly and has finite resolution, the output is well defined and is easily calculated. Mathematically, it can be shown that this can be expressed in terms of the 'Fourier transform in the limit' of the function $\cos(\omega_0 t)$, and this transform can be shown to be

$$\cos(\omega_0 t) \leftrightarrow \pi\delta(\omega + \omega_0) + \pi\delta(\omega - \omega_0). \tag{A.20}$$

The output is obtained by convolving this transform (with respect to frequency) with a function representing the response of the spectrum analyzer for a single-frequency input. Equation (A.20) thus represents a prescription for calculating the output of any linear spectrum analyzer with finite resolution, for an input $\cos(\omega_0 t)$. It can also be regarded as the output in the limit when the resolution falls to zero.

Many of the Fourier transform formulae, and all those involving the delta function, are valid only in this limiting sense. The formal justification is based on generalized functions, as described for example by Papoulis [1, p. 269].

Some particular transforms

In the following, a and ω_0 are real constants:

$$\sin(\omega_0 t) \leftrightarrow j\pi[\delta(\omega + \omega_0) - \delta(\omega - \omega_0)], \tag{A.21}$$

$$\exp(-at^2) \leftrightarrow \sqrt{\pi/a}\exp[-\omega^2/(4a)], \quad \text{for } a > 0, \tag{A.22}$$

$$\exp(\pm jat^2) \leftrightarrow \sqrt{\pi/a}\exp[\pm j(\pi - \omega^2/a)/4], \quad \text{for } a > 0, \tag{A.23}$$

$$\delta(t - a) \leftrightarrow \exp(-ja\omega), \tag{A.24}$$

$$\delta(\omega - \omega_0) \leftrightarrow \frac{1}{2\pi}\exp(j\omega_0 t). \tag{A.25}$$

The function rect(x) is unity for $|x| < 1/2$ and zero for other x, and for $a > 0$ it gives the transforms

$$\text{rect}(t/a) \leftrightarrow a\,\text{sinc}(a\omega/2); \quad \text{rect}(\omega/a) \leftrightarrow \frac{a}{2\pi}\text{sinc}(at/2), \tag{A.26}$$

where $\text{sinc}(x) \equiv (\sin x)/x$.

The function $\text{sgn}(x) = 1$ for $x > 0$ and $\text{sgn}(x) = -1$ for $x < 0$, and it gives the transforms

$$1/t \leftrightarrow -j\pi\text{sgn}(\omega); \quad 1/\omega \leftrightarrow j\text{sgn}(t)/2. \tag{A.27}$$

The step function $U(x) = 1$ for $x > 0$ and $U(x) = 0$ for $x < 0$, so that

$$U(x) = \frac{1}{2} + \frac{1}{2}\,\text{sgn}(x). \tag{A.28}$$

Using eqs (A.25) and (A.27) we have

$$U(t) \leftrightarrow \pi\delta(\omega) + \frac{1}{j\omega}. \tag{A.29}$$

Using the modulation theorems, eqs (A.7) and (A.8), we have

$$U(\pm t)\exp(\mp j\omega_0 t) \leftrightarrow \pi\delta(\omega \pm \omega_0) \mp \frac{j}{\omega \pm \omega_0}. \tag{A.30}$$

Adding the transforms of eq. (A.30) gives

$$\exp(-j\omega_0|t|) \leftrightarrow \pi\delta(\omega + \omega_0) + \pi\delta(\omega - \omega_0) + \frac{2j\omega_0}{\omega^2 - \omega_0^2}. \tag{A.31}$$

An infinite train of delta functions transforms into a function of the same form:

$$\sum_{n=-\infty}^{\infty} \delta(t - na) \leftrightarrow \frac{2\pi}{|a|}\sum_{m=-\infty}^{\infty} \delta(\omega - 2\pi m/a), \tag{A.32}$$

From this it follows that

$$\sum_{n=-\infty}^{\infty} e^{jnat} = \frac{2\pi}{|a|}\sum_{m=-\infty}^{\infty} \delta(t - 2\pi m/a), \tag{A.33}$$

which is proved by transforming both sides, using eqs (A.25) and (A.32).

A.2 LINEAR FILTERS

We consider a two-port device with an input waveform $v(t)$ and an output waveform $g(t)$. These waveforms represent physical quantities measured at the two ports, and their precise meaning must be assigned before applying the relations below to a practical device. They usually refer to either voltages or currents. The input $v(t)$ is often defined as the voltage which a specified waveform generator would produce across a matched load.

The term *linear* means that the device obeys superposition with regard to different input waveforms. Suppose that an input waveform $v_1(t)$ gives an output waveform $g_1(t)$, while an input waveform $v_2(t)$ gives an output waveform $g_2(t)$. The device is linear if an input waveform $v_1(t) + v_2(t)$ gives an output waveform $g_1(t) + g_2(t)$, irrespective of the forms of $v_1(t)$ and $v_2(t)$. The device is *time invariant* if an input $v(t - \tau)$ gives an output $g(t - \tau)$ for any input waveform $v(t)$ and for any delay τ. The term 'linear filter' used here refers to a device that is both linear and time invariant. Most surface-wave devices can be taken to be linear filters provided the power of the input waveform is not too high. This is usually a good approximation in practice, though non-linear effects are sometimes significant (Section 4.2 in Chapter 4 and Section 7.6.3 in Chapter 7).

It can be assumed here that the input and output waveforms are real. A consequence of linearity and time invariance is that, if the input $v(t)$ is a real sinusoid, then the output $g(t)$ is also a real sinusoid, with the same frequency but generally with a different amplitude and phase. Thus, if $v(t) = \cos \omega_0 t$ for $\omega_0 \geq 0$, then

$$g(t) = A(\omega_0) \cos[\omega_0 t + \phi(\omega_0)], \tag{A.34}$$

where $A(\omega_0)$ and $\phi(\omega_0)$ are functions of frequency but not of time. Since the filter is linear, the input can be written as a sum of two complex exponentials and the output can be regarded as the sum of the responses to these exponentials. We define the *frequency response* $H(\omega)$ such that if $v(t) = \exp(j\omega t)$, then

$$g(t) = H(\omega) \exp(j\omega t). \tag{A.35}$$

This is consistent with eq. (A.34) if

$$H(\pm\omega) = A(\omega) \exp[\pm j\phi(\omega)]. \tag{A.36}$$

This defines the response for positive and negative frequencies. Clearly the negative-frequency components are given by conjugates, so that $H(-\omega) = H^*(\omega)$.

Now consider a general real input waveform $v(t)$, as in Fig. A.1. The Fourier transform of $v(t)$ is $V(\omega)$, which is also called the 'spectrum' of $v(t)$.

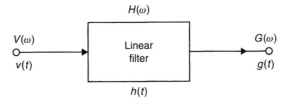

Figure A.1 Linear filter, showing notation for input and output waveforms.

By eq. (A.3), $v(t)$ can be written as a sum of complex exponentials with coefficients $V(\omega)/(2\pi)$, and for each component eq. (A.35) can be used. The corresponding outputs are summed to give the total output $g(t)$. This has a spectrum $G(\omega)$ given by eq. (A.2), and hence

$$G(\omega) = H(\omega)V(\omega). \tag{A.37}$$

Thus the output spectrum can be obtained for any input waveform once $H(\omega)$ is known. The output waveform $g(t)$ can be obtained by transforming $G(\omega)$. Alternatively, $g(t)$ can be obtained from the *impulse response* $h(t)$, defined as the inverse Fourier transform of the frequency response:

$$h(t) \leftrightarrow H(\omega). \tag{A.38}$$

Applying the convolution theorem of eq. (A.13) to eq. (A.37) gives

$$g(t) = v(t) * h(t) = \int_{-\infty}^{\infty} v(\tau)h(t-\tau)d\tau. \tag{A.39}$$

Hence, the output waveform is the convolution of the input waveform with the device impulse response. In the particular case $v(t) = \delta(t)$, the output is equal to $h(t)$, by eq. (A.16). Hence $h(t)$ is the output waveform when a delta function is applied to the input. As expected, $h(t)$ is real; this follows from the symmetry of $H(\omega)$, eq. (A.36). A practical constraint is that $h(t) = 0$ for $t < 0$, since otherwise the device is not causal.

An ideal delay line delays the input waveform by an amount τ_0, say, without distortion. In this case $g(t) = v(t - \tau_0)$, and the device responses are

$$h(t) = \delta(t - \tau_0); \quad H(\omega) = \exp(-j\omega\tau_0).$$

An important conclusion is that if the phase of $H(\omega)$ is proportional to ω, this causes no distortion of the waveform.

If the input waveform is a random noise waveform, the analysis must be treated statistically [4, p. 312]. We assume here that the noise is stationary, that is, its statistical properties are independent of time. It follows that the output noise will also be stationary, and a power spectral density can be defined for both the input and the output noise waveforms. The power of the input noise waveform is denoted by P_i, defined by

$$P_i = E\{[v(t)]^2\}, \tag{A.40}$$

where $E\{\ \}$ denotes the statistical expectation value and $v(t)$ is now a statistical ensemble of functions. The expectation value is independent of t because the noise is stationary. A similar definition gives the output noise power P_o, with $v(t)$ replaced by $g(t)$. The input noise has a spectral density $N_i(\omega)$, and the output noise has a power spectral density $N_o(\omega)$. These spectra are assumed to exist only for $\omega \geq 0$. They are related to the noise powers by

$$P_i = \frac{1}{2\pi} \int_0^\infty N_i(\omega)d\omega; \quad P_o = \frac{1}{2\pi} \int_0^\infty N_o(\omega)d\omega, \tag{A.41}$$

and the two spectral densities are related by

$$N_o(\omega) = N_i(\omega)|H(\omega)|^2. \tag{A.42}$$

Physically, $N_o(\omega)$ can be taken as the power per Hz of bandwidth for the output noise spectral components in the immediate vicinity of frequency ω, and $N_i(\omega)$ is interpreted similarly. This interpretation follows from eqs (A.41 and A.42), taking the linear filter to be a narrow-band device.

In many cases the input waveform is taken to be *white* noise, which has a spectral density $N_i(\omega)$ independent of frequency. Strictly, this implies that the noise power P_i will be infinite. In practice, the spectral density will decay at high frequencies, and the power will be finite. White noise can be defined such that its spectral density is constant up to some high frequency, beyond which the filter response $H(\omega)$ can be taken to be negligible. Equations (A.41) and (A.42) can then be used, with $N_i(\omega)$ independent of ω.

A.3 MATCHED FILTERING

We now consider a waveform $s(t)$ of finite length, accompanied by wide-band noise. If this is applied to a linear filter, the output signal-to-noise ratio will depend on the response of the filter. It is shown here that the output signal-to-noise ratio is maximized if the filter is designed such that its impulse response has a specific form, essentially the time reverse of $s(t)$. Such a filter is a matched filter [5–8]. The terminology refers to the fact that the filter response is matched to the input waveform. Matched filters are used in pulse compression radar systems (Chapter 7, Section 7.1), where the waveform $s(t)$ represents the signal received after reflection from a target. They are also used in spread-spectrum communications (Chapter 7, Section 7.6). It is assumed here that the noise at the filter input is white, though this is not always the case.

The filter output waveform is a linear sum of a waveform $g(t)$, due to the input waveform $s(t)$, plus random noise due to the noise applied at the input. The power of the output noise is P_o. The output signal power is $[g(t)]^2$, and since this varies with time its maximum value is used when defining the output signal-to-noise power ratio (SNR). Denoting this by SNR_o, we define

$$SNR_o = \frac{[g(t)]^2_{max}}{P_o}. \tag{A.43}$$

For an oscillatory waveform this refers to *peak* signal power, not the r.m.s value. The noise power P_o refers to average power. If $H(\omega)$ is the frequency response of the filter and $S(\omega)$ is the spectrum of the input waveform $s(t)$, the output waveform has spectrum $G(\omega) = S(\omega)H(\omega)$, from eq. (A.37). The output signal power is therefore

$$[g(t)]^2 = \left| \frac{1}{2\pi} \int_{-\infty}^{\infty} S(\omega)H(\omega)e^{j\omega t}d\omega \right|^2. \tag{A.44}$$

The input noise is taken to be stationary and white, with spectral density $N_i(\omega)$ per Hz of bandwidth, so that $N_i(\omega)$ is independent of frequency. From eq. (A.41), the output noise power is

$$P_i = \frac{N_i}{4\pi} \int_{-\infty}^{\infty} |H(\omega)|^2 d\omega. \tag{A.45}$$

We now apply Schwartz's inequality which states that, if $A(\omega)$ and $B(\omega)$ are complex functions of ω, then

$$\left| \int_a^b A^*(\omega)B(\omega)d\omega \right|^2 \leq \int_a^b |A(\omega)|^2 d\omega \int_a^b |B(\omega')|^2 d\omega'. \tag{A.46}$$

The equality applies only when $B(\omega) = kA(\omega)$, with k an arbitrary constant. Assuming that $[g(t)]^2$ is maximized at some time $t = t_0$, the output SNR is given by eqs (A.43)–(A.45) with $t = t_0$. Using eq. (A.46), with $A^*(\omega) = S(\omega)$ $\exp(j\omega t_0)$ and $B(\omega) = H(\omega)$, we find

$$\text{SNR}_o \leq \frac{1}{\pi N_i} \int_{-\infty}^{\infty} |S(\omega)|^2 d\omega. \tag{A.47}$$

The energy E_s of the input signal is given by

$$E_s = \int_{-\infty}^{\infty} [s(t)]^2 dt = \int_{-\infty}^{\infty} |S(\omega)|^2 d\omega/(2\pi), \tag{A.48}$$

where the equality of the two forms follows from eq. (A.11). We thus have

$$\text{SNR}_o \leq 2E_s/N_i. \tag{A.49}$$

Thus the maximum value of SNR_o is simply $2E_s/N_i$. This depends only on the signal energy and the noise spectral density, and not on the form of the input signal. The maximum SNR is obtained when the equality applies, that is, when $B(\omega) = kA(\omega)$, and the filter response is then

$$H(\omega) = kS^*(\omega)\exp(-j\omega t_0). \tag{A.50}$$

A filter satisfying this criterion is a matched filter, matched to the waveform $s(t)$. Its impulse response $h(t)$ is the inverse Fourier transform of $H(\omega)$, and from standard Fourier analysis we have

$$h(t) = ks(t_0 - t). \tag{A.51}$$

This is simply the time reverse of $s(t)$, delayed by an amount t_0. The delay in $h(t)$ is of course arbitrary except for causality, that is, $h(t)$ must be zero for $t < 0$.

As for any linear filter, the output is given by the convolution $g(t) = s(t)*h(t)$. Here, this can conveniently be written in terms of the correlation function, $c(t)$, of $s(t)$. This is defined as the convolution of $s(t)$ with $s(-t)$, so that

$$c(t) \equiv s(t) * s(-t) = \int_{-\infty}^{\infty} s(\tau)s(\tau - t)d\tau. \tag{A.52}$$

Using eq. (A.51), we find

$$g(t) = kc(t - t_0), \tag{A.53}$$

so that $g(t)$ is essentially the delayed correlation function of $s(t)$. Setting $\tau = \tau' + t$ in eq. (A.52) we find that $c(-t) = c(t)$, so that $c(t)$ is symmetric about $t = 0$. Hence $g(t)$ is symmetric about the time $t = t_0$ when its power is maximized. The spectrum of the output waveform is $G(\omega) = S(\omega)H(\omega)$, and from eq. (A.50) we have

$$G(\omega) = k|S(\omega)|^2 \exp(-j\omega t_0) = \frac{1}{k}|H(\omega)|^2 \exp(-j\omega t_0). \qquad \text{(A.54)}$$

A.4 NON-UNIFORM SAMPLING

The theory of uniform sampling was described in Chapter 6, Section 6.2. It was shown that a bandpass waveform can be recovered from its sampled version by low-pass filtering, provided the sampling rate is high enough. Here we show that a similar result applies if the sample spacing varies. This is needed for analysis of chirp filters. The analysis is based on that of Tancrell and Holland [9] and others [10, 11].

We first establish a relation concerning delta functions. Consider a function $u(t)$ which has one zero at $t = t_0$, and whose differential $\dot{u}(t)$ is non-zero at this point. We then have [2, p. 95]

$$\delta[u(t)] = \frac{\delta(t - t_0)}{|\dot{u}(t)|}. \qquad \text{(A.55)}$$

This can be proved by substituting $\delta[u(t)]$ for $\delta(t - a)$ in eq. (A.15) and taking u as the independent variable.

Now consider sampling a finite-length continuous waveform $v(t)$. We consider a general case first, and later specialize by taking $v(t)$ to be a chirp waveform. The sampled version of $v(t)$ is $v_s(t)$, a sequence of delta functions at times t_n, given by

$$v_s(t) = v(t) \sum_{n=-\infty}^{\infty} \delta(t - t_n). \qquad \text{(A.56)}$$

The sampling times t_n are to have non-uniform spacing. It is assumed that their values can be obtained from a smooth monotonic non-linear function $\theta(t)$, by solving the equation

$$\theta(t_n) = n\Delta, \quad n = 0, \pm 1, \pm 2, \ldots, \qquad \text{(A.57)}$$

where $\Delta > 0$. The times t_n thus correspond to uniform increments of $\theta(t)$. With $u(t) = \theta(t) - n\Delta$, we obtain from eq. (A.55)

$$\sum_{n=-\infty}^{\infty} \delta(t - t_n) = |\dot{\theta}(t)| \sum_{n=-\infty}^{\infty} \delta[\theta(t) - n\Delta]. \qquad (A.58)$$

Here the delta functions on the right can be expressed as a sum of complex exponentials using eq. (A.33), giving

$$v_s(t) = v(t) \sum_{n=-\infty}^{\infty} \delta(t - t_n) = v(t) \frac{|\dot{\theta}(t)|}{\Delta} \sum_{m=-\infty}^{\infty} \exp[j2\pi m\theta(t)/\Delta]. \qquad (A.59)$$

This shows that the sampled waveform can be expressed as a fundamental, with $m = 0$, plus a series of 'harmonics' with other m. The fundamental has the same form as the original waveform $v(t)$ expect for an amplitude distortion produced by the term $|\dot{\theta}(t)|$. Each 'harmonic' is essentially the original waveform multiplied by a chirp waveform. If the original waveform is band-limited and has finite length, and if the increment Δ is small enough, the frequency band occupied by the fundamental will not overlap the bands occupied by the harmonics. The fundamental component may then be obtained from the sampled waveform by using a low-pass filter to reject the harmonics.

In surface-wave chirp filters, the waveform $v(t)$ to be sampled is itself a chirp waveform. Sampling is nearly always done in synchronism with the waveform, that is, at corresponding points in each cycle. This implies that the function $\theta(t)$ is also the time-domain phase of the waveform, apart from an additive constant. Thus the original chirp waveform can be written as

$$v(t) = a(t) \cos[\theta(t) + \phi_0], \qquad (A.60)$$

where ϕ_0 is an arbitrary constant. The envelope $a(t)$ will have finite length, and it is taken to be a smooth function such that $v(t)$ is a bandpass waveform. Thus, the spectrum of $v(t)$ is non-zero only for $\omega_1 < |\omega| < \omega_2$, where ω_1 and ω_2 are two positive frequencies. We assume that an integer number of samples is taken in each cycle, as done in practical device design. This number is denoted by S_e, and in practice $S_e \geq 2$. The increment Δ is therefore $2\pi/S_e$, and the sampling times t_n are the solutions of $\theta(t_n) = 2\pi n/S_e$. From eqs (A.59) and (A.60), the sampled waveform is

$$
v_s(t) = \frac{S_e a(t)|\dot{\theta}(t)|}{2\pi} \left[\cos[\theta(t) + \phi_0] \right.
$$

$$
\left. + \sum_{m=1}^{\infty} \{\cos[(mS_e + 1)\theta(t) + \phi_0] + \cos[(mS_e - 1)\theta(t) - \phi_0]\} \right].
$$

$$(A.61)$$

We are mainly interested in the fundamental component of $v_s(t)$, which will be written as $\tilde{v}_s(t)$. For $S_e > 2$ the fundamental is obtained by omitting the terms dependent on m, giving

$$
\tilde{v}_s(t) = \frac{1}{2\pi} S_e |\dot{\theta}(t)| a(t) \cos[\theta(t) + \phi_0], \quad \text{for } S_e > 2, \qquad (A.62)
$$

which is essentially the original waveform, eq. (A.60), multiplied by $|\dot{\theta}(t)|$. The harmonic terms that have been omitted have the same form except that the phase $\theta(t)$ in the cosine is replaced by $M\theta(t)$, with the integer $M \geq 2$. For $S_e = 2$ the term in eq. (A.61) involving $(mS_e - 1)\theta(t)$ contributes to the fundamental when $m = 1$, and we find

$$
\tilde{v}_s(t) = \frac{1}{\pi} S_e |\dot{\theta}(t)| a(t) \cos[\theta(t)] \cos \phi_0, \quad \text{for } S_e = 2. \qquad (A.63)
$$

If ϕ_0 is a multiple of 2π, this is essentially the original waveform multiplied by $|\dot{\theta}(t)|$; in this case, the sampling times are at the maxima and minima of the original waveform. For other ϕ_0 a constant phase term is introduced, and this is generally inconsequential. However, ϕ_0 must not be an odd multiple of $\pi/2$ because the samples are then at the zeros of the original waveform, and this gives $\tilde{v}_s(t) = 0$.

A.5 SOME PROPERTIES OF BANDPASS WAVEFORMS

The impulse response of a surface-wave filter is always a bandpass waveform, so that its spectrum is zero at $\omega = 0$. This section gives some properties of bandpass waveforms, relating to the design of surface-wave transducers discussed in Chapter 6, Section 6.2.

The bandpass waveform is $v(t)$, with spectrum $V(\omega)$. We define $\hat{V}(\omega)/2$ as the positive-frequency part of the spectrum, shifted downward in frequency by an amount ω_r, where ω_r is a reference frequency within the band of $V(\omega)$. The positive-frequency part of $V(\omega)$ is $\hat{V}(\omega - \omega_r)/2$, and this is zero for $\omega < 0$. Since $v(t)$ is real, the negative-frequency part is the conjugate of this, so the total spectrum is

$$V(\omega) = \hat{V}(\omega - \omega_r)/2 + \hat{V}^*(-\omega - \omega_r)/2. \qquad (A.64)$$

The complex envelope of the waveform $v(t)$ is denoted by $\hat{v}(t)$. This is defined as the Fourier transform of $\hat{V}(\omega)$. Using the shifting theorem, eq. (A.6), the transform of $\hat{V}(\omega - \omega_r)$ is $\hat{v}(t)\exp(j\omega_r t)$. This gives the positive-frequency part of $v(t)$. The negative-frequency part is the complex conjugate, so we have

$$v(t) = \frac{1}{2}\hat{v}(t)\exp(j\omega_r t) + \frac{1}{2}\hat{v}^*(t)\exp(-j\omega_r t). \qquad (A.65)$$

The complex envelope is written as

$$\hat{v}(t) = \hat{a}(t)\exp[j\hat{\theta}(t)], \qquad (A.66)$$

where $\hat{a}(t)$ and $\hat{\theta}(t)$ are the amplitude and phase. Substituting into eq. (A.65), the bandpass waveform can be written as

$$v(t) = \hat{a}(t)\cos[\omega_r t + \hat{\theta}(t)]. \qquad (A.67)$$

Linear phase in the frequency domain

In surface-wave transducer design, it is often a requirement that the spectrum $V(\omega)$ should have a phase linear with frequency. Here, we consider what constraints this places on $v(t)$. If $V(\omega)$ has linear phase, then $\hat{V}(\omega)$ also has linear phase. We take its phase to be $\theta_c - \omega t_0$, where θ_c and t_0 are constants. Thus $\hat{V}(\omega) = |\hat{V}(\omega)|\exp[j(\theta_c - \omega t_0)]$, and we have

$$|\hat{V}(\omega)| = \hat{V}(\omega)\exp[-j(\theta_c - \omega t_0)] \leftrightarrow \exp(-j\theta_c)\hat{v}(t + t_0), \qquad (A.68)$$

where the transform has been done using the shifting theorem, eq. (A.6). The left side of eq. (A.68) is real, so its transform must conjugate if the sign of t is reversed. This gives $\hat{v}(t_0 - t) = \hat{v}^*(t_0 + t)\exp(2j\theta_c)$. Using eq. (A.66) for $\hat{v}(t)$, we find

$$\hat{a}(t_0 - t)/\hat{a}(t_0 + t) = \exp\{j[2\theta_c - \hat{\theta}(t_0 + t) - \hat{\theta}(t_0 - t)]\}. \qquad (A.69)$$

Here the left side is real, so the right side must be ± 1 and it can be written as $\exp(jn\pi)$. Hence $\hat{a}(t_0 - t) = \pm\hat{a}(t_0 + t)$, so that $\hat{a}(t)$ is either symmetric or antisymmetric about $t = t_0$. The phase $\hat{\theta}(t)$ equals a constant $\theta_c - n\pi/2$ plus a function antisymmetric about time t_0.

Amplitude modulation

Consider an amplitude-modulated waveform, with linear time-domain phase, given by

$$v(t) = \hat{a}(t)\cos(\omega_r t + \theta_c), \tag{A.70}$$

where θ_c is a constant and ω_r is the carrier frequency. The form of this waveform places constraints on its spectrum $V(\omega)$. The amplitude $\hat{a}(t)$ is real and the complex envelope is $\hat{v}(t) = \hat{a}(t)\exp(j\theta_c)$. Hence $\hat{V}(\omega) = \hat{A}(\omega)\exp(j\theta_c)$, where $\hat{A}(\omega)$ is the transform of $\hat{a}(t)$. Since $\hat{a}(t)$ is real we have $\hat{A}(-\omega) = \hat{A}^*(\omega)$, and hence

$$\hat{V}(-\omega) = \hat{V}^*(\omega)\exp(2j\theta_c). \tag{A.71}$$

For positive frequencies, we have $V(\omega) = \hat{V}(\omega - \omega_r)/2$. Using eq. (A.71) we find

$$V(\omega_r - \omega) = V^*(\omega_r + \omega)\exp(2j\theta_c), \quad \text{for } |\omega| < \omega_r. \tag{A.72}$$

If $V(\omega) = A(\omega)\exp[j\phi(\omega)]$, we find, for $|\omega| < \omega_r$,

$$\phi(\omega_r - \omega) + \phi(\omega_r + \omega) = 2\theta_c - n\pi,$$

and

$$A(\omega_r - \omega) = A(\omega_r + \omega)e^{jn\pi} = \pm A(\omega_r + \omega).$$

Thus, the amplitude $A(\omega)$ is symmetric or antisymmetric about the carrier frequency ω_r. The phase $\phi(\omega)$ equals a constant $\theta_c - n\pi/2$, plus a function antisymmetric about ω_r.

Sampling of amplitude-modulated waveforms

In Chapter 6, Section 6.2 it was shown that a band-limited waveform can be sampled uniformly in such a way that the original waveform can be recovered by low-pass filtering. In general, this requires the sampling frequency ω_s to be above the Nyquist frequency $2\omega_2$, where ω_2 is such that the original waveform has negligible energy for $\omega > \omega_2$. Here it is shown that, for an amplitude-modulated waveform, the sampling frequency may be *below* the Nyquist frequency.

Suppose that $v(t)$ is a bandpass waveform whose spectrum $V(\omega)$ is negligible except for $\omega_1 < |\omega| < \omega_2$. Sampling $v(t)$ at times $t = n\tau_s$ gives a sampled

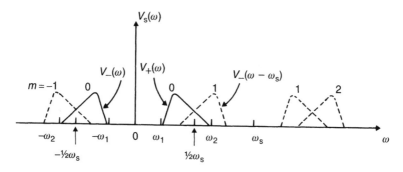

Figure A.2 Effect of sampling a bandpass waveform, with sampling frequency below the Nyquist frequency.

waveform $v_s(t)$, so that

$$v_s(t) = v(t) \sum_{n=-\infty}^{\infty} \delta(t - n\tau_s).$$

The spectrum of $v_s(t)$ is $V_s(\omega)$, and from eq. (6.16) this is given by

$$V_s(\omega) = \frac{\omega_s}{2\pi} \sum_{m=-\infty}^{\infty} V(\omega - m\omega_s), \qquad (A.73)$$

with $\omega_s = 2\pi/\tau_s$. If the sampling frequency ω_s is below the Nyquist frequency, there will generally be some overlap of the original spectrum $V(\omega)$ and the image spectra due to the sampling. This is shown in Fig. A.2, where ω_s is assumed to be between ω_1 and ω_2 and the terms with $m \neq 0$ are shown by broken lines.

Define $U(\omega)$ as the part of the spectrum $V_s(\omega)$ below frequency ω_s, so that $U(\omega) = V_s(\omega)$ for $|\omega| < \omega_s$ and $U(\omega) = 0$ for $|\omega| > \omega_s$. This function can be obtained from $V_s(\omega)$ by low-pass filtering. Defining $V_+(\omega)$ as the positive-frequency part of $V(\omega)$, and $V_-(\omega)$ as the negative-frequency part, we have

$$V(\omega) = V_+(\omega) + V_-(\omega). \qquad (A.74)$$

Using eq. (A.73), the positive-frequency part of $U(\omega)$ is given by

$$\begin{aligned} 2\pi U(\omega)/\omega_s &= V_+(\omega) + V_-(\omega - \omega_s) \\ &= V_+(\omega) + V_+^*(\omega_s - \omega), \quad \text{for } \omega > 0. \end{aligned} \qquad (A.75)$$

Here the second form follows because $v(t)$ is real, implying that $V_-(\omega) = V_+^*(-\omega)$. The original waveform $v(t)$ can be recovered from the sampled waveform if $U(\omega)$ is proportional to the original spectrum $V(\omega)$. We therefore consider the case

$$2\pi U(\omega)/\omega_s = KV_+(\omega), \quad \text{for } \omega > 0, \tag{A.76}$$

where K is a constant which may be complex. For consistency with eq. (A.75) there must be a constraint on $V_+(\omega)$, and therefore on the waveform $v(t)$. Writing $\omega = \omega_s/2 - \Delta\omega$ we find

$$(K - 1)V_+(\omega_s/2 - \Delta\omega) = V_+^*(\omega_s/2 + \Delta\omega). \tag{A.77}$$

By setting $\Delta\omega = 0$ we see that $|K - 1| = 1$. We can set $K - 1 = \exp(-2j\alpha_c)$, where the constant α_c is the phase of $V_+(\omega_s/2)$, and eq. (A.77) becomes

$$V_+(\omega_s/2 - \Delta\omega) = V_+^*(\omega_s/2 + \Delta\omega)\exp(2j\alpha_c). \tag{A.78}$$

This equation has the form of eq. (A.72), with ω_r replaced by $\omega_s/2$ and θ_c replaced by α_c. Hence, $v(t)$ must be an amplitude-modulated waveform, with carrier frequency $\omega_s/2$. Comparison with eq. (A.77) shows that

$$v(t) = \hat{a}(t)\cos(\omega_s t/2 + \alpha_c), \tag{A.79}$$

where the envelope $\hat{a}(t)$ is not determined. This shows that, if an amplitude-modulated waveform is sampled using a sampling frequency ω_s equal to twice the carrier frequency, the sampled waveform has the same spectrum as the original in the region $|\omega| < \omega_s$, apart from a complex constant. From eq. (A.76), the constant is $K\omega_s/(2\pi)$, and K is given by

$$K = 2\exp(-j\alpha_c)\cos\alpha_c. \tag{A.80}$$

If the sampled waveform is applied to an ideal low-pass filter which rejects components with frequencies $|\omega| > \omega_s$, the resulting waveform $u(t)$ is the transform of $U(\omega)$. This is found to be

$$u(t) = (\omega_c/\pi)\hat{a}(t)\cos(\omega_s t/2)\cos\alpha_c. \tag{A.81}$$

This is the same as the original waveform $v(t)$, eq. (A.79), except for changes of amplitude and phase. The angle α_c must not be an odd multiple of $\pi/2$, since this gives $u(t) = 0$. The reason for this is that the sampling points were taken to be at $t = n\tau_s = 2n\pi/\omega_s$, and for $\alpha_c = \pi/2$ the waveform is zero at these points. The waveform is usually sampled at the maxima and minima, in which case $\alpha_c = n\pi$ and $K = 2$.

A.6 HILBERT TRANSFORMS

In Chapter 5, Section 5.5.1 it was shown that a uniform non-reflective transducer has conductance approximately given by $G_a(\omega) \approx [(\sin X)/X]^2$, where
$X = \pi N_p(\omega - \omega_0)/\omega_0$ and a constant has been omitted. Here we derive the corresponding susceptance $B_a(\omega)$, which is the Hilbert transform of $G_a(\omega)$. We
need to convolve $G_a(\omega)$ with $-1/(\pi\omega)$. Since the Fourier transform of $1/\omega$ is
$j\,\text{sgn}(t)/2$, we can transform $G_a(\omega)$ to the time domain, multiply by $-j\,\text{sgn}(t)$
and then transform back.

Consider first the function $G(\omega) = [\text{sinc}(a\omega/2)]^2$, with Fourier transform $g(t)$.
The transform of $\text{sinc}(a\omega/2)$ is $(1/a)\,\text{rect}(t/a)$, by eq. (A.26). From the convolution theorem, the transform of $G(\omega)$ is $g(t) = (1/a^2)\,\text{rect}(t/a)*\text{rect}(t/a)$. This
function is easily seen to be a triangle of length $2a$ and height $1/a$, so that
$g(t) = (a - |t|)/a^2$ for $|t| < a$, and $g(t) = 0$ for other t. Defining $B(\omega)$ as the
Hilbert transform of $G(\omega)$, the Fourier transform of $B(\omega)$ is $b(t) = -j\,\text{sgn}(t)g(t)$,
giving

$$b(t) = -j[a\,\text{sgn}(t) - t]/a^2 \quad \text{for } |t| < a, \tag{A.82}$$

with $b(t) = 0$ for other t. This is transformed back to the ω domain to give $B(\omega)$,
using integration by parts for the second term. This gives

$$B(\omega) = G(\omega)^* \, [-1/(\pi\omega)] = 2[\sin(a\omega) - a\omega)]/(a^2\omega^2). \tag{A.83}$$

Using eq. (A.12) we find that the Hilbert transform of $G(\omega - \omega_0)$ is

$$G(\omega - \omega_0)^* \, [-1/(\pi\omega)] = B(\omega - \omega_0). \tag{A.84}$$

Substituting $\omega - \omega_0$ for ω in eq. (A.83), and setting $a = 2\pi N_p/\omega_0$, we find that
$G_a(\omega) = G(\omega - \omega_0) = [(\sin X)/X]^2$, and

$$B_a(\omega) = B(\omega - \omega_0) = \frac{\sin(2X) - 2X}{2X^2}, \tag{A.85}$$

which is the required result.

Numerical transformation

The above transform provides a convenient method for evaluating $B_a(\omega)$
from a general conductance $G_a(\omega)$. Numerically, $G_a(\omega)$ will normally be
expressed at frequencies ω_n which have constant spacing $\Delta\omega$, say. The
continuous function can be approximated as $G_a(\omega) \approx \sum_n G_a(\omega_n)\,\text{sinc}^2 X_n$,

where $X_n = \pi(\omega - \omega_n)/\Delta\omega$. In this series, the nth term is equal to $G_a(\omega)$ when $\omega = \omega_n$, and it is zero for all other data frequencies ω_m. Hence, the series is exactly correct at the data points. For each term the Hilbert transform is given by eq. (A.85), so B_a is obtained by a simple substitution.

REFERENCES

1. A. Papoulis. *The Fourier Integral and Its Applications*, McGraw-Hill, 1962.
2. R.M. Bracewell. *The Fourier Transform and Its Applications*, McGraw-Hill, 1965.
3. A. Papoulis. *Systems and Transforms with Applications in Optics*, McGraw-Hill, 1968.
4. A. Papoulis. *Signal Analysis*, McGraw-Hill, 1977.
5. J.R. Klauder, A.C. Price, S. Darlington and W.J. Albersheim. 'The theory and design of chirp radars', *Bell Syst. Tech. J.*, **39**, 745–808 (1960).
6. M.I. Skolnik. *Introduction to Radar Systems*, McGraw-Hill, 1962.
7. C.E. Cook and M. Bernfeld. *Radar Signals*, Academic Press, 1967.
8. G.L. Turin. 'An introduction to matched filters', *IRE Trans.*, **IT-6**, 311–329 (1960).
9. R.H. Tancrell and M.G. Holland. 'Acoustic surface wave filters', *Proc. IEEE*, **59**, 393–409 (1971).
10. W.R. Smith, H.M. Gerard and W.R. Jones. 'Analysis and design of dispersive interdigital surface-wave transducers', *IEEE Trans.*, **MTT-20**, 458–471 (1972).
11. C. Atzeni and L. Masotti. 'Linear signal processing by acoustic surface-wave transversal filters', *IEEE Trans.*, **MTT-21**, 505–519 (1973).

Appendix B

RECIPROCITY

The theory of elastic piezoelectric materials can be use to derive a general reciprocity relation, with a wide range of applications. This appendix discusses the application of reciprocity to transducers consisting of electrodes on the surface of a piezoelectric solid. The main consequence is that the processes of launching and receiving surface waves are mathematically related. We first consider a generalized geometry, and then later discuss surface waves on a half-space.

Throughout this appendix it is assumed that the surface is force free, so that the electrodes cause no mechanical loading.

B.1 GENERAL RELATION FOR A MECHANICALLY FREE SURFACE

We consider a homogeneous insulating piezoelectric material, governed by the piezoelectric equations and Newton's laws of motion. Any disturbance in the material gives rise to an elastic displacement $\mathbf{u}$, stress $\mathbf{T}$, potential Φ and electric displacement $\mathbf{D}$, all functions of coordinates and of time. These are assumed to satisfy the constitutive relations of the material and to be proportional to $\exp(j\omega t)$, with frequency ω considered as a constant. A second solution satisfying the same constraints is denoted by the functions $\mathbf{u}'$, $\mathbf{T}'$, Φ' and $\mathbf{D}'$. These solutions are related at each point by the *real reciprocity relation*, written as

$$\mathrm{div}[\{\mathbf{u} \cdot \mathbf{T}'\} - \{\mathbf{u}' \cdot \mathbf{T}\} + \Phi\mathbf{D}' - \Phi'\mathbf{D}] = 0, \qquad (B.1)$$

where the vector $\{\mathbf{u} \cdot \mathbf{T}'\}$ is defined such that its x_j component is given by

$$\{\mathbf{u} \cdot \mathbf{T}'\}_j = \sum_{i=1}^{3} u_i T'_{ij}, \quad j = 1, 2, 3. \qquad (B.2)$$

Equation (B.1) is valid provided there are no mechanical or electrical sources within the material, and in particular this means that there are no free charges.

The derivation, given by Auld [1, p. 153], uses the piezoelectric constitutive relations with Maxwells's equations and Newton's laws of mechanics. The electric field is taken to be quasi-static, so that $\mathbf{E} = -\text{grad } \Phi$. The equation is valid if the material is lossy.

The solid is assumed to be enclosed by a mechanically free surface S, with a vacuum in the space outside this surface. The integral of eq. (B.1) over the volume V of the solid is related to an integral over the surface S by the divergence theorem

$$\int_V \text{div } \mathbf{A} \, dV = \int_S \mathbf{A} \cdot \mathbf{n} \, dS,$$

where $\mathbf{n}$ is the outward-directed normal to the surface, with magnitude unity. For a mechanically free surface, with no forces, it can be shown [1] that $\{\mathbf{u} \cdot \mathbf{T}'\} = \{\mathbf{u}' \cdot \mathbf{T}\} = 0$. Thus, integration of eq. (B.1) over the volume yields

$$\int_S (\Phi \mathbf{D}' - \Phi' \mathbf{D}) \cdot \mathbf{n} \, dS = 0. \tag{B.3}$$

Note that no electrical boundary condition has been applied at this stage.

The normal component of displacement at the surface ($\mathbf{D} \cdot \mathbf{n}$ or $\mathbf{D}' \cdot \mathbf{n}$) can be related to the charge density on a set of electrodes on the surface. The free charges must however be outside the surface because eq. (B.1) is valid only if there are no sources. To comply with this, the electrodes are assumed to be separated from the surface by an infinitesimal gap. Thus $\mathbf{D} \cdot \mathbf{n}$ is equal to $-\sigma$, where σ is the charge density on the adjacent electrode. If σ' is the charge density corresponding to the displacement $\mathbf{D}'$, eq. (B.3) becomes

$$\int_S (\Phi \sigma' - \Phi' \sigma) \, dS = 0. \tag{B.4}$$

In this equation σ and σ' are the charge densities on the sides of the electrodes adjacent to the piezoelectric. There will also be charges on the vacuum side. The reciprocity argument can be applied to the vacuum as well as the piezoelectric, so that eq. (B.4) is valid if σ and σ' are charges on the vacuum side. It follows that the equation is also valid if σ and σ' are the total charge densities, including both sides. In the following description, this is taken to be the case.

B.2   RECIPROCITY FOR TWO-TERMINAL TRANSDUCERS

For the type of transducer considered here, each electrode is connected to one of two terminals. Transducers with more than two terminals may be analyzed by using results for two-terminal transducers, in conjunction with superposition. The resistivity of the electrodes is assumed to be negligible.

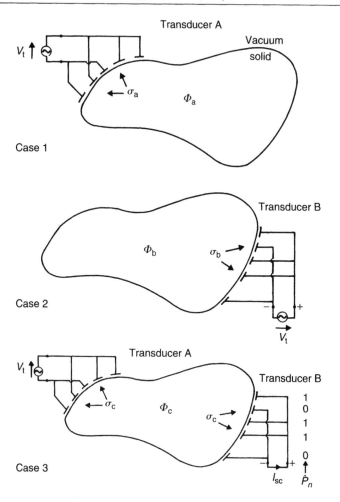

Figure B.1 Two-terminal transducers: reciprocity.

We consider a piezoelectric solid with three different transducer configurations, as shown in Fig. B.1. In case 1, transducer A is placed on the surface and a voltage V_t is applied across it. The voltage is proportional to $\exp(j\omega t)$. A charge density σ_a appears on the electrodes and the potential, specified everywhere, is Φ_a. The transducer may radiate acoustic waves of various types.

In case 2, transducer B is placed on the surface and a voltage V_t is applied, with transducer A absent. The charge density on the electrodes is σ_b and the potential is Φ_b. In case 3 both transducers are present. A voltage V_t is applied to transducer A, while transducer B is shorted and produces a current I_{sc}.

The potential is Φ_c and the charge density is σ_c, which includes charges on both transducers.

It is assumed that the launching process can be analyzed, so that the functions Φ_a, σ_a, Φ_b and σ_b can be found. Reciprocity enables the current I_{sc} in the receiving transducer to be expressed in terms of these functions. Using reciprocity in the form of eq. (B.4) we have

$$\int_S \Phi_c \sigma_b \, dS = \int_S \Phi_b \sigma_c \, dS \tag{B.5}$$

and

$$\int_S \Phi_a \sigma_b \, dS = \int_S \Phi_b \sigma_a \, dS \tag{B.6}$$

where it is assumed that any fields generated by the leads connecting the transducers can be ignored. Now, in case 3, Φ_c has the same value at all points on the electrodes of transducer B, because this transducer is shorted. In case 2, σ_b is zero everywhere except on the electrodes of transducer B. We therefore have

$$\int_S \Phi_c \sigma_b = \text{const.} \int_S \sigma_b \, dS = 0,$$

since the total charge on transducer B is zero. Equation (B.5) therefore gives

$$\int_S \Phi_b \sigma_c \, dS = 0. \tag{B.7}$$

It is now assumed that, for case 3, the presence of transducer B does not affect the charge density on transducer A. Thus, on transducer A we have $\sigma_c = \sigma_a$. This is usually justified in practice. The coupling between the two transducers is of two types, electrostatic and acoustic. Electrostatic coupling causes a 'feedthrough' signal to appear at the terminals of the receiving transducer. This signal is assumed to be small by design. Hence, resulting perturbations of the charge density on the *launching* transducer due to the presence of the *receiving* transducer can be ignored. Acoustic coupling can occur because transducer B generates acoustic waves when the waves from transducer A are incident on it. However, in cases that we are concerned with this effect is usually small when the receiver is shorted, as it is here. In any case, the result is to produce multiple-transit signals which are easily identified because they have differing delays.

Thus, assuming the charge density on transducer A is not affected by the presence of transducer B, we have $\sigma_c = \sigma_a$ on transducer A. Define σ_{cb} as the part of σ_c on the electrodes of transducer B, so that $\sigma_{cb} = 0$ except on transducer B. Then $\sigma_c = \sigma_a + \sigma_{cb}$, and eq. (B.7) gives

$$\int_S \Phi_b \sigma_{cb}\, dS = -\int_S \Phi_b \sigma_a\, dS. \tag{B.8}$$

To find the current I_{sc} produced by transducer B the electrodes are labeled $n = 1, 2, \ldots$, with areas S_n. The polarity of electrode n is designated by $\hat{P}_n$, with value 1 if it is connected to the positive terminal, and 0 if it is connected to the negative terminal, as in Fig. B.1. When a voltage V_t is applied, as in case 2, the potential of electrode n is $(\hat{P}_n - 1/2)\,V_t$, apart from a constant which does not affect the charge density. In case 3, the total charge on electrodes connected to the positive terminal is

$$Q_+ = \sum_n \hat{P}_n \int_{S_n} \sigma_{cb}\, dS,$$

and the total charge on electrodes connected to the negative terminal is

$$Q_- = \sum_n (1 - \hat{P}_n) \int_{S_n} \sigma_{cb}\, dS,$$

where S_n is the area of electrode n. The current is $I_{sc} = j\omega Q_+$, and since the total charge is zero we have $Q_+ = (Q_+ - Q_-)/2$, giving

$$I_{sc} = j\omega \sum_n \int_{S_n} (\hat{P}_n - 1/2)\sigma_{cb}\, dS. \tag{B.9}$$

Now, for case 2 we have $\Phi_b = (\hat{P}_n - 1/2)V_t$ on electrode n of transducer B. Also, σ_{cb} is zero except on these electrodes so, in eq. (B.9), $(\hat{P}_n - 1/2)$ may be replaced by Φ_b/V_t. This gives

$$I_{sc} = \frac{j\omega}{V_t} \int_S \Phi_b \sigma_{cb}\, dS.$$

Finally, using eqs (B.8) and (B.6) we have

$$V_t I_{sc} = -j\omega \int_S \Phi_b \sigma_a\, dS = -j\omega \int_S \Phi_a \sigma_b\, dS. \tag{B.10}$$

This is the required result, giving the output I_{sc} of the receiving transducer in terms of functions obtained by analysis of launching transducers. Analysis of the launching process also gives the impedance of transducer B, so the output voltage and current produced by this transducer can be calculated for any load impedance.

A convenient modification of eq. (B.10) is obtained by introducing the functions $\rho_a = \sigma_a/V_t$ and $\rho_b = \sigma_b/V_t$. Here ρ_a is the charge density on transducer A when unit voltage is applied, with no other electrodes present. Equation (B.10) can be written as

$$I_{sc} = -j\omega \int_S \Phi_b \rho_a \, dS = -j\omega \int_S \Phi_a \rho_b \, dS. \qquad (B.11)$$

B.3 SYMMETRY OF THE GREEN'S FUNCTION

If the charge density σ is specified at all points on a surface S enclosing the solid, this determines the potential Φ everywhere, apart from a constant term which is ignored here because it does not affect the charge density. This linear relationship can be expressed in terms of a Green's function $G_1(\mathbf{r}; \mathbf{r}')$, defined as the potential at $\mathbf{r}$ due to the charge density at $\mathbf{r}'$, so that

$$\Phi(\mathbf{r}) = \int_S G_1(\mathbf{r}; \mathbf{r}')\sigma(\mathbf{r}')dS'. \qquad (B.12)$$

The subscript distinguishes this from the Green's function $G(x, \omega)$ for a half-space (Chapter 3). Using an argument given by Auld [1, p. 366] we show that the Green's function is symmetrical. The equations involve values of the potential $\Phi(\mathbf{r})$ only at the surface, so in the following the points $\mathbf{r}$ and $\mathbf{r}'$ are both on the surface.

Consider two solutions, one with potential $\Phi_1(\mathbf{r})$ and charge density $\sigma_1(\mathbf{r})$ and the other with potential $\Phi_2(\mathbf{r})$ and charge density $\sigma_2(\mathbf{r})$. Using reciprocity in the form of eq. (B.4) we have

$$\int_S \Phi_1(\mathbf{r})\sigma_2(\mathbf{r}) \, dS = \int_S \Phi_2(\mathbf{r})\sigma_1(\mathbf{r}) \, dS.$$

Using eq. (B.12) for the two potentials gives

$$\iint_S \sigma_2(\mathbf{r})G_1(\mathbf{r}; \mathbf{r}')\sigma_1(\mathbf{r}') \, dS' \, dS = \iint_S \sigma_1(\mathbf{r})G_1(\mathbf{r}; \mathbf{r}')\sigma_2(\mathbf{r}')dS' \, dS.$$

If we interchange $\mathbf{r}$ and $\mathbf{r}'$ in the integral on the right, it becomes the same as the integral on the left except that $G_1(\mathbf{r}; \mathbf{r}')$ is replaced by $G_1(\mathbf{r}'; \mathbf{r})$. Since the integrals must be equal for any choice of $\sigma_1(\mathbf{r})$ and $\sigma_2(\mathbf{r})$, we conclude that the two forms of the Green's function must be equal, that is

$$G_1(\mathbf{r}'; \mathbf{r}) = G_1(\mathbf{r}; \mathbf{r}'). \qquad (B.13)$$

The Green's function is therefore symmetrical.

B.4   RECIPROCITY FOR SURFACE EXCITATION OF A HALF-SPACE

We now consider a half-space with its plane force-free surface normal to the z-axis, and assume that there are no variations in the y-direction, so that the surface potential and charge density are functions of x only. The surface potential is denoted by $\phi(x)$ and the charge density, which exists only at the surface, is $\sigma(x)$. Equation (B.12) thus becomes

$$\phi(x) = \int_{-\infty}^{\infty} G_1(x; x')\sigma(x')\mathrm{d}x'. \tag{B.14}$$

For a half-space the relation between $\phi(x)$ and $\sigma(x)$ must be unchanged if the origin for the x-axis is displaced, and hence $G_1(x; x')$ depends only on the distance between x and x'. This is expressed by defining a new Green's function $G(x)$, such that

$$G_1(x; x') = G(x - x'), \tag{B.15}$$

and hence

$$\phi(x) = \int_{-\infty}^{\infty} G(x - x')\sigma(x')\mathrm{d}x' = G(x)^*\sigma(x),$$

so that $G(x)$ is the same as the Green's function introduced in Chapter 3, Section 3.4. Comparing eqs (B.13) and (B.15) shows that $G(x)$ is symmetrical:

$$G(-x) = G(x). \tag{B.16}$$

It follows that the Fourier transform of $G(x)$, denoted $\overline{G}(\beta)$, is also symmetrical, so that $\overline{G}(-\beta) = \overline{G}(\beta)$. In addition, the effective permittivity $\varepsilon_s(\beta)$ equals $1/[\,|\beta|\,\overline{G}(\beta)]$ from eq. (3.33), and hence

$$\varepsilon_s(-\beta) = \varepsilon_s(\beta). \tag{B.17}$$

B.5   RECIPROCITY FOR SURFACE-WAVE TRANSDUCERS

In general, a transducer on the surface of a half-space may generate surface waves and bulk waves, and it will also produce an electrostatic potential. For such cases, reciprocity may be used in the form of eq. (B.11). However, for devices in which the coupling between transducers is predominantly due to surface waves some additional formulae can be derived.

We first consider reception of surface waves by a two-terminal transducer, as shown on Fig. B.2. This represents the same situation as case 3 of Fig. B.1, but with the transducers on the plane surface of a half-space. The aperture W is

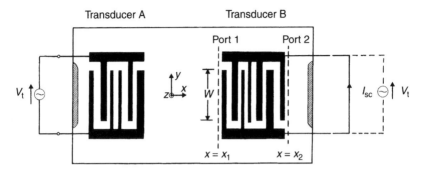

Figure B.2 General two-terminal transducers on a plane surface.

assumed to be large, so that variations in the y-direction can be ignored. From eq. (B.11), the short-circuit current produced by transducer B is

$$I_{sc} = -j\omega W \int_{-\infty}^{\infty} \phi_a(x)\rho_b(x)dx. \tag{B.18}$$

Here $\phi_a(x)$ is the surface potential produced by transducer A when a voltage V_t is applied to it, assuming transducer B to be absent. The function $\rho_b(x)$ is the charge density on transducer B for unit applied voltage, with transducer A absent. Generally, these functions also depend on the frequency ω.

In general, $\phi_a(x)$ includes bulk-wave and electrostatic terms, but these decay with distance. We assume that transducer A is generating surface waves and that the transducer separation is large, so that in the region of transducer B the bulk-wave and electrostatic terms are negligible in comparison with the surface-wave term. Thus, in the region of transducer B, $\phi_a(x)$ has the form

$$\phi_a(x) = \phi_0 \exp s(-jk_f x), \tag{B.19}$$

where ϕ_0 is a constant and k_f is the free-surface wavenumber for surface waves, taken to be real. This is substituted into eq. (B.18). If $\overline{\rho}_b(\beta)$ is the Fourier transform of $\rho_b(x)$, this gives

$$I_{sc} = -j\omega W \phi_0 \overline{\rho}_b(k_f). \tag{B.20}$$

This can also be expressed in terms of the surface-wave amplitude at the input port, port 1, of transducer B. The port is defined as the line at $x = x_1$ near the left edge of the transducer, as in Fig. B.2. The exact location of this line is immaterial. Defining ϕ_{i1} as the potential of the incident wave at this port, eq. (B.19) gives $\phi_{i1} = \phi_0 \exp(-jk_f x_1)$, and the output current is

$$I_{sc} = -j\omega W \phi_{i1} \overline{\rho}_b(k_f) \exp(jk_f x_1). \tag{B.21}$$

This relation is valid even if the transducer couples to bulk waves; it is only necessary to assume that the incident wave has no bulk-wave or electrostatic terms.

The same method can be used to deduce the current produced by transducer B when surface waves are incident from the right instead of from the left. In this case eq. (B.18) is still valid, but transducer A must be taken to be on the right of transducer B. The input port of transducer B is now port 2, at $x = x_2$, and the potential of the incident wave at this port is ϕ_{i2}. The current produced by transducer B is

$$I_{sc} = -j\omega W \phi_{i2} \bar{\rho}_b(-k_f) \exp(-jk_f x_2). \tag{B.22}$$

Relation between launching and reception

The amplitudes of the surface waves generated by an isolated transducer are derived in Section B.6 below, assuming the wave to be a piezoelectric Rayleigh wave. If $\sigma(x)$ is the charge density on the transducer, with Fourier transform $\bar{\sigma}(\beta)$, the potential of the wave launched in the $-x$-direction is

$$\phi_s(x) = j\Gamma_s \bar{\sigma}(k_f) \exp(jk_f x), \tag{B.23}$$

where Γ_s is defined in Section B.6. Now suppose that a voltage V_t is applied to the two-terminal transducer B of Fig. B.2, with transducer A absent. For unit applied voltage the charge density is $\rho_b(x)$, with Fourier transform $\bar{\rho}_b(\beta)$, so that $\bar{\sigma}(k_f) = V_t \bar{\rho}_b(k_f)$. The transducer generates a surface wave with potential $\phi_s(x)$, and we define ϕ_{s1} as the potential at port 1, so that $\phi_{s1} = \phi_s(x_1)$. We thus have

$$\phi_{s1} = j\Gamma_s V_t \bar{\rho}_b(k_f) \exp(jk_f x_1) \tag{B.24}$$

for the wave launched at port 1. For surface waves incident on port 1 the current produced, when the transducer is shorted, is given by eq. (B.21). Comparing with eq. (B.24) we have

$$\left[\frac{I_{sc}}{\phi_{i1}} \right]_{\text{receive}} = -\frac{\omega W}{\Gamma_s} \left[\frac{\phi_{s1}}{V_t} \right]_{\text{launch}}. \tag{B.25}$$

This equation relates the reception and launching of surface waves at port 1. The reader is reminded that the derivation assumed the electrodes to have negligible mechanical loading and negligible resistivity. The equation is valid if the transducer couples to bulk waves as well as surface waves. If so, ϕ_{s1} is the surface-wave component of the potential at port 1 for the launching process, and for reception it is assumed that the incident wave is a surface wave, with no bulk-wave component. The same relation applies for port 2, with suitable changes of the variables.

A generalization to include the effects of mechanical loading is given in Appendix D, Section D.2.

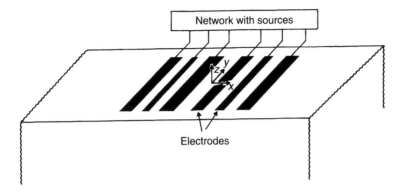

Figure B.3 General transducer on a half-space.

B.6 SURFACE-WAVE GENERATION

This section derives the formula for the amplitude of surface waves generated by a transducer, given in eq. (B.23). The formula follows from the nature of the effective permittivity $\varepsilon_s(\beta)$, and therefore assumes that there is no mechanical loading. We also assume that the wave is a piezoelectric Rayleigh wave, so that $\varepsilon_s(\beta)$ is zero at $\beta = k_f$, the free-surface wavenumber, and it is real when β is close to k_f.

Consider a set of electrodes on the surface, connected to some electrical network that includes at least one current or voltage source, as in Fig. B.3. The electrode edges are parallel to the y-axis, and the fields are assumed to be invariant in the y-direction. The surface potential $\phi(x)$ and $\sigma(x)$ have Fourier transforms $\overline{\phi}(\beta)$ and $\overline{\sigma}(\beta)$, and these are related by

$$\overline{\phi}(\beta) = \frac{\overline{\sigma}(\beta)}{|\beta|\,\varepsilon_s(\beta)}, \tag{B.26}$$

as in Chapter 3, Section 3.2, eq. (3.19).

In general the potential $\phi(x)$ includes contributions due to surface waves, bulk waves and electrostatic effects. However, the surface-wave contribution may be identified by noting that this is the only term that does not decay in amplitude for large positive or negative values of x, remote from the transducer electrodes. Since the surface is unmetallized outside the transducer, the potential for large positive or negative x must have the form $\exp(-jk_f|x|)$. The Fourier transform of such a function must be infinite at $\beta = \pm k_f$, and hence the right side of eq. (B.26) must be infinite at these points. Now, the charge density is localized in a finite region of x, occupied by the transducer electrodes, and it follows that its transform $\overline{\sigma}(\beta)$ cannot be infinite for any β. The poles of eq. (B.26) therefore arise from zeros of $\varepsilon_s(\beta)$ at $\beta = \pm k_f$.

To evaluate the surface-wave potential, it is assumed that the total potential $\phi(x)$ can be written as a sum of two surface-wave terms, representing waves of constant amplitude radiated away from the transducer, plus some potential $\phi_1(x)$ which is localized, so that it vanishes for $x \rightarrow \pm\infty$. The exact forms assumed for the surface-wave terms are not consequential, though they must be defined such that they are harmonic at points remote from the transducer, and they must represent waves propagating away from the transducer. Assuming that the origin for x is within the transducer, a suitable form for the potential is

$$\phi(x) = \phi_- U(-x)e^{jk_fx} + \phi_+ U(x)e^{-jk_fx} + \phi_1(x), \tag{B.27}$$

where the constants ϕ_- and ϕ_+ are the amplitudes of the two surface-wave potentials. $U(x)$ is the step function, equal to unity for $x > 0$ and zero for $x < 0$. From Appendix A, eq. (A.30), the transform of the first term is given by

$$U(-x)\exp(jk_fx) \leftrightarrow \pi\delta(\beta - k_f) + j/(\beta - k_f).$$

This is infinite at $\beta = k_f$. The second term has a similar transform (eq. (A.30)), and is infinite at $\beta = -k_f$. Using these and eq. (B.26), the charge density has a Fourier transform

$$\begin{aligned}\overline{\sigma}(\beta) &= |\beta|\varepsilon_s(\beta)\overline{\phi}(\beta)\\ &= |\beta|\varepsilon_s(\beta)[\phi_-\pi\delta(\beta - k_f) + j\phi_-/(\beta - k_f)\\ &\quad + \phi_+\pi\delta(\beta + k_f) - j\phi_+/(\beta + k_f) + \overline{\phi}_1(\beta)],\end{aligned}$$

where $\overline{\phi}_1(\beta)$ is the transform of $\phi_1(x)$.

We now evaluate this function at $\beta = k_f$, noting that at this point $\varepsilon_s(\beta) = 0$. In the square bracket, the two ϕ_+ terms and the term $\overline{\phi}_1(\beta)$ can be omitted because at $\beta = k_0$ they are finite or zero. In addition $\varepsilon_s(\beta)\delta(\beta - k_f)$ is zero for all β. Thus for β close to k_f we only need to consider the remaining term

$$\overline{\sigma}(\beta) = j\phi_-|\beta|\varepsilon_s(\beta)/(\beta - k_f). \tag{B.28}$$

Noting that $\varepsilon_s(k_f) = 0$, we replace $\varepsilon_s(\beta)$ by the linear term of its Taylor expansion:

$$\varepsilon_s(\beta) \approx -(\beta - k_f)/(k_f\Gamma_s), \tag{B.29}$$

where the constant Γ_s is defined by

$$\frac{1}{\Gamma_s} = -k_f\left[\frac{d\varepsilon_s(\beta)}{d\beta}\right]_{k_f},$$

as in Chapter 3, Section 3.3. Substituting eq. (B.29) into eq. (B.28) and evaluating at $\beta = k_f$ gives

$$\phi_- = j\Gamma_s\overline{\sigma}(k_f).$$

Similarly, by evaluating $\overline{\sigma}(\beta)$ at $\beta = -k_f$, we find that $\phi_+ = j\Gamma_s\overline{\sigma}(-k_f)$. Thus, for locations remote from the transducer, such that the localized potential $\phi_1(x)$ in eq. (B.27) is negligible, the total potential is given by

$$\phi(x) = j\Gamma_s\overline{\sigma}(k_f)\exp(jk_fx), \quad \text{for } x \ll 0,$$
$$= j\Gamma_s\overline{\sigma}(-k_f)\exp(-jk_fx), \quad \text{for } x \gg 0, \quad (B.30)$$

which represents surface waves traveling away from the transducer. An alternative proof of this equation can be obtained by contour integration, using the theory of analytic functions [2, 3].

An alternative interpretation is to define a surface-wave potential $\phi_s(x)$ for all values of x on the free surface outside the transducer. We can thus write

$$\phi_s(x) = j\Gamma_s\overline{\sigma}(\mp k_f)\exp(\mp jk_fx) \quad (B.31)$$

taking the upper signs for $x > 0$ and the lower signs for $x < 0$. At points remote from the transducer this is equal to the total potential $\phi(x)$. For points close to the transducer the surface-wave potential can be taken to be given by eq. (B.31), though the total potential $\phi(x)$ also includes the term $\phi_1(x)$, which is due to electrostatic effects and, possibly, bulk-wave excitation.

REFERENCES

1. B.A. Auld. *Acoustic Fields and Waves in Solids*, Vol. 2, Krieger, 1990.
2. R.F. Milsom, N.H.C. Reilly and M. Redwood. 'Analysis of generation and detection of surface and bulk acoustic waves by interdigital transducers', *IEEE Trans.*, **SU-24**, 147–166 (1977).
3. D. Royer and E. Dieulesaint. 'Elastic Waves in Solids', *Generation, Acousto-optic Interaction and Applications*, Vol. 2, Springer, 2000, p. 83.

Appendix C

ELEMENTAL CHARGE DENSITY FOR REGULAR ELECTRODES

In Chapter 5, Section 5.4 it was shown that the properties of transducers with regular electrodes can conveniently be expressed in terms of a function $\bar{\rho}_f(\beta)$, defined as the Fourier transform of an elemental charge density $\rho_f(x)$. The main purpose of this appendix is to demonstrate the validity of analytic expression for $\bar{\rho}_f(\beta)$. However, before doing this Section C.1 summarizes some properties of Legendre functions that are needed in this appendix and elsewhere.

C.1 SOME PROPERTIES OF LEGENDRE FUNCTIONS

Properties of Legendre functions are given in many texts, including that of Erdelyi [1], which gives all the properties quoted here.

The Legendre function with variable x and degree v is written as $P_v(x)$. For surface-wave analysis x and v are real, and x is in the range $-1 < x < 1$. The function can be evaluated using the expansion

$$P_v(x) = \sum_{m=0}^{\infty} a_m, \quad \text{for } |x| \leq 1, \tag{C.1}$$

where $a_0 = 1$ and

$$a_m = \frac{(m - 1 - v)(m + v)(1 - x)}{2m^2} a_{m-1}.$$

Some plots of $P_v(x)$, regarded as functions of v, are shown in Fig. C.1. The recursion relation is

$$vP_v(x) = (2v - 1)xP_{v-1}(x) - (v - 1)P_{v-2}(x), \tag{C.2}$$

and the symmetry relation is

$$P_v(x) = P_{-v-1}(x). \tag{C.3}$$

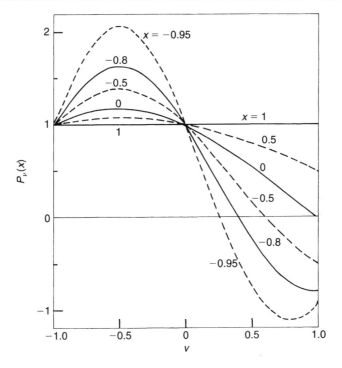

Figure C.1 Legendre functions $P_\nu(x)$.

Thus $P_\nu(x)$ is symmetric about $\nu = -1/2$. From the series of eq. (C.1) it can be seen that $P_\nu(-1)$ is infinite if ν is not an integer. For $\nu = -1/2$, $P_\nu(x)$ is related to the elliptic integral by

$$P_{-1/2}(x) = 2K(m)/\pi, \quad m = \sqrt{(1-x)/2}, \tag{C.4}$$

where $K(m)$ is the complete elliptic integral of the first kind. The Mehler–Dirichlet formula is

$$P_\nu(\cos \Delta) = \frac{1}{\pi\sqrt{2}} \int_{-\Delta}^{\Delta} \frac{\exp[j(\nu + 1/2)\phi]}{\sqrt{\cos \phi - \cos \Delta}} d\phi, \quad \text{for } 0 < \Delta < \pi. \tag{C.5}$$

For positive and negative values of x and ν, the Legendre functions are related by

$$P_\nu(x)P_{-\nu}(-x) + P_\nu(-x)P_{-\nu}(x) = \frac{2 \sin \pi\nu}{\pi\nu}, \tag{C.6}$$

as noted by Bløtekjaer *et al.* [2]. To prove this relation, first use the differentiation formulae [1] to show that the left side is independent of x. For $x = 0$, $P_\nu(x)$ can

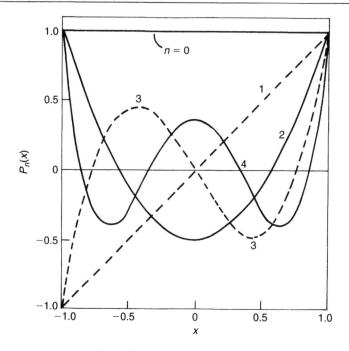

Figure C.2 Legendre polynomials $P_\nu(x)$.

be expressed in terms of the gamma function, and the properties of this function can be used to derive eq. (C.6) for $x = 0$, completing the proof.

For the particular case when ν is an integer, the series of eq. (C.1) truncates, so that $P_\nu(x)$ becomes a polynomial. This is called the Legendre polynomial, written as $P_n(x)$. The above relations are valid for $P_n(x)$, with ν replaced by the integer n. Thus,

$$P_0(x) = 1, \quad P_1(x) = x. \tag{C.7}$$

For larger n, $P_n(x)$ may be obtained from the recursion relation, eq. (C.2), giving

$$P_2(x) = (3x^2 - 1)/2, \quad P_3(x) = (5x^3 - 3x)/2, \tag{C.8}$$

and so on. Some Legendre polynomials are shown in Fig. C.2. The polynomials are orthogonal over the interval $-1 < x < 1$, and they have the symmetry

$$P_n(-x) = (-1)^n P_n(x). \tag{C.9}$$

The Legendre function $P_{-\nu}(\cos\Delta)$ is expressed as a sum of the polynomials by Dougall's expansion

$$P_{-\nu}(\cos\Delta) = \frac{\sin(\pi\nu)}{\pi} \sum_{-\infty}^{\infty} \frac{S_n P_n(-\cos\Delta)}{\nu + n}, \tag{C.10}$$

where

$$S_n = 1 \quad \text{for } n \geq 0; \qquad S_n = -1 \quad \text{for } n < 0. \tag{C.11}$$

Also, for $0 < \Delta < \pi$,

$$\sum_{n=-\infty}^{\infty} P_n(\cos\Delta)e^{jn\theta} = \frac{(-1)^m\sqrt{2}e^{-j\theta/2}}{\sqrt{\cos\theta - \cos\Delta}}, \quad \text{for } |\theta - 2m\pi| < \Delta$$

$$= 0, \qquad\qquad \text{for } \Delta < |\theta - 2m\pi| \leq \pi \quad \text{(C.12)}$$

This is valid for any value of θ; the integer m is chosen such that $|\theta - 2m\pi| \leq \pi$. Another series valid for $0 < \Delta < \pi$ is

$$\sum_{n=-\infty}^{\infty} S_n P_n(\cos\Delta)e^{jn\theta} = \frac{j(-1)^m\sqrt{2}\,\mathrm{sgn}(\theta - 2m\pi)e^{-j\theta/2}}{\sqrt{\cos\Delta - \cos\theta}},$$

$$\text{for } \Delta < |\theta - 2m\pi| < \pi,$$

$$= 0, \quad \text{for } |\theta - 2m\pi| < \Delta, \tag{C.13}$$

where S_n is given by eq. (C.11), $\mathrm{sgn}(x) = 1$ for $x > 0$ and $\mathrm{sgn}(x) = -1$ for $x < 0$.

Combining eq. (C.12) with the Mehler–Dirichlet formula, eq. (C.5), we have, for $0 < \Delta < \pi$,

$$\int_{-\Delta}^{\Delta} \sum_{n=-\infty}^{\infty} P_{n+m}(\cos\Delta)e^{-j(n+v)\theta}d\theta = 2\pi P_{m-v}(\cos\Delta). \tag{C.14}$$

This equation relates to the current flowing into an electrode of an interdigital transducer.

C.2 ELEMENTAL CHARGE DENSITY

The elemental charge density $\rho_f(x)$ is defined by considering an infinite array of regular electrodes on a half-space, with width a and pitch p, as in Fig. 5.8. One of the electrodes, centered at $x = 0$, has unit voltage applied, while all other electrodes have zero voltage. The charge density is then $\rho_f(x)$. This section shows that the Fourier transform of $\rho_f(x)$ is

$$\overline{\rho}_f(\beta) = (\varepsilon_0 + \varepsilon_p)\frac{2\sin(\pi s)}{P_{-s}(-\cos\Delta)}P_n(\cos\Delta), \quad \text{for } n \leq \frac{\beta p}{2\pi} \leq n+1, \quad \text{(C.15)}$$

where $\Delta = \pi a/p$, $s = (\beta p)/(2\pi) - n$, and ε_p is a function of the permittivities of the half-space material, as in eq. (3.3). The parameter s is in the range $0 \leq s \leq 1$.

To verify eq. (C.15), we show that the associated potential satisfies Laplace's equation and that the boundary conditions are satisfied. Thus, the charge density must be zero in the gaps between the electrodes, and the potential on each electrode must be uniform and equal to the applied voltage. In the x-domain, the charge density $\rho_f(x)$ is given by the inverse Fourier transform of $\overline{\rho}_f(\beta)$, so that

$$\rho_f(x) = \int_{-\infty}^{\infty} \overline{\rho}(\beta) \exp(j\beta x) d\beta / (2\pi). \tag{C.16}$$

In view of eq. (C.15), this can be integrated from $\beta = 2\pi n/p$ to $\beta = 2\pi(n+1)/p$, and then summed with respect to n. The summation has the form of eq. (C.12). Defining a normalized x-coordinate by $\theta = 2\pi x/p$, the result is

$$\rho_f(x) = \frac{\varepsilon_0 + \varepsilon_p}{p} \frac{2\sqrt{2}(-1)^m}{\sqrt{\cos\theta - \cos\Delta}} \Gamma(\theta, \Delta), \quad \text{for } |x - mp| < a/2,$$

$$= 0, \qquad\qquad\qquad \text{for } a/2 < |x - mp| < p/2, \tag{C.17}$$

where the function $\Gamma(\theta, \Delta)$ is given by

$$\Gamma(\theta, \Delta) = \int_0^1 \frac{\sin(\pi s)\cos[(s - 1/2)\theta]}{P_{-s}(-\cos\Delta)} ds. \tag{C.18}$$

Thus, $\rho_f(x)$ is zero in the gaps between electrodes, as required. Its form is shown in Fig. 5.8, for $a/p = 1/2$.

The associated electric field at the surface has an x-component $E_x(x)$, with Fourier transform $\overline{E}_x(\beta)$. If the potential at the surface is $\phi(x)$, with transform $\overline{\phi}(\beta)$, we have $\overline{E}_x(\beta) = -j\beta\overline{\phi}(\beta)$. From the electrostatic analysis in Chapter 3, Section 3.1, this is related to the charge density by

$$\overline{E}_x(\beta) = -j\,\text{sgn}(\beta)\overline{\rho}_f(\beta)/(\varepsilon_0 + \varepsilon_p). \tag{C.19}$$

Using eq. (C.15) we have

$$\overline{E}_x(\beta) = -\frac{2j\sin(\pi s)}{P_{-s}(-\cos\Delta)} S_n P_n(\cos\Delta), \quad \text{for } n \leq \frac{\beta p}{2\pi} \leq n + 1, \tag{C.20}$$

with S_n defined by eq. (C.11). Transforming to the x-domain involves a sum with the form of eq. (C.13), giving the result

$$E_x(x) = \frac{2\sqrt{2}(-1)^m\,\text{sgn}(\theta - 2m\pi)}{p\sqrt{\cos\Delta - \cos\theta}} \Gamma(\theta, \Delta), \quad \text{for } a/2 < |x - mp| \leq p/2,$$

$$= 0, \qquad\qquad\qquad \text{for} |x - mp| < a/2. \tag{C.21}$$

This shows that on each electrode we have $E_x(x) = 0$ and the potential is uniform, as required.

To show that the potential $\phi(x)$ is correct, it is sufficient to evaluate it at the electrode centers $x = mp$. In the β-domain we have $\overline{\phi}(\beta) = j\overline{E}_x(\beta)/\beta$, with $\overline{E}_x(\beta)$ given by eq. (C.20). Transforming to the x-domain, the integral involves a summation over n, and for $x = mp$ this has the form of Dougall's expansion, eq. (C.10). This gives $\phi(mp) = 1$ for $m = 0$ and $\phi(mp) = 0$ for $m \neq 0$. Hence the potentials are correct. This completes the proof, since the boundary conditions are satisfied and Laplace's equation is satisfied in the form of eq. (C.19).

C.3 NET CHARGES ON ELECTRODES

The total charge per unit length on the electrode centered at $x = mp$ is denoted by Q_m, so that

$$Q_m = \int_{mp-a/2}^{mp+a/2} \rho_f(x)dx. \tag{C.22}$$

These quantities are useful when calculating the capacitance of a transducer. On substituting for $\rho_f(x)$ from eq. (C.17), the x-integral is seen to have the form of the Mehler–Dirichlet formula, eq. (C.5), and Q_m is found to be

$$Q_m = 2(\varepsilon_0 + \varepsilon_p) \int_0^1 \frac{\sin(\pi s)\cos(2\pi m s)}{P_{-s}(-\cos \Delta)} P_{-s}(\cos \Delta)ds. \tag{C.23}$$

If the metallization ratio $a/p = 1/2$ we have $\cos \Delta = 0$ and the Legendre functions cancel. This gives the simple formula

$$Q_m = \frac{4(\varepsilon_0 + \varepsilon_p)}{\pi(1 - 4m^2)}, \quad \text{for } a/p = 1/2. \tag{C.24}$$

For uniform transducers, the capacitance is given by a sum of the Q_n. The summation can be done using the formula

$$N \sum_{m=-\infty}^{\infty} \cos[2\pi x(mN + n)] = \cos(2\pi nx) \sum_{i=-\infty}^{\infty} \delta(x - i/N), \tag{C.25}$$

where $N \neq 0$. This follows from eq. (A.33), or by Fourier transformation of both sides. Using eq. (C.23) for Q_m we find that, for $N \geq 1$,

$$\sum_{m=-\infty}^{\infty} Q_{mN+n} = \frac{2(\varepsilon_0 + \varepsilon_p)}{N} \sum_{i=0}^{N} \frac{\sin(\pi v)\cos(2\pi n v)}{P_{-v}(-\cos \Delta)} P_{-v}(\cos \Delta), \tag{C.26}$$

with $v = i/N$.

Another summation formula is useful for analysis of multistrip couplers, and this again follows from eqs (A.33) and (C.23). The formula is

$$\sum_{m=-\infty}^{\infty} Q_m e^{-jkmp} = 2(\varepsilon_0 + \varepsilon_p)\frac{\sin(\pi\mu)}{P_{-\mu}(-\cos\Delta)}P_{-\mu}(\cos\Delta). \qquad (C.27)$$

Here μ is defined such that $kp = 2\pi(i + \mu)$ and i is the integer part of $kp/(2\pi)$, so that $0 \le \mu \le 1$.

REFERENCES

1. A. Erdelyi. *Higher Transcendental Functions*, McGraw-Hill, 1953.
2. K. Bløtekjaer, K.A. Ingebrigtsen and H. Skeie. 'Acoustic surface waves in piezoelectric materials with periodic metal strips on the surface', *IEEE Trans.*, **ED-20**, 1139–1146 (1973).

Appendix D

P-MATRIX RELATIONS

The *P*-matrix [1] is a type of scattering matrix commonly used to describe the behavior of surface acoustic wave (SAW) gratings or transducers. It has wide applicability, and here we summarize many useful properties which do not depend on the physics involved. It is assumed that there is only one type of acoustic wave present. For example, in a Rayleigh-wave device the coupling to bulk waves is not considered. We also assume that the wave amplitude is uniform in the transverse direction, thus excluding waveguiding and diffraction. For transducers, this also implies that there is no apodization.

D.1 GENERAL RELATIONS

We consider a surface-wave transducer with acoustic ports on either side. The ports are reference positions to specify where the incident and outgoing waves are measured. Their precise locations are unimportant, but they are conveniently placed near the transducer edges. For a transducer with acoustic symmetry, the ports are positioned symmetrically so that the symmetry is maintained. The wave amplitudes will be specified in terms of the accompanying surface potential, ϕ_s. The amplitude is denoted by A, defined to have the same phase as ϕ_s and such that $|A|^2/2$ equals the wave power. In terms of the theory in Chapter 3, Section 3.3, A is given by $A = \phi_s \sqrt{\omega W/(2\Gamma_s)}$, where W is the beam width and $\Gamma_s = (\Delta v/v)/\varepsilon_\infty$. Other definitions of A are valid for the *P*-matrix theory, provided they give the same power relation and they exhibit the symmetry when appropriate.

Define A_i as the amplitudes of waves incident on the transducer, and A_t as the amplitudes of waves leaving. These are also subscripted 1 or 2 to refer to the acoustic ports, as in Fig. D.1. The electrical port, with current I and voltage V, is port 3. These variables are related by

$$\begin{bmatrix} A_{t1} \\ A_{t2} \\ I \end{bmatrix} = \begin{bmatrix} P_{11} & P_{12} & P_{13} \\ P_{21} & P_{22} & P_{23} \\ P_{31} & P_{32} & P_{33} \end{bmatrix} \cdot \begin{bmatrix} A_{i1} \\ A_{i2} \\ V \end{bmatrix}. \tag{D.1}$$

397

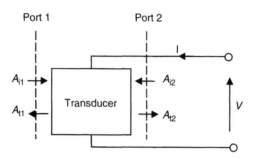

Figure D.1 Definition of variables for *P*-matrix.

Here P_{33} is the transducer admittance, as seen when no acoustic waves are incident. For reflecting gratings the same matrix is valid, but the P_{ij} with i or j equal to 3 are not needed. They may be set to zero if necessary.

Reciprocity

A SAW transducer obeys the reciprocity relations

$$P_{21} = P_{12}; \quad P_{31} = -2P_{13}; \quad P_{32} = -2P_{23}. \tag{D.2}$$

A proof of these relations is given later in this appendix.

Power conservation

Here we assume that there is no power loss, a condition which is often well approximated in practice. If $V = 0$, the power of the waves leaving the transducer must be equal to the power of the incident waves. Using reciprocity, this gives the relations

$$|P_{11}|^2 + |P_{12}|^2 = |P_{22}|^2 + |P_{12}|^2 = 1, \tag{D.3}$$

$$P_{11}P_{12}^* + P_{22}^*P_{12} = 0, \tag{D.4}$$

where the asterisk denotes a complex conjugate. To prove these, note that power conservation gives $|A_{t1}|^2 + |A_{t2}|^2 = |A_{i1}|^2 + |A_{i2}|^2$. Setting $A_{i1} = 0$ or $A_{i2} = 0$ gives eq. (D.3). For eq. (D.4), set $A_{i2}A_{i1}^*$ to be either real or imaginary, and subtract the results. Equation (D.3) also shows that $|P_{22}| = |P_{11}|$, that is, the reflection coefficients are of equal magnitude.

If a voltage is applied and there are no acoustic inputs ($A_{i1} = A_{i2} = 0$), the power of the waves generated is equal to the power taken from the source, which is $|V|^2 G_a/2$. This gives

$$G_a \equiv \text{Re}\{P_{33}\} = |P_{13}|^2 + |P_{23}|^2. \tag{D.5}$$

If input signals are applied at the electrical port and at one of the acoustic ports, power conservation gives

$$P_{11}P_{13}^* + P_{12}P_{23}^* + P_{13} = 0, \tag{D.6}$$

$$P_{22}P_{23}^* + P_{12}P_{13}^* + P_{23} = 0. \tag{D.7}$$

Symmetrical lossless transducer

A symmetrical transducer is taken to be one that gives $P_{22} = P_{11}$ and $P_{23} = P_{13}$. This condition applies if the transducer has symmetrical geometry and the substrate behaves symmetrically (i.e. the N-SPUDT effect is absent). Setting $P_{22} = P_{11}$ in eq. (D.4) gives

$$P_{12}/P_{11} = -P_{12}^*/P_{11}^*. \tag{D.8}$$

This shows that P_{12}/P_{11} is imaginary, so that P_{11} and P_{12} are in phase quadrature with each other. This applies, for example, to a symmetrical transducer (short circuit or open circuit), a symmetrical grating or a single strip. In addition,

$$|P_{11} \pm P_{12}| = 1. \tag{D.9}$$

To show this, multiply $(P_{11} \pm P_{12})$ by its conjugate and then use eqs (D.3) and (D.4).

Now consider the acoustic reflection coefficient of a symmetrical loss-less transducer connected to a matched load. The load admittance will be P_{33}^*, so we have $I/V = -P_{33}^*$. Assuming that $A_{i2} = 0$, eq. (D.1) gives $A_{t1}/A_{i1} = P_{11} + P_{13}^2/G_a$. Using $P_{23} = P_{13}$, eq. (D.5) gives $G_a = 2|P_{13}|^2$ and eq. (D.7) gives $P_{13} = -P_{13}^*(P_{11} + P_{12})$. From these, the reflection coefficient is found to be $A_{t1}/A_{i1} = (P_{11} - P_{12})/2$. This has magnitude 1/2, by eq. (D.9). Hence, a matched symmetrical transducer has a high reflection coefficient of $-6\,\mathrm{dB}$. The transducer voltage is found to be $V = P_{13}A_{i1}/G_a$, and from this we find that the power delivered to the load is half of the incident surface-wave power. The conversion coefficient is therefore $-3\,\mathrm{dB}$. This also applies for a launching transducer – when a voltage is applied, half of the available power emerges as surface waves in each direction, by symmetry.

Non-reflective transducer: triple-transit

A non-reflective transducer is one that does not reflect when it is shorted, so that $P_{11} = P_{22} = 0$. Assuming no loss, eq. (D.3) gives $|P_{12}| = 1$, and using this in eq. (D.6) gives

$$|P_{13}| = |P_{23}|. \tag{D.10}$$

This shows that a non-reflective transducer is bidirectional, that is, it generates waves of equal amplitude in the two directions when a voltage is applied. This is true even if the geometry is not symmetric, as in the case of a chirp transducer, for example. If the transducer is connected to a load with admittance Y_L we have $I/V = -Y_L$. Taking $P_{11} = 0$, the power reflection and conversion coefficients for a wave incident on port 1 are found to be

$$R_p \equiv |A_{t1}/A_{i1}|^2 = G_a^2/|P_{33} + Y_L|^2, \tag{D.11}$$

$$C_p \equiv P_L/P_s = 2G_aG_L/|P_{33} + Y_L|^2, \tag{D.12}$$

where $G_L = \mathrm{Re}\{Y_L\}$. P_s and P_L are respectively the incident wave power and the power delivered to the load. These coefficients are found to be the same for a launching transducer when a source with admittance Y_L is used.

To illustrate the triple-transit problem, consider applying a sinusoidal signal to a device consisting of two identical transducers, with identical source and load admittances. The power of the main output signal, relative to the available input power, is C_p^2. The power of the triple-transit signal, relative to the main signal, is R_p^2. Assuming that the transducers have inductors to tune out the imaginary part of the admittance P_{33}, the term $P_{33} + Y_L$ becomes $G_a + G_L$. When the device is matched ($G_L = G_a$) each transducer gives $C_p = 1/2$ and $R_p = 1/4$, so the insertion loss is 6 dB and the triple-transit suppression is 12 dB. To improve the latter, we can increase G_L. For $G_L \gg G_a$ we find $C_p = 2G_a/G_L$ and $R_p = G_a^2/G_L^2 = C_p^2/4$. Expressed in decibels, we have

Triple-transit suppression $\approx 2 \times$ insertion loss $+ 12\,\mathrm{dB}$, for $G_L \gg G_a$.

$$\tag{D.13}$$

Hence, good triple-transit suppression can be obtained at the expense of high insertion loss. On the other hand, if $G_L \ll G_a$ we have $R_p \approx 1$, giving high loss and poor triple-transit suppression.

D.2 CASCADING FORMULAE

Change of port position

The formulae here can be used to analyze a wide variety of devices. We first consider the effect of changing a port position. Suppose port 1 is moved away from the transducer by a distance d, changing the *P*-matrix from P_{ij} to P'_{ij}. It is easily seen that the matrices are related by

$$P'_{11}/P_{11} = \exp(-2jkd); \quad P'_{12}/P_{12} = \exp(-jkd); \quad P'_{13}/P_{13} = \exp(-jkd),$$

$$\tag{D.14}$$

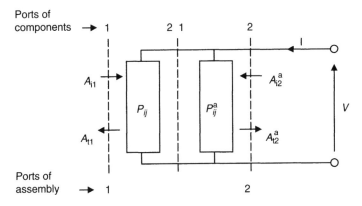

Figure D.2 Cascading of adjacent components.

where k is the wavenumber for waves between the two port locations. The reciprocity relations, eq. (D.2), apply, and the remaining components P_{22} and P_{33} are unaffected. Corresponding relations apply for port 2. These equations can be used to shift the ports to locations which are convenient for cascading.

Cascading of two components

Consider two adjacent transducers with the same aperture and with P-matrices P_{ij} and P_{ij}^a, as in Fig. D.2. For simplicity, it is assumed here that the transducers interact only via the surface waves, excluding for example electrostatic coupling. Each transducer has port 1 at the left and port 2 at the right. The bus bars are common, so the transducer voltages are the same. Port 2 of the left transducer is taken to be coincident with port 1 of the right transducer, using the method of eq. (D.14) to adjust either of these port positions. The common port is indicated by the central broken line. The left broken line is port 1 of the left transducer, where the wave amplitudes are A_{i1} and A_{t1}. The right broken line is port 2 of the right transducer, where the amplitudes are A_{i2}^a and A_{t2}^a. The P-matrix of the assembly is P_{ij}'. This is defined as in eq. (D.1), with port 1 at the left, so that

$$[A_{t1}, A_{t2}^a, I]^T = [P_{ij}'] \cdot [A_{i1}, A_{i2}^a, V]^T, \tag{D.15}$$

where superscript T indicates a transpose. Using continuity of the wave amplitudes in the central gap, the combined P-matrix is as follows. Reciprocity is applied, and we define $D = 1 - P_{22}P_{11}^a$.

$$P_{11}' = P_{11} + P_{11}^a P_{12}^2/D; \quad P_{12}' = P_{21}' = P_{12}^a P_{12}/D;$$

$$P_{22}' = P_{22}^a + P_{22}(P_{12}^a)^2/D;$$

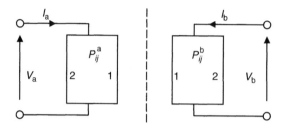

Figure D.3 Y-matrix for a two-component device.

$$P'_{13} = -P'_{31}/2 = P_{13} + P_{12}(P^a_{11}P_{23} + P^a_{13})/D;$$

$$P'_{23} = -P'_{32}/2 = P^a_{23} + P^a_{12}(P^a_{13}P_{22} + P_{23})/D;$$

$$P'_{33} = P_{33} + P^a_{33} - 2P_{23}(P^a_{11}P_{23} + P^a_{13})/D - 2P^a_{13}(P_{22}P^a_{13} + P_{23})/D.$$

$$(D.16)$$

These equations can be used repeatedly to give the P-matrix of more complex assemblies.

Y-matrix of a two-port device

Now consider two components with P-matrices P^a_{ij} and P^b_{ij}, as in Fig. D.3. These 'components' may be assemblies of many transducers or gratings, analyzed by eqs (D.16). Each component has port 1 facing the other component, and these ports are coincident. All transducers are assumed to have the same aperture. The admittance matrix Y_{ij} is defined by

$$I_a = Y_{11}V_a + Y_{12}V_b; \quad I_b = Y_{21}V_a + Y_{22}V_b. \quad (D.17)$$

The Y_{ij} are given by

$$Y_{11} = P^a_{33} - 2P^b_{11}(P^a_{13})^2/(1 - P^a_{11}P^b_{11});$$

$$Y_{12} = Y_{21} = -2P^a_{13}P^b_{13}/(1 - P^a_{11}P^b_{11});$$

$$Y_{22} = P^b_{33} - 2P^a_{11}(P^b_{13})^2/(1 - P^a_{11}P^b_{11}). \quad (D.18)$$

This matrix gives a full description of the electrical behavior of the device.

Scattering matrix

The response of a two-port device is often expressed by the scattering matrix S_{ij}. We imagine the device to be connected between two transmission lines, with characteristic impedance R_1 at port 1 and R_2 at port 2. The matrix S_{ij}

gives the transmission and reflection of waves incident on either port [2]. From the Y-matrix, and noting the reciprocity condition $Y_{21} = Y_{12}$, the transmission coefficient is found to be

$$S_{12} = S_{21} = \frac{2Y_{12}\sqrt{R_1 R_2}}{R_1 R_2 Y_{12}^2 - (1 + Y_{11}R_1)(1 + Y_{22}R_2)}. \tag{D.19}$$

The insertion loss is $20\log|S_{12}|$ dB, giving the ratio of transmitted power to available input power. The reflection coefficient at port 1 is

$$S_{11} = \frac{(1 - R_1 Y_{11})(1 + R_2 Y_{22}) + R_1 R_2 Y_{12}^2}{(1 + R_1 Y_{11})(1 + R_2 Y_{22}) - R_1 R_2 Y_{12}^2}. \tag{D.20}$$

S_{22} is given by the same formula with indices 1 and 2 interchanged.

For eqs (D.19) and (D.20), it is assumed that a device described by the Y_{ij} is connected directly to the transmission lines. In practice, there are often extra components such as tuning inductors, impedance-matching components or strays. The Y-matrix needs to be modified to allow for these components before using eq. (D.19) or (D.20).

Cascading of identical transducers

Here we summarize results of an analysis given elsewhere [3]. It assumed that N identical unapodized transducers are arranged in sequence and connected to one pair of lossless bus bars, as in Fig. D.4. The P-matrix of each transducer is denoted by p_{ij}, defined as in eq. (D.1) and assumed to obey the reciprocity relations, eq. (D.2). To cascade these transducers, we define a transmission matrix relating variables on the right to those on the left. This is expressed as

$$\begin{bmatrix} c_{n+1} \\ b_{n+1} \\ V \\ I_{n+1} \end{bmatrix} = \begin{bmatrix} t_{11} & t_{12} & t_{13} & 0 \\ t_{21} & t_{22} & t_{23} & 0 \\ 0 & 0 & 1 & 0 \\ t_{31} & t_{32} & t_{33} & 1 \end{bmatrix} \cdot \begin{bmatrix} c_n \\ b_n \\ V \\ I_n \end{bmatrix} \tag{D.21}$$

where c_n and b_n refer to waves propagating to the right and left, respectively. From the P-matrix of eq. (D.1), the elements in eq. (D.21) are

$$t_{11} = p_{21} - p_{22}p_{11}/p_{12}; \quad t_{12} = p_{22}/p_{12};$$

$$t_{13} = p_{23} - p_{22}p_{13}/p_{12}; \quad t_{21} = -p_{11}/p_{12};$$

$$t_{22} = 1/p_{12}; \quad\quad\quad\quad\quad t_{23} = -p_{13}/p_{12};$$

$$t_{31} = p_{31} - p_{32}p_{11}/p_{12}; \quad t_{32} = p_{32}/p_{12};$$

$$t_{33} = p_{33} - p_{32}p_{13}/p_{12}. \tag{D.22}$$

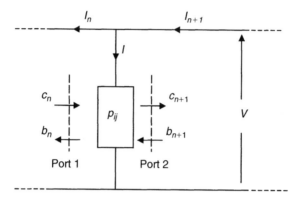

Figure D.4 Parameters for cascading identical transducers.

For an infinite array of gratings or shorted transducers, a grating mode has the form $c_{n+1} = c_n \exp(-j\theta)$ and $b_{n+1} = b_n \exp(-j\theta)$. Using eq. (D.21) with $V = 0$, we find that

$$2\cos\theta = t_{11} + t_{22}. \tag{D.23}$$

Here, θ is essentially the wavenumber of the grating mode. If the repeat distance of the transducers is L, we have $\theta = \gamma L$, where γ is the wavenumber of the grating mode.

To cascade N transducers, we first diagonalize the upper left part of the transmission matrix. Define new variables $u_n = c_n + \alpha b_n$ and $v_n = c_n + \beta b_n$. The required values of α and β are found to be

$$\alpha = (e^{j\theta} - t_{11})/t_{21}; \quad \beta = (e^{-j\theta} - t_{11})/t_{21}. \tag{D.24}$$

With these definitions, eq. (D.21) gives

$$\begin{bmatrix} u_{n+1} \\ v_{n+1} \\ V \\ I_{n+1} \end{bmatrix} = \begin{bmatrix} e^{j\theta} & 0 & R_{13} & 0 \\ 0 & e^{-j\theta} & R_{23} & 0 \\ 0 & 0 & 1 & 0 \\ R_{31} & R_{32} & R_{33} & 1 \end{bmatrix} \cdot \begin{bmatrix} u_n \\ v_n \\ V \\ I_n \end{bmatrix} \tag{D.25}$$

where the R_{ij} are to be defined. When $V = 0$ the wave amplitudes are related by the top left 2×2 part of this matrix. For $V = 0$, an array of N transducers is easily cascaded because the terms $\exp(\pm j\theta)$ are simply replaced by $\exp(\pm jN\theta)$.

The acoustic components P_{11}, P_{12} and P_{22} of the array are obtained by back-substituting, giving

$$P_{11} = \frac{p_{11} \sin N\theta}{\sin N\theta - p_{12} \sin (N-1)\theta};$$

$$P_{12} = P_{21} = \frac{p_{12} \sin \theta}{\sin N\theta - p_{12} \sin (N-1)\theta};$$

$$P_{22} = \frac{p_{22} \sin N\theta}{\sin N\theta - p_{12} \sin (N-1)\theta}. \tag{D.26}$$

Other terms of the array P-matrix are found to be [3].

$$P_{31} = (1 - P_{12})K_1 + P_{11}K_2; \quad P_{32} = -P_{22}K_1 + (P_{12} - 1)K_2, \tag{D.27}$$

with $P_{13} = -P_{31}/2, P_{23} = -P_{32}/2$ and

$$K_1 = \frac{p_{31}(p_{21} - 1) - p_{32}p_{11}}{p_{11}p_{22} - (p_{12} - 1)^2}; \quad K_2 = \frac{p_{32}(1 - p_{12}) + p_{31}p_{22}}{p_{11}p_{22} - (p_{12} - 1)^2}. \tag{D.28}$$

The admittance of the array is

$$P_{33} = NR'_{33} - P_{23}K_1 + P_{13}K_2, \tag{D.29}$$

where

$$R'_{33} = t_{33} - \alpha_1 R_{31} - \beta_1 R_{32};$$
$$\alpha_1 = (t_{13} + \alpha t_{23})/[\exp(j\theta) - 1]; \quad \beta_1 = (t_{13} + \beta t_{23})/[\exp(-j\theta) - 1];$$
$$R_{31} = (t_{32} - \beta t_{31})/(\alpha - \beta); \quad R_{32} = -(t_{32} - \alpha t_{31})/(\alpha - \beta).$$

General cascading scheme

The following method is more flexible than the method of eqs (D.16) and (D.17), though rather more complex.

Figure D.5 illustrates the process. Each transducer is connected to either the top bus bars, with voltage V and currents I_n, or the lower bus bars with voltage V' and currents I'_n. In the case shown, transducer n is connected to the top bars, and transducer $n+1$ is connected to the lower bars. The ports between adjacent transducers (broken lines) are assumed to be coincident. For a transducer

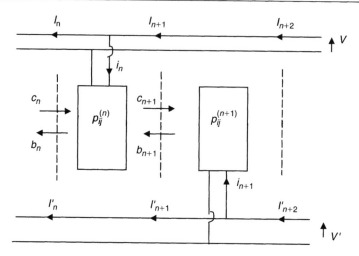

Figure D.5 Cascading of transducers with arbitrary connections to two sets of bus bars.

connected to the upper bus bar, we can use eq. (D.21) to obtain a transmission matrix

$$
\begin{bmatrix}
c_{n+1} \\
b_{n+1} \\
V \\
I_{n+1} \\
V' \\
I'_{n+1}
\end{bmatrix}
=
\begin{bmatrix}
t_{11} & t_{12} & t_{13} & 0 & 0 & 0 \\
t_{21} & t_{22} & t_{23} & 0 & 0 & 0 \\
0 & 0 & 1 & 0 & 0 & 0 \\
t_{31} & t_{32} & t_{33} & 1 & 0 & 0 \\
0 & 0 & 0 & 0 & 1 & 0 \\
0 & 0 & 0 & 0 & 0 & 1
\end{bmatrix}
\begin{bmatrix}
c_n \\
b_n \\
V \\
I_n \\
V' \\
I'_n
\end{bmatrix}
\tag{D.30}
$$

If the transducer is connected to the lower bus bars, the matrix is

$$
\begin{bmatrix}
c_{n+1} \\
b_{n+1} \\
V \\
I_{n+1} \\
V' \\
I'_{n+1}
\end{bmatrix}
=
\begin{bmatrix}
t_{11} & t_{12} & t_{13} & 0 & 0 & 0 \\
t_{21} & t_{22} & t_{23} & 0 & 0 & 0 \\
0 & 0 & 1 & 0 & 0 & 0 \\
0 & 0 & 0 & 1 & 0 & 0 \\
0 & 0 & 0 & 0 & 1 & 0 \\
t_{31} & t_{32} & 0 & 0 & t_{33} & 1
\end{bmatrix}
\begin{bmatrix}
c_n \\
b_n \\
V \\
I_n \\
V' \\
I'_n
\end{bmatrix}
\tag{D.31}
$$

In these equations, the t_{ij} are as in eq. (D.22). The t_{ij} will be different for each transducer if the designs are different. These matrices can be multiplied to cascade up the N transducers, giving a device transmission matrix. In addition, the input waves c_1 and b_N are zero, and the currents at the left, I_1 and I'_1, can be set to zero. This gives a set of simultaneous equations which can be solved to find I_N and I'_N in terms of the voltages V and V', thus giving the device Y-matrix.

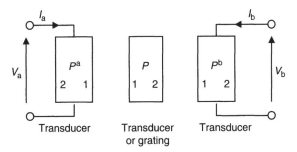

Figure D.6 Device analysis for reciprocity.

This method has considerable flexibility because the transducers can be connected to the bus bars in any sequence. For example, the method can be used for interdigitated interdigital transducers (IIDTs) and for complex forms of longitudinally-coupled resonator (LCR) filters.

Reciprocity proofs

To prove the reciprocity relations given earlier (eq. (D.2)), consider the device of Fig. D.6. This has two transducers plus a central component. The latter may be a grating or a shorted transducer, for example. The transducers have P-matrices P_{ij}^a and P_{ij}^b, and the central component has P-matrix P_{ij}. Numbers indicate the acoustic ports. Without assuming reciprocity or power conservation, it is found that

$$Y_{12} \equiv \left[I_a/V_b\right]_{V_a=0} = -P_{31}^a P_{12} P_{13}^b/D$$

$$Y_{21} \equiv \left[I_b/V_a\right]_{V_b=0} = -P_{31}^b P_{21} P_{13}^a/D, \qquad \text{(D.32)}$$

where $D = \alpha\beta X - P_{12}P_{21}/X$, $\alpha = P_{11}^a + P_{22}/X$, $\beta = P_{11}^b + P_{11}/X$ and $X = P_{12}P_{21} - P_{11}P_{22}$. We now use the fact that $Y_{21} = Y_{12}$. Assume first that the two outer transducers are identical, so that $P_{ij}^a = P_{ij}^b$. Then $Y_{12} = -P_{31}^a P_{12} P_{13}^a/D$, and Y_{21} is the same except that P_{12} is replaced by P_{21}. Since $Y_{21} = Y_{12}$, this shows that $P_{21} = P_{12}$.

Now assume that the central component is absent, so that $P_{21} = P_{12} = 1$. In this case the equation $Y_{21} = Y_{12}$ gives $P_{31}^b/P_{13}^b = P_{31}^a/P_{13}^a$. This shows that P_{31}/P_{13} is a universal quantity, the same for all transducers. In Appendix B, Section B.5, reciprocity was deduced in the form of eq. (B.25), and this gives $P_{31}/P_{13} = -2$. Equation (B.25) was derived for a transducer with negligible mechanical loading. Since this ratio is universal, we can now say that $P_{31}/P_{13} = -2$ for all transducers. The argument for P_{32}/P_{23} is the same.

A transducer scattering matrix S_{ij}^{t} can be derived from the P_{ij}. The above reciprocity conditions can be shown to lead to symmetry of this matrix [1], so that $S_{ji}^{t} = S_{ij}^{t}$.

REFERENCES

1. G. Tobolka. 'Mixed matrix representation of SAW transducers', *IEEE Trans.*, **SU-26**, 426–428 (1979).
2. S. Ramo, J.R. Whinnery and T. Van Duzer. *Fields and Waves in Communication Electronics*, Wiley, 1994.
3. D.P. Morgan. 'Cascading formulas for identical transducer *P*-matrices', *IEEE Trans. Ultrason. Ferroelect. Freq. Contr.*, **43**, 985–987 (1996).

Appendix E

ELECTRICAL LOADING IN AN ARRAY OF REGULAR ELECTRODES

This appendix gives a simplified account of the work of Bløtekjaer *et al.* [1, 2], analyzing an infinite array of regular electrodes taking account of electrical loading. The effective permittivity described in Chapter 3 is used, and solutions are obtained by using Floquet's theorem. The analysis makes use of Legendre function expansions to ensure that the boundary conditions are met. Stop bands are predicted for frequencies such that the electrode pitch is a multiple of the half-wavelength, and the electrode reflection coefficient is deduced. The theory also gives expressions for the change in wave velocity due to the presence of the electrodes. These results are used in the analysis of gratings and transducers in Chapter 8. Here, the electrodes are considered to be thin enough so that mechanical loading is negligible, but the effects of this loading are considered in Chapter 8.

The surface is taken to be in the x–y plane, with wave propagation in the x-direction and with the electrodes parallel to the y-axis. The fields are assumed to be invariant with y. The frequency ω is assumed to be below the sampling frequency $\omega_s = 2\pi v/p$, which is the frequency at which the electrode pitch p equals the wavelength. Here v is the wave velocity. The range $\omega < \omega_s$ covers most cases of interest, including the first stop band which occurs at $\omega = \omega_s/2$. The analysis for higher frequencies is more complex [1, 2], and it is considered briefly elsewhere [3].

E.1 GENERAL SOLUTION FOR LOW FREQUENCIES

For a piezoelectric half-space the surface potential $\phi(x)$ and charge density $\sigma(x)$ are related by the effective permittivity $\varepsilon_s(\beta)$, which gives the relation between their Fourier transforms $\overline{\phi}(\beta)$ and $\overline{\sigma}(\beta)$. For the present analysis it is

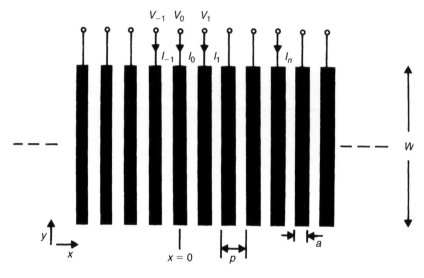

Figure E.1 Infinite array of regular electrodes.

convenient to use the electric field $E_x(x)$ at the surface, rather than the potential. The transform of this field is $\overline{E}_x(\beta) = -\mathrm{j}\beta\overline{\phi}(\beta)$. Thus, with $\varepsilon_s(\beta)$ defined by eq. (3.18) we have

$$\overline{\sigma}(\beta) = \mathrm{j}\varepsilon_s(\beta)\mathrm{sgn}(\beta)\overline{E}_x(\beta) \qquad (\text{E.1})$$

with $\mathrm{sgn}(\beta) = 1$ for $\beta > 0$ and $\mathrm{sgn}(\beta) = -1$ for $\beta < 0$. The solution must satisfy this equation and the boundary conditions that $E_x(x) = 0$ on the electrodes and $\sigma(x) = 0$ on the unmetallized regions between the electrodes.

The electrode configuration is shown in Fig. E.1. The electrodes are centered at $x = mp$, and have width a. For the electrode centered at $x = mp$ the voltage is V_m and the current entering the electrode is I_m. We consider solutions in which these take the form

$$V_m = V_0 \exp(-\mathrm{j}\kappa mp); \quad I_m = I_0 \exp(-\mathrm{j}\kappa mp) \qquad (\text{E.2})$$

where a term $\exp(\mathrm{j}\omega t)$ is implicit. The ratio I_m/V_m is called the *harmonic admittance*, also discussed in Chapter 8, Section 8.3.1. These equations are unaffected if $2\pi/p$ is added to κ. We restrict κ to the range $0 \leq \kappa \leq 2\pi/p$ without affecting the generality of the solution.

With the voltages given by eq. (E.2), the electric field can be expected to have the property $E_x(x+p) = E_x(x)\exp(-\mathrm{j}\kappa p)$, and it follows that $E_x(x)\exp(\mathrm{j}\kappa x)$ is a periodic function with period p. Thus, the field can be written in the form

$$E_x(x) = \sum_{n=-\infty}^{\infty} E_n e^{-j2\pi nx/p} e^{-j\kappa x} \qquad (E.3)$$

and similarly the charge density has the form

$$\sigma(x) = \sum_{n=-\infty}^{\infty} \sigma_n e^{-j2\pi nx/p} e^{-j\kappa x} \qquad (E.4)$$

where the coefficients E_n and σ_n are to be determined. These equations have the form of Floquet's theorem, which can be proved for a wave in an unbounded medium [4] and is assumed to apply for a surface wave.

In the β domain, each term in eqs (E.3) and (E.4) has the form of a delta function located at $\beta = -(\kappa + 2\pi n/p)$. The ratio of E_n to σ_n is thus given by eq. (E.1) for this value of β. Since $\kappa \leq 2\pi/p$ we have $\mathrm{sgn}(-\kappa - 2\pi n/p) = -S_n$, with

$$S_n = 1 \quad \text{for } n \geq 0; \quad S_n = -1 \quad \text{for } n < 0. \qquad (E.5)$$

Noting also that $\varepsilon_s(-\beta) = \varepsilon_s(\beta)$, we have

$$\sigma_n/E_n = -jS_n\varepsilon_s(\kappa + 2\pi n/p). \qquad (E.6)$$

Assuming that the wave is a piezoelectric Rayleigh wave, $\varepsilon_s(\beta)$ is given by Ingebrigtsen's approximation, eq. (3.28). It is shown for Y–Z lithium niobate in Fig. 3.2. This function is almost constant except when β is close to $\pm k_f$, the free-surface wavenumber. We write $\varepsilon_s(\beta) = \varepsilon_s(\infty)$ for all β except for values near $\pm k_f$. We also assume that the frequency is restricted such that

$$0 \leq k_f \leq 2\pi/p \qquad (E.7)$$

which implies that $\omega \leq \omega_s$. From this restriction and the assumption $0 < \kappa < 2\pi/p$ it is found that $\kappa + 2\pi n/p$ can be close to $+k_f$ only if $n = 0$ or -1, so that eq. (E.6) becomes

$$\sigma_n/E_n = -jS_n\varepsilon_s(\kappa + 2\pi n/p) \quad \text{for } n = 0, -1 \qquad (E.8a)$$

$$= -jS_n\varepsilon_\infty \qquad \qquad \text{for other } n \qquad (E.8b)$$

using the notation $\varepsilon_\infty \equiv \varepsilon_s(\infty)$. If k_f is close to $2\pi/p$ an extra case $n = 1$ must be included in eq. (E.8a), but here we assume that k_f is not near $2\pi/p$.

The solution is found by using properties of Legendre functions, described in Appendix C. Consider the functions

$$E'(x) = \sum_{n=-\infty}^{\infty} S_{n-r}P_{n-r}(\cos \Delta)e^{-j2\pi nx/p}e^{-j\kappa x}$$

$$\sigma'(x) = -j\varepsilon_\infty \sum_{n=-\infty}^{\infty} P_{n-r}(\cos \Delta)e^{-j2\pi nx/p}e^{-j\kappa x} \tag{E.9}$$

where $\Delta = \pi a/p$, and a and p are respectively the width and pitch of the electrodes. Equation (C.13) shows that $E'(x)$ is zero at the electrode locations, irrespective of the integer r. Equation (C.12) shows that $\sigma'(x)$ is zero in the gaps between electrodes, for any r. If $\varepsilon_s(\beta)$ were independent of β, eqs (E.8), would be satisfied by eqs (E.9), with $r = 0$. However, because $\varepsilon_s(\beta)$ has different values for $n = 0$ and -1, it is necessary to add terms with different r values. It is found that terms with $r = -1, 0$ and 1 are sufficient, so that the field and charge density are given by

$$E_x(x) = \sum_{n=-\infty}^{\infty} \sum_{r=-1}^{1} \alpha_r S_{n-r}P_{n-r}(\cos \Delta)e^{-j2\pi nx/p}e^{-j\kappa x} \tag{E.10}$$

$$\sigma(x) = -j\varepsilon_\infty \sum_{n=-\infty}^{\infty} \sum_{r=-1}^{1} \alpha_r P_{n-r}(\cos \Delta)e^{-j2\pi nx/p}e^{-j\kappa x} \tag{E.11}$$

where the coefficients α_r are to be determined. These equations have the same form as eqs (E.3) and (E.4). Writing $P_n(\cos \Delta)$ as P_n for brevity, the coefficients are related by

$$\frac{\sigma_n}{E_n} = -j\varepsilon_\infty \frac{\alpha_{-1}P_{n+1} + \alpha_0 P_n + \alpha_1 P_{n-1}}{\alpha_{-1}S_{n+1}P_{n+1} + \alpha_0 S_n P_n + \alpha_1 S_{n-1}P_{n-1}} \tag{E.12}$$

and this is required to satisfy eqs (E.8). From eq. (E.5) we have $S_{n-1} = S_n = S_{n+1}$ if $n \neq 0$ or -1, and it follows that eq. (E.8b) is already satisfied. Hence the ratios of the α_r are determined by eq. (E.8a). Define quantities A_0 and A_{-1} by

$$A_j = \frac{\varepsilon_s(\kappa + 2\pi j/p) + \varepsilon_\infty}{\varepsilon_s(\kappa + 2\pi j/p) - \varepsilon_\infty} \quad \text{for } j = 0, -1. \tag{E.13}$$

Substituting eq. (E.12) into eq. (E.8a) gives

$$\frac{\alpha_1}{\alpha_0} = \frac{A_{-1} + \cos \Delta}{A_0 A_{-1} - \cos^2 \Delta}; \quad \frac{\alpha_{-1}}{\alpha_0} = \frac{A_0 + \cos \Delta}{A_0 A_{-1} - \cos^2 \Delta} \tag{E.14}$$

where use has been made of the relations $P_{-1} = P_0 = 1$ and $P_{-2} = P_1 = \cos \Delta$.

The effective permittivity is taken to be given by Ingebrigtsen's approximation, eq. (3.28), so that

$$\varepsilon_s(\beta) = \varepsilon_\infty \frac{\beta^2 - k_f^2}{\beta^2 - k_m^2} \tag{E.15}$$

where k_f and k_m are the wavenumbers for a free surface and a metallized surface, respectively. Using eq. (E.13) gives

$$A_0 = (2\kappa^2 - k_f^2 - k_m^2)/(k_m^2 - k_f^2) \tag{E.16}$$

$$A_{-1} = [2(\kappa - 2\pi/p)^2 - k_f^2 - k_m^2]/(k_m^2 - k_f^2). \tag{E.17}$$

Electrode voltages and currents

The voltages V_m are found by integrating the field to give the potential, evaluating this at the electrode centers $x = mp$. Using eq. (E.3) for the field, the voltages are $V_m = V_0 \exp(-j\kappa mp)$, with V_0 given by

$$V_0 = -\frac{jp}{2\pi} \sum_{n=-\infty}^{\infty} \frac{E_n}{n + s} \tag{E.18}$$

where

$$s = \kappa p/(2\pi) \tag{E.19}$$

and $0 \le s \le 1$ because of the constraint on κ. The coefficients E_n are identified by comparing eq. (E.3) with eq. (E.10), and the summation over n can be done using Dougall's expansion, eq. (C.10). This leads to

$$V_0 = -\frac{jp}{2\sin(\pi s)} \sum_{r=-1}^{1} (-1)^r \alpha_r P_{r+s-1}(-\cos\Delta). \tag{E.20}$$

The current I_m entering the electrode centered at $x = mp$ is found by integrating the charge density over the electrode and differentiating with respect to time. The integral over x has limits $x = mp \pm a/2$, and the electrode length, in the y-direction, is taken as W. Using eq. (E.4) for $\sigma(x)$ gives $I_m = I_0 \exp(-j\kappa mp)$, with

$$I_0 = \frac{j\omega Wp}{2\pi} \int_{-\Delta}^{\Delta} \sum_{n=-\infty}^{\infty} \sigma_n e^{-j(n+s)\theta} d\theta \tag{E.21}$$

where $\theta = 2\pi(x - mp)/p$ and $\Delta = \pi a/p$. The coefficients σ_n are evaluated by comparing eqs (E.4) and (E.11). Using also eq. (C.14) of Appendix C, eq. (E.21) gives

$$I_0 = \omega W p \varepsilon_\infty \sum_{r=-1}^{1} \alpha_r P_{r+s-1}(\cos \Delta). \tag{E.22}$$

Equations (E.20) and (E.22) give the voltages and currents in terms of the α_r, which are related to κ by eqs (E.14), (E.16) and (E.18). If the ratio I_0/V_0 is specified, the equations determine the values of the α_r and the value of κ, and hence the field and charge density, eqs (E.10) and (E.11), are determined.

This type of approach was also considered by Danicki [5], and Hashimoto [6] has extended it to include the mechanical effects in electrodes of finite thickness, analyzed using the finite-element method (FEM). This generally requires more than three values of r, the number depending on the numerical accuracy required. The method also applies for arbitrary forms of $\varepsilon_s(\beta)$, including leaky wave propagation. However, here we consider only Rayleigh-type waves and the electrodes are taken to have infinitesimal thickness, so that mechanical effects are absent. Consequently, it is found that three values of r are sufficient.

E.2   PROPAGATION OUTSIDE THE STOP BAND

For free-surface propagation the wavenumber is $\pm k_f$. Strong coupling to a surface wave is therefore expected only if one or more of the wavenumbers $\kappa + 2\pi n/p$ in the Floquet expansion is close to $\pm k_f$. Usually, at most one of the wavenumbers is close to $\pm k_f$. However, for the particular case $k_f \approx \pi/p$ we can have $\kappa \approx k_f$ and $\kappa - 2\pi/p \approx -k_f$, so that coupling occurs for $n = 0$ and for $n = -1$. In this case two surface waves propagating in opposite directions are coupled by the electrodes, giving rise to a stop band. In this section it is assumed that the frequency is not close to this value. The stop band will be considered in Section E.3.

It is assumed that, if there is coupling to surface waves, it occurs for $\kappa \approx k_f$. The alternative case, when $\kappa \approx -k_f$, gives essentially the same solution. In view of this, we take $\varepsilon_s(\kappa - 2\pi/p) = \varepsilon_\infty$, and eq. (E.13) gives $A_{-1} = \infty$. Equations (E.14) then give $\alpha_{-1} = 0$ and

$$\alpha_0/\alpha_1 = A_0 = (2\kappa^2 - k_f^2 - k_m^2)/(k_m^2 - k_f^2) \tag{E.23}$$

where A_0 has been obtained from eq. (E.16). Thus, the α_{-1} term in the electric field and charge density, eqs (E.10) and (E.11), is not required here. This can also be deduced more directly by noting that eq. (E.8b) must be valid for all $n \neq 0$. We now consider cases with different electrical terminations.

Shorted electrodes

If the electrodes are all connected together the voltages V_m must be zero, so $V_0 = 0$. Thus, noting that $\alpha_{-1} = 0$, eq. (E.20) gives

$$\alpha_0/\alpha_1 = P_s(-\cos \Delta)/P_{s-1}(-\cos \Delta). \tag{E.24}$$

Using eq. (E.23) this determines the propagation constant κ, which for this case is denoted by k_{sc}. The result is

$$k_{sc}^2 = k_f^2 + \frac{1}{2}(k_m^2 - k_f^2)\left[1 + \frac{P_s(-\cos \Delta)}{P_{s-1}(-\cos \Delta)}\right] \tag{E.25}$$

Since k_m is close to k_f we see that k_{sc} is also close to k_f. Since the Legendre functions vary slowly with s, defined in eq. (E.19), we can use $s \approx k_f p/(2\pi)$, and the right side of eq. (E.25) becomes independent of k_{sc}. The wave velocity is $v_{sc} = \omega/k_{sc}$, and using $k_m \approx k_f$ we find

$$v_{sc} \approx v_f + \frac{1}{2}(v_m - v_f)[1 + P_s(-\cos \Delta)/P_{s-1}(-\cos \Delta)] \tag{E.26}$$

where $v_f = \omega/k_f$ and $v_m = \omega/k_m$.

Open-circuit electrodes

With the electrodes disconnected electrically we have $I_0 = 0$ and eq. (E.22) gives $\alpha_0/\alpha_1 = -P_s(\cos \Delta)/P_{s-1}(\cos \Delta)$. Using eq. (E.23) gives κ, which for this case is denoted by k_{oc}, giving

$$k_{oc}^2 = k_f^2 + \frac{1}{2}(k_m^2 - k_f^2)\left[1 - \frac{P_s(\cos \Delta)}{P_{s-1}(\cos \Delta)}\right]. \tag{E.27}$$

As before, the wavenumbers are all quantitatively similar and we can set $s \approx k_f p/(2\pi)$. The velocity is $v_{oc} = \omega/k_{oc}$, given by

$$v_{oc} \approx v_f + \frac{1}{2}(v_m - v_f)[1 - P_s(\cos \Delta)/P_{s-1}(\cos \Delta)]. \tag{E.28}$$

The velocities v_{sc} and v_{oc} are shown as functions of frequency in Fig. 8.7.

With $s \approx k_f p/(2\pi)$, the difference $k_{sc}^2 - k_{oc}^2$ can be obtained from eqs (E.25) and (E.27). Using eq. (C.6), and noting that the wavenumbers are all similar, this gives

$$k_{sc} - k_{oc} = \frac{\Delta v/v}{p} \frac{2\sin(\pi s)}{P_{-s}(-\cos \Delta)P_{-s}(\cos \Delta)}. \tag{E.29}$$

This equation can also be developed from the quasi-static theory of Chapter 5, as shown elsewhere [7].

General terminations: harmonic admittance

Still using the assumption that the frequency is not in a stop band, consider the solution for general values of the ratio I_0/V_0. Noting that $\alpha_{-1} = 0$, eqs (E.20) and (E.22) give

$$\frac{I_0}{V_0} = 2j\omega W \varepsilon_\infty \sin(\pi s) \frac{P_{s-1}(\cos \Delta)}{P_{s-1}(-\cos \Delta)} \frac{X_+}{X_-} \tag{E.30}$$

where

$$X_\pm = 1 \pm \frac{\alpha_1}{\alpha_0} \frac{P_s(\pm\cos \Delta)}{P_{s-1}(\pm\cos \Delta)} \tag{E.31}$$

and $s = \kappa p/(2\pi)$. The ratio α_1/α_0 is related to κ by eq. (E.23). Hence, eqs (E.30) and (E.31) give the relation between I_0/V_0 and κ. When κ is close to k_f the ratio of Legendre functions in eq. (E.31) is given by eqs (E.25) and (E.27), leading to

$$X_+/X_- \approx (\kappa^2 - k_{oc}^2)/(\kappa^2 - k_{sc}^2). \tag{E.32}$$

The solutions for shorted or open-circuit electrodes occur when this ratio is infinite or zero, giving $\kappa = k_{sc}$ or k_{oc}.

The ratio of I_0/V_0 is the harmonic admittance, discussed in Chapter 8, Section 8.3.1. Equation (E.30) gives a pole at $\kappa = k_{sc}$. Section 8.3.1 of Chapter 8 shows that I_0/V_0 is symmetric about $\kappa = \pi/p$, so there should be another pole at $\kappa = 2\pi/p - k_{sc}$. This is not predicted by eq. (E.30) because the analysis assumed coupling only to forward-propagating waves. We can account for this by calculating I_0/V_0 for $\kappa < \pi/p$ and then using the symmetry. Figure E.2 shows the result obtained from eq. (E.30), for *ST–X* quartz. This is quite similar to accurate calculations [8] though bulk waves are not included here, as shown by the fact that I_0/V_0 is purely imaginary. In addition the analysis here does not include electrode reflections, which make the function much more complicated when the frequency is within the stop band. Accurate methods use finite-element (FEM) analysis to account for mechanical effects [8].

The quasi-static theory of Section 5.1 in Chapter 5 can be developed to give the formula

$$\frac{I_0}{V_0} = j\omega W \left[C(\kappa) - \frac{2\Gamma_s k_f}{p} \sum_{m=-\infty}^{\infty} \frac{[\overline{\rho}_f(\beta_m)]^2}{k_f^2 - \beta_m^2} \right] \tag{E.33}$$

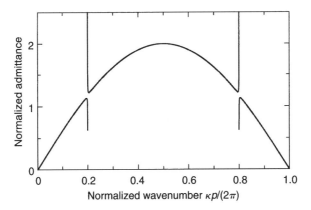

Figure E.2 Harmonic admittance for *ST–X* quartz from first-order theory, eq. (E.30), with $a/p = 1/2$. The frequency is $0.2f_s$, where $f_s = v/p$ is the sampling frequency. The function plotted is the normalized form $(I_0/V_0)/(j\omega\, W \varepsilon_\infty)$.

where $\beta_m = \kappa + 2\pi m/p$, $C(\kappa)$ is the function of eq. (C.27), and $\overline{\rho}_f(\beta)$ is defined in eq. (5.59). Equation (E.33) is derived in Chapter 5 of the previous book [7]. The terms in the summation give poles in the admittance, but for the κ range of Fig. E.2 there are only two poles, and the other terms have little effect. Equation (E.33) then becomes equivalent to eq. (E.30).

E.3 STOP BANDS

When k_f is close to π/p it is no longer valid to ignore the α_{-1} term in the equations of Section E.1. It is then found that the propagation constant κ can be complex, giving a stop band. In this section we consider the stop bands for shorted and open-circuit electrodes. It is sufficient to assume that $k_f \approx \pi/p$, so that $\omega \approx \omega_B = \pi v/p$, the Bragg frequency. The results for other k_f are given in Section E.2.

Shorted electrodes

In this case V_0, given by eq. (E.20), is zero. Since $k_f \approx \pi/p$ and κ must be close to k_f we have $s = \kappa p/(2\pi) \approx 1/2$. The Legendre functions vary slowly with s, so we can set $s = 1/2$ in these functions. Equation (E.20) gives

$$(\alpha_1 + \alpha_{-1})/\alpha_0 = P_{-1/2}(-\cos \Delta)/P_{1/2}(-\cos \Delta).$$

The left side of this equation is related to κ by eqs (E.14), (E.16) and (E.17). Noting that κ, k_f and k_m are all close to π/p, eqs (E.16) and (E.17) give

$$A_0 \approx (2\kappa - k_f - k_m)/(k_m - k_f) \tag{E.34a}$$

and

$$A_{-1} \approx (4\pi/p - 2\kappa - k_f - k_m)/(k_m - k_f). \tag{E.34b}$$

The solution is now found by using eqs (E.14). It is convenient to define the functions

$$F_{\pm}(\Delta) = \mp\cos(\Delta) + P_{1/2}(\pm\cos \Delta)/P_{-1/2}(\pm\cos \Delta). \tag{E.35}$$

Note also that $\Delta v/v \equiv (v_f - v_m)/v_f \approx p(k_m - k_f)/\pi$. It is then found that

$$(\kappa - \pi/p)^2 \approx (\omega - \omega_1)(\omega - \omega_{sc})/v_f^2 \tag{E.36}$$

where

$$\omega_1 = \frac{\pi v_f}{p}\left[1 + \frac{1}{2}\frac{\Delta v}{v}(\cos \Delta - 1)\right] \tag{E.37}$$

and ω_{sc} is given by

$$\omega_1 - \omega_{sc} = \frac{\pi v_f}{p}\frac{\Delta v}{v}F_-(\Delta). \tag{E.38}$$

The function $F_-(\Delta)$ is positive, so that $\omega_1 > \omega_{sc}$. Equation (E.36) shows that κ is complex for $\omega_{sc} < \omega < \omega_1$, so the frequencies ω_{sc} and ω_1 give the edges of the stop band. These frequencies are shown in Fig. E.3, as functions of the metallization ratio.

Open-circuit electrodes

For this case we have $I_0 = 0$, and with $s = 1/2$ eq. (E.22) gives

$$(\alpha_1 + \alpha_{-1})/\alpha_0 = -P_{-1/2}(\cos\Delta)/P_{1/2}(\cos\Delta). \tag{E.39}$$

Using eqs (E.34) and (E.14), the solution for κ is given by

$$(\kappa - \pi/p)^2 \approx (\omega - \omega_1)(\omega - \omega_{oc})/v_f^2 \tag{E.40}$$

where ω_1 is given by eq. (E.37) and ω_{oc} is given by

$$\omega_{oc} - \omega_1 = \frac{\pi v_f}{p}\frac{\Delta v}{v}F_+(\Delta) \tag{E.41}$$

with $F_+(\Delta)$ given by eq. (E.35). Here $\omega_{oc} > \omega_1$, so κ is complex for frequencies in the range $\omega_1 < \omega < \omega_{oc}$. The frequency ω_{oc} is included in Fig. E.3. The functions $F_{\pm}(\Delta)$, which are positive and proportional to the widths of the stop bands, are shown in Fig. E.4.

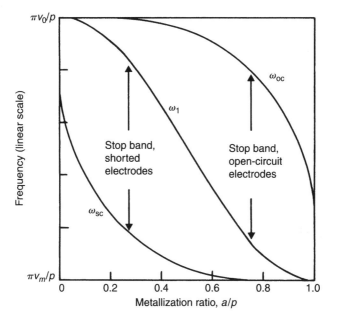

Figure E.3 Frequencies of the stop-band edges.

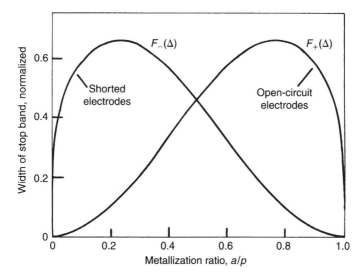

Figure E.4 Width of stop band, as a function of metallization ratio.

Electrode reflection coefficients

The stop bands can be considered to arise from reflections caused by individual electrodes. The electrode reflection coefficients vary slowly with frequency and they are significant mainly within the stop bands, which are quite narrow. Consequently, it is often sufficient to evaluate them at the Bragg frequency ω_B and to regard them as constants. This is done here. The values are related to the stop-band widths, as shown using coupling-of-modes (COM) theory in Chapter 8, Section 8.3.2.

Note first that one stop-band edge for shorted electrodes, at $\omega = \omega_1$, is the same as a stop-band edge for open-circuit electrodes. From Section 8.3.2 of Chapter 8, it follows that the electrode behavior is symmetrical. The electrode reflection coefficients r_{s1} and r_{s2}, for waves incident from either side, are the same and we write $r_{s1} = r_{s2} = r_s$. Equation (8.3) shows that r_s is imaginary. From eq. (8.55) the stop-band width is $\Delta\omega = 2|r_s|v/p$, and this applies for both shorted and open-circuit electrodes. Using eqs (E.38) and (E.41) for these widths, the reflection coefficients are

$$r_s = \pm j \frac{\pi}{2} \frac{\Delta v}{v} F_\pm(\Delta) \quad \text{for } \omega = \omega_B \qquad (E.42)$$

taking the upper signs for open-circuit electrodes and the lower signs for shorted electrodes. In this equation, the signs have been chosen so that r_s/j is negative for shorted electrodes, and positive for open-circuit electrodes. This choice follows from the fact that the upper stop-band edge of the shorted grating is also the lower stop-band edge of the open-circuit grating, as shown in Chapter 8, Section 8.3.2. Equation (E.42) agrees with eq. (8.15) in Chapter 8, Section 8.1.4.

Another derivation of r_s can be obtained by extending the Green's function theory of Section 5.1.1 in Chapter 5 to include higher-order terms [9]. This gives r_s as a function of frequency. For frequencies below the sampling frequency $\omega_s = 2\pi v/p$, the result is

$$r_s = -\frac{j}{2} \frac{\Delta v}{v} kp \left[P_{2s}(\cos\Delta) \mp \frac{P_s(\pm\cos\Delta)}{P_{-s}(\pm\cos\Delta)} P_{-2s}(\cos\Delta) \right] \qquad (E.43)$$

taking the upper (lower) signs for open-circuit (shorted) electrodes, as before. Datta and Hunsinger [10] deduced this formula earlier, and gave a form valid for all frequencies.

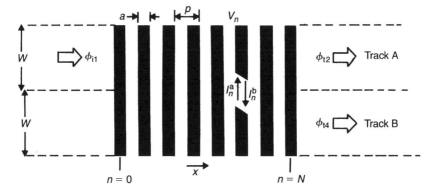

Figure E.5 Simple multistrip coupler.

E.4 THEORY OF THE MULTISTRIP COUPLER

As mentioned in Chapter 5, Section 5.6.4, the multistrip coupler (m.s.c.) consists of a set of parallel disconnected electrodes spanning two surface-wave tracks [11]. It is used to couple the waves from one track to the other. Figure E.5 shows the arrangement, where an incident wave at the left has surface potential ϕ_{i1} and there are output waves with potentials ϕ_{t2} and ϕ_{t4}. Both tracks have width W. In each track the ratio I_0/V_0 is given by the analysis of Section E.2, assuming that the frequency is not in a stop band. For electrode n the currents flowing into the segments at the central boundary are I_n^a and I_n^b, and since the electrode is disconnected we have $I_n^b = -I_n^a$.

The coupler is conveniently analyzed by considering two modes. A *symmetric* mode has the wave amplitude the same in the two tracks. This must give $I_n^b = I_n^a$, which implies that both currents are zero. This corresponds to open-circuit electrodes, so the wavenumber must be k_{oc}. The wave amplitudes in the two tracks have the form $\phi_n^a = \phi_n^b = A \exp(-jk_{oc}np)$, where A is some constant. An *antisymmetric* mode has opposite wave amplitudes in the two tracks, and this gives zero electrode voltages. This corresponds to shorted electrodes, giving wavenumber k_{sc}. The wave amplitudes have the form $\phi_n^a = -\phi_n^b = B \exp(-jk_{sc}np)$, where B is a constant. We now add these two modes, taking $B = A$, giving total amplitudes

$$\phi_n^a = 2A \cos[\pi n/(2N_c)]e^{-jknp}; \quad \phi_n^b = 2jA \sin[\pi n/(2N_c)]e^{-jknp} \quad (E.44)$$

where $k = (k_{oc} + k_{sc})/2$. The parameter N_c is given by $pN_c = \pi/(k_{sc} - k_{oc})$. Equation (E.44) shows that for $n = 0$ the wave power is entirely in track A, while for $n = N_c$ the power is entirely in track B. Thus N_c is the number of strips needed to transfer the power from one track to the other. This is given by eq. (E.29) above. It varies quite slowly with frequency, and a typical value is $N_c \approx 3/(\Delta v/v)$. Hence, efficient coupling is obtainable only for strongly piezoelectric materials such as lithium niobate, which give large $\Delta v/v$ values.

Taking the first and last electrodes of the m.s.c. to be at $n = 0$ and $n = N$, the transfer function for waves transferred from track A to track B is

$$\phi_N^b / \phi_0^a = j \sin[\pi N/(2N_c)] \exp(-jkNp).$$

The amplitude of this function varies quite slowly with frequency, so it causes little distortion when an m.s.c. is incorporated in a surface-wave device.

There are several variants of the basic coupler [11]. For example, the tracks may have different widths, the tracks may have different electrode pitches (enabling efficient coupling even if the track widths are different), and there may be an inactive central region enabling the active tracks to be separated (Fig. 5.17). These cases can all by analyzed using a variant of the above method [7], with I_0/V_0 given by eq. (E.30). In addition, a unidirectional transducer can be formed from a coupler bent into a 'U' shape, with a conventional transducer between the two arms of the 'U'. In all cases, the m.s.c. is normally designed to operate below the Bragg frequency, that is, the electrode pitch p is made less than half the wavelength. The stop band is then outside the region of interest, as assumed above.

REFERENCES

1. K. Bløtekjaer, K.A. Ingebrigtsen and H. Skeie. 'Acoustic surface waves in piezoelectric materials with periodic metal strips on the surface', *IEEE Trans.*, **ED-20**, 1139–1146 (1973).
2. K. Bløtekjaer, K.A. Ingebrigtsen and H. Skeie. 'A method for analysing waves in structures consisting of metal strips on dispersive media', *IEEE Trans.*, **ED-20**, 1133–1138 (1973).
3. D.P. Morgan. *Surface-Wave Devices for Signal Processing*, Elsevier, 1991, Appendix D.
4. C. Elachi. 'Waves in active and passive periodic structures: a review', *Proc. IEEE*, **64**, 1666–1698 (1976).
5. E. Danicki. 'Spectral theory for IDTs', *IEEE Ultrason. Symp.*, **1**, 213–222 (1994).
6. K.-Y. Hashimoto. *Surface Acoustic Wave Devices in Telecommunications – Modelling and Simulation*, Springer, 2000, Appendix B.
7. D.P. Morgan. Surface-Wave Devices for Signal Processing, Elsevier, 1991, Chapter 5.
8. P. Ventura, J.M. Hodé, M. Solal, J. Desbois and J. Ribbe. 'Numerical methods for SAW propagation characterization', *IEEE Ultrason. Symp.*, **1**, 175–186 (1998).
9. D.P. Morgan. Surface-Wave Devices for Signal Processing, Elsevier, 1991, Section E.2.
10. S. Datta and B.J. Hunsinger. 'An analytical theory for the scattering of surface acoustic waves by a single electrode in a periodic array on a piezoelectric substrate', *J. Appl. Phys.*, **51**, 4817–4823 (1980).
11. F.G. Marshall, C.O. Newton and E.G.S. Paige. 'Surface acoustic wave multistrip components and their applications', *IEEE Trans.*, **MTT-21**, 216–265 (1973).

INDEX

Printed and bound by CPI Group (UK) Ltd, Croydon, CR0 4YY

14/05/2025

01871570-0001